Space, Number, and Geometry
from Helmholtz to Cassirer

Archimedes
NEW STUDIES IN THE HISTORY AND PHILOSOPHY
OF SCIENCE AND TECHNOLOGY

VOLUME 46

EDITOR

Archimedes has three fundamental goals; to further the integration of the histories of science and technology with one another: to investigate the technical, social and practical histories of specific developments in science and technology; and fi nally, where possible and desirable, to bring the histories of science and technology into closer contact with the philosophy of science. To these ends, each volume will have its own theme and title and will be planned by one or more members of the Advisory Board in consultation with the editor. Although the volumes have specific themes, the series itself will not be limited to one or even to a few particular areas. Its subjects include any of the sciences, ranging from biology through physics, all aspects of technology, broadly construed, as well as historically-engaged philosophy of science or technology. Taken as a whole, *Archimedes* will be of interest to historians, philosophers, and scientists, as well as to those in business and industry who seek to understand how science and industry have come to be so strongly linked.

More information about this series at http://www.springer.com/series/5644

Francesca Biagioli

Space, Number, and Geometry from Helmholtz to Cassirer

 Springer

Francesca Biagioli
Zukunftskolleg
University of Konstanz
Konstanz, Germany

ISSN 1385-0180   ISSN 2215-0064   (electronic)
Archimedes
ISBN 978-3-319-81116-1   ISBN 978-3-319-31779-3   (eBook)
DOI 10.1007/978-3-319-31779-3

Printed on acid-free paper

This Springer imprint is published by Springer Nature
The registered company is Springer International Publishing AG Switzerland

To my parents

Acknowledgments

This book is a reworked version of the PhD thesis, *Spazio, numero e geometria. La sfida di Helmholtz e il neokantismo di Marburgo: Cohen e Cassirer* (Space, Number, and Geometry. Helmholtz's Challenge and Marburg Neo-Kantianism: Cohen and Cassirer), I defended at the University of Turin on February 24, 2012. I am thankful to my supervisor Massimo Ferrari for his helpful comments on my dissertation and to the members of my thesis committee, Luciano Boi, Paolo Pecere, and especially Renato Pettoello for valuable advice and suggestions. I am especially indebted to Jeremy Gray for his suggestions and for his insightful comments on a previous version of my work. Some of the chapters are a reworked version of previously published articles. I wish to thank all the colleagues and friends with whom I had the opportunity to discuss these contributions and topics related to other parts of the book at conferences, during my doctoral studies, and during my subsequent researches at several institutions: the University of Paderborn, the New Europe College in Bucharest, the Mediterranean Institute for Advanced Research of the Aix-Marseille University, and the Centre for History and Philosophy of Science at the University of Leeds. I am thankful for the supportive environment at my current institution, the University of Konstanz, where I am employed as a postdoctoral fellow of the Zukunftskolleg – Marie Curie affiliated with the Departement of Philosophy. In particular, I wish to thank Eric Audureau, Julien Bernard, Silvio Bozzi, Henny Blomme, Paola Cantù, Andrea Casà, Gabriella Crocco, Christian Damböck, Michael Demo, Vincenzo De Risi, Elena Ficara, Steven French, Giovanni Gellera, Marco Giovanelli, Michael Heidelberger, Don Howard, David Hyder, Alexandru Lesanu, Pierre Livet, Winfried Lücke, Igor Ly, Samuel Marcone, Nadia Moro, Philippe Nabonnand, Matthias Neuber, Anca Oroveanu, Gheorghe Pascalau, Volker Peckhaus, Henning Peucker, Helmut Pulte, David Rowe, Thomas Ryckman, Oliver Schlaudt, Dirk Schlimm, Jean Seidengart, Wolfgang Spohn, Michael Stöltzner. I am thankful to David McCarty and Jeremy Gray for both comments and stylistic suggestions. I also wish to remark that the current book is my own work, and no one else is responsible for any mistakes in it.

Note on Translations

I have slightly modified existing translations so as to conform with the original sources. When not otherwise indicated, all translations are my own.

Contents

Introduction

This book offers a reconstruction of the debate on non-Euclidean geometry in neo-Kantianism between the second half of the nineteenth century and the first decades of the twentieth century. Kant famously characterized space and time as a priori forms of intuitions, which lie at the foundation of mathematical knowledge. The success of his philosophical account of space was due not least to the fact that Euclidean geometry was widely considered to be a model of certainty at his time. However, such later scientific developments as non-Euclidean geometries and Einstein's general theory of relativity called into question the certainty of Euclidean geometry and posed the problem of reconsidering space as an open question for empirical research. The transformation of the concept of space from a source of knowledge to an object of research can be traced back to a tradition, which includes mathematicians such as Carl Friedrich Gauss, Bernhard Riemann, Richard Dedekind, Felix Klein, and Henri Poincaré, and which found one of its clearest expressions in Hermann von Helmholtz's epistemological works.

Although Helmholtz formulated compelling objections to Kant, I reconstruct different strategies for a philosophical account of the transformation of the concept of space in the neo-Kantian movement. I believe that the neo-Kantian reception of Helmholtz might shed some light on his relationship to Kant and, thereby, on Helmholtz's approach to the problem of determining the geometry of space. Helmholtz was one of the first scientists to provide a physical interpretation of non-Euclidean geometry based on the empirical origin of the notion of a rigid body. Helmholtz's definition is rooted in his psychological theory of spatial perception: he defined as "rigid" those bodies whose motion is not accompanied by remarkable changes in shape and size, and this observation can, therefore, be reproduced by corresponding movements of our body. Helmholtz maintained that the generalized formulation of this fact, namely, the free mobility of rigid bodies, lies at the foundation of geometry. He used the same principle to obtain a characterization of space as a manifold of constant curvature. Therefore, a choice has to be made among special cases of such a manifold, which may be either Euclidean or non-Euclidean according to the actual structure of space. Helmholtz contrasted his conception of geometry with the older view that geometrical axioms are evident truths. In that case, not only would geometrical axioms have no need of any justification, but they could not be subject to revision for the purposes of measurement. Helmholtz seemed to attribute such a view to Kant or sometimes, more specifically, to Kant's followers. Nevertheless Helmholtz emphasized, especially in his later writings, the possibility of consistently generalizing the Kantian notion of form of spatial intuition, so as to include both Euclidean and non-Euclidean cases.

The Kantian aspect of Helmholtz's approach has long been neglected after Moritz Schlick, in his comments on the centenary edition of Helmholtz's *Epistemological Writings* from 1921, used Helmholtz's argument to rule out the assumption of a pure intuition in Kant's sense and, therefore, the synthetic a priori character of mathematics. Helmholtz's theory of spatial perception showed that what Kant called the form of intuition could be reduced to the qualitative impressions experienced by a perceiving subject and combined by her in series. According to Schlick, Helmholtz's argument suggests that the theory of space should be clearly distinguished from geometry as the study of abstract structures. Schlick's reading enabled him to relate Helmholtz to his own project of a scientific empiricism, whose goal was to take into account the formalist approach to the foundations of geometry – exemplified for Schlick by the work of David Hilbert – on the one hand, and the philosophical consequences of Einstein's general relativity, on the other. Therefore, Schlick argued for a conventional rather than empirical origin of geometry. Following Poincaré, Schlick acknowledged the possibility of giving a more precise definition of rigidity as one of the properties left unchanged by the set of idealized operations used to represent spatial motion. Such a set univocally determines geometrical properties, insofar as it forms a group. Given the possibility of identifying and classifying geometries according to the group under consideration, the choice of geometrical hypotheses is conventional and can be subject to revision in order to establish a univocal coordination in physics. At the same time, Schlick maintained that spatiotemporal coincidences provide us with objective knowledge about a mind-independent reality. He used the idea of establishing a univocal coordination between various sense impressions to explain how the formation of an intuitive concept of space offers the ground for acquiring knowledge about the structure of nature, once the qualities of sensations are neglected and the consideration is restricted to quantitative relations.

More recent scholarship, initiated by Michael Friedman, called into question Schlick's reading of Helmholtz. Schlick's sharp distinction between intuitive space, geometry, and physics presupposed a very different scientific context from that in which Helmholtz lived. However, the main problem is that by restricting Helmholtz's consideration to intuitive space, Schlick failed to appreciate the significance of Helmholtz's theory of spatial perception regarding the problem of identifying the characteristics of physical space. In order to overcome this problem, Friedman reconsiders the Kantian structure of Helmholtz's argument for the applicability of mathematics in physics. Helmholtz's derivation of the properties of space from the lawful order of appearance found in spatial perception enabled him to consider the free mobility of rigid bodies as a precondition of spatial measurement. Friedman interprets Helmholtz's definition of space as a consistent development of Kant's view that mathematical principles are constitutive of the objects of experience for the following reasons. Firstly, there is evidence that Helmholtz himself saw a connection between his epistemological views and Kant's *Critique of Pure Reason*. Helmholtz defended the aprioricity of the principle of causality even in some of his earliest writings. In his later epistemological papers, he made it clear that the objectivity of measurements depends not on the choice of some particular objects as standards, but on the repeatability of measuring procedures and, therefore, ultimately on the rational demand that there is some regularity of nature.

Secondly, Friedman compares Helmholtz's idea of defining geometric notions by studying all the possible perspectives on the space of a moving subject to Kant's account of motion. Although Kant disregarded motion as an empirical factor in the definition of space, he also distinguished the motion of an object in space from the description of a space, which he called "a pure act of the successive synthesis of the productive imagination." The successive character of such a synthesis suggests that at least the germ of Helmholtz's kinematical conception of geometry can be traced back to Kant himself.

The same idea finds a powerful expression in Klein's and Poincaré's later classifications of geometries in group-theoretical terms. However, Friedman, unlike Schlick, does not consider this an argument against a Kantian view of mathematics. On the contrary, the analogy with Kant regards precisely Kant's argument for the synthetic a priori character of mathematics: not only do mathematical concepts make the definition of physical objects first possible, but all judgments of mathematics (including geometry) are synthetic because of the temporal aspect of the mathematical reasoning. It was for this reason that Kant identified the a priori part of mechanics with kinematics or the general theory of motion in the *Metaphysical Foundations of Natural Science*.

Following Friedman's interpretation, Thomas Ryckman maintains that appreciating Helmholtz's reliance on a basically Kantian theory of space would provide us with a consistent reading of his definition of rigid bodies as opposed to Schlick's. Whereas Schlick replaced Helmholtz's definition with geometrical conventionalism, free mobility, as a restriction imposed by the form of spatial intuition, presents itself as a constitutive property of solid bodies used as measuring standards, and not simply as a stipulation which could be arbitrarily varied. Regarding this view, the main disagreement with Kant lies in the fact that Helmholtz acknowledged the possibility of specifying the structure of physical space in different ways. Whereas Helmholtz does not deny the aprioricity of some of the principles formulated by Kant, the line between the a priori and the empirical part of physical theory is drawn somewhere else, and the specific structure of space appears to belong to the empirical part. Friedman and Ryckman attach particular importance to this development, because, from a Kantian perspective, this would enable one to refurbish the Kantian theory of the a priori without being committed to the aprioricity of Euclidean geometry. Friedman calls this a relativized and historicized conception of the a priori, because it follows that the constitutive role of a priori principles in Kant's sense can only be reaffirmed relative to specific theories. Although Helmholtz himself did not advocate such a conception, his generalization of Kant's form of spatial intuition, along with his new, kinematical approach to geometry, can be considered a decisive step in this direction.

I largely agree with this reading. However, my studies in German neo-Kantianism have led me to approach these topics in a different way. Instead of proposing a consistent reading of Helmholtz's epistemological writings from a Kantian perspective, I focus on their reception in neo-Kantianism and on the debate about the Kantian theory of space and the foundations of geometry. Besides his scientific contributions, Helmholtz played an important role in nineteenth-century philosophy, because he was the first renowned scientist to make a plea for a return to Kant to bridge the gap between post-Kantian idealism and the sciences. As in the current debate about the relativized a priori, some of the leading figures of the neo-Kantian movement were especially interested in a comparison with Helmholtz to handle the question of which aspects of Kant's philosophy are in agreement with later scientific development, including non-Euclidean geometry, and which ones should be refurbished or even rejected by someone who is willing to take such developments into account. However, this should not obscure the fact that the philosophical roots of this debate go back to the earlier objections against the necessity of the representation of space formulated by Johann Friedrich Herbart and to the controversy between Friedrich Adolf Trendelenburg and Kuno Fischer about the status of the forms of intuition. Cohen's rejoinders to these objections were the background for both the earlier phase of the neo-Kantian debate on the foundations of geometry in the nineteenth century and for Cassirer's later reading of Helmholtz in continuity with the group-theoretical treatment of geometry.

In order to introduce the reader to Cassirer's approach to the problems posed by Helmholtz, it was necessary to provide a brief account of Cassirer's insights into the history of mathematics in the nineteenth century. The problem with such an account is that Cassirer considered the comprehensive development of mathematical method from ancient Greek mathematics to the latest mathematical researches of his time. A more careful look at his epistemological works shows that his interest lay, more specifically, in the transformation of mathematics from a discipline that defined itself by referring to a specific domain of objects (i.e., as the science of quantity) to the study of mathematical structures. Cassirer expressed this change of perspective by saying that the unity of mathematics is found no more in its object than in its method. However general this remark may be, Cassirer's recurring examples show that, from his earliest writings, he bore in mind the methodology of some mathematicians in particular. I refer to such examples as the analysis of the continuum by Richard Dedekind, his definition of number, and Klein's classification of geometries into elliptic, hyperbolic, and parabolic. Klein's proof that these geometries are equivalent to the three classical cases of manifolds of constant curvature was one of the first examples of a group-theoretical treatment of geometry, and it was Cassirer's starting point for his reconstruction of the group-theoretical analysis of space from Helmholtz to Poincaré.

The sections of the book which are devoted to the aforementioned episodes in the history of mathematics are far from being conclusive about the methodological issues at stake. My aim is to emphasize the relevance of these examples to Cassirer's view that there is a comparable tendency in the history of science to conceive of objects in terms of structures and to clarify the notion of objectivity in terms of stability of intra-theoretical relations. The historical aspect of Cassirer's approach enabled him to follow the view that Kant's form of spatial intuition deserved a generalization and, at the same time, to restrict the consideration of hypotheses to the cases that could be given a physical interpretation according to the best available theoretical framework of his time. In other words, Cassirer looked at Helmholtz for the opposite reason to Schlick and presupposed a completely different approach to mathematics than Schlick's formalistic view.

Although this later phase of the debate has been largely discussed in the literature and Cassirer's stance has been reconsidered, especially by Ryckman, I believe that a more comprehensive contextualization of the debate since its beginning might shed some light on aspects which have been neglected and lend plausibility to some of the philosophical theses under consideration. It is not always acknowledged that the Marburg School of neo-Kantianism undertook what is known today as a relativization of the a priori. The reception of Helmholtz played an important role in the extension of this conception of the a priori to geometry. However, I believe that this contributed to the development of a preexisting idea. One should not forget that Helmholtz's relationship to Kant was problematic, and it was only in the reception that some of the philosophical tensions in his epistemological works were resolved in one way or another. One of the risks of a Kantian interpretation of Helmholtz in particular is to obscure the empiricist aspect of his approach, that is, the reason why Helmholtz distanced himself from Kant on several occasions. The problem that clearly emerges in the early debate about Helmholtz's psychological interpretation of the forms of intuition is that the Kantian notion of a possible experience in general cannot be reduced to individual experiences. Therefore, Cohen sharply distinguished transcendental philosophy from any direction in psychology by identifying the Kantian notion of experience with the domain of scientific knowledge. Nevertheless, owing to his reliance on the history of science for the inquiry into the conditions of knowledge, Cohen and the Marburg School of neo-Kantianism agreed with an important consequence of Helmholtz's approach: the system of experience cannot be delimited once and for all, but it must be left open to further generalizations. Notwithstanding the fact that Cohen and especially Cassirer looked at the history of mathematics to find examples of such generalizations, the thesis about the structure of experience is not restricted to mathematical objects. The same model of formation of mathematical concepts must apply to physics in order for mathematical principles to make experience possible. This point of agreement with empiricism is understandable, if one considers that the idea that conceptual transformations in cultural history reflect and integrate a systematical order of ideas was quite common in nineteenth-century philosophy. However, this aspect does not always receive enough attention in current debates in the philosophy of science because of the abandonment of this idea in the analytic tradition.

A second problem is that Helmholtz's disagreement with Kant about the status of mathematics depends on a deeper disagreement about the approach to the problems concerning measurement. In addition to the theory of spatial perception, Helmholtz's argument against the restriction of the form of spatial intuition to Euclidean geometry is that analytic methods made it possible to define space as a specific case according to the general theory of manifolds. Following this approach, which goes back to Riemann, Helmholtz's goal was to obtain the definition of space from the simplest relations that can occur between spatial magnitudes, beginning with congruence or equivalence. This required him to introduce the free mobility of rigid bodies to account for the fact that magnitudes, in order to be comparable, must be homogeneous or divisible into equal parts. Helmholtz believed that the same principle was implicit in Euclid's method of proof by ruler and compass constructions. However, Euclid's reliance on intuitive construction prevented him from considering the law of homogeneity in its generality. By contrast, Helmholtz defined spatial magnitudes as a specific case of relations that can be expressed by numbers. The problem with Kant's philosophy of geometry lies not so much in the fact that he could not know about non-Euclidean geometry, but in the fact that he seemed to bear in mind the Euclidean method of proof for the characterization of the objects of geometry. According to Kant, any part of space is homogeneous, because constructions in pure intuition, including infinite division, necessarily take place in one and the same space. This was Kant's argument for the introduction of a distinction between general concepts and pure intuitions, because, according to the syllogistic logic of Kant's time, infinite terms could not be subsumed under one concept. Construction in pure intuition appeared as the only means to capture the idea of an indefinite repetition of some operation.

The definition of mathematics as synthetic a priori does not follow straightforwardly from the use of intuitive or synthetic methods in mathematics, because this definition presupposes Kant's proof that experience is made possible by the application of the category of quantity to the manifold of intuition. However, intuitive constructions as opposed to syllogistic inferences play an important role in the definition of space as pure intuition and, therefore, in the argument for the synthetic a priori status of mathematics as a whole. Indeed, Helmholtz distanced himself both from Kant's view of the method of geometry and from the view that geometry, and mathematics in general, is synthetic a priori. Helmholtz rather called the knowledge of space obtained by his combination of analytic methods and empirical observations "physical geometry," and contrasted this with Kant's allegedly "pure geometry" grounded in spatial intuition.

Helmholtz's disagreement with Kant regarding the mathematical method and the theory of measurement has been emphasized in the literature by Olivier Darrigol and David Hyder. Although they do not refer to the neo-Kantian movement in this connection, I profit from their remarks to reconsider one of the most controversial aspects of the neo-Kantian interpretation of Kant. It is well known that both schools of neo-Kantianism distanced themselves from Kant's assumption of pure intuitions and reinterpreted the notions of space and time as conceptual constructions. As Helmut Holzhey and Massimo Ferrari showed, this debate in the Marburg School of neo-Kantianism was related to the revival of Leibniz in the second half of the nineteenth century and led to a reinterpretation of Kant's theory in continuity with Leibniz's definition of space and time as orders of coexistence and succession, respectively. This is sometimes considered to be incompatible with a synthetic conception of mathematics and, therefore, with the view that mathematical principles are constitutive of the objects of experience. However, it is not least because of this reading of Kant that Helmholtz received a lot of attention, especially by those neo-Kantians who were looking for a reformulation of the argument for the synthetic character of mathematics after important changes of methodology in nineteenth-century geometry. In particular, Cohen agreed with Helmholtz on the role of analytic methods in geometry, and the former was one of the first philosophers to emphasize the connection between this subject and Helmholtz's theory of measurement. Owing to these aspects of Helmholtz's view, Cohen referred to Helmholtz to support his own attempt to reformulate the Kantian argument for the applicability of mathematical principles by pointing out the defining role of mathematical concepts in physics. In Cohen's view, what can be established a priori is not the truth of some statements, but the conceptual classification of all possible cases that may occur in experiment. He maintained that the history of the exact sciences suggests that the unifying power of mathematics in physics increases in the measure that the definition of mathematical concepts is made independent of external elements, such as intuitions, and carried out by purely conceptual means. Although Cohen's stance led him to avoid reference to intuitions (whether empirical or pure), he considered this to be a basically Kantian argument, insofar as Kant himself disallowed specific ontological assumptions in order to address the question of the conditions for making generally valid judgments about empirical objects.

To sum up, owing to important points of disagreement with Kant, I do not believe that Helmholtz's epistemological writings can receive a consistent interpretation in terms of a Kantian epistemology. Nevertheless, I do believe that the transformation of the theory of the a priori undertaken by Cohen in a philosophical context in the first place enabled him to appreciate those aspects of Helmholtz's approach that indeed admit a Kantian interpretation. In particular, it seems to me implausible to trace back a relativized conception of the a priori to Helmholtz himself. Even if one focuses on Helmholtz's generalization of the form of spatial intuition, it is noteworthy that he only gradually acknowledged the possibility of a connection with the Kantian theory of space. Even then, his considerations remained problematic, because he did not address the question about the status of a priori knowledge. The first attempt to clarify the relation between the notion of the a priori and that of necessity is found in Cohen, who interpreted the aprioricity of mathematics as not only compatible, but even inherently related to the hypothetical character of mathematical inferences. Therefore, the only kind of necessity which can be attributed to a priori knowledge is relative necessity. However, it was especially Cassirer who emphasized the connection between Cohen's theory of the a priori and the hypothetical character of geometry emerging from nineteenth-century inquiries into the foundations of geometry, on the one hand, and from Einstein's use of Riemannian geometry in general relativity, on the other. Cassirer used the group-theoretical classification of geometry as one of the clearest example of the anticipatory role of mathematics in the formulation of physical hypotheses.

The first chapter provides a general introduction to Helmholtz's relationship to Kant, with a special focus on the aforementioned problems of the Kantian reading of Helmholtz's epistemology. The following chapters offer a reconstruction of the debate about the Kantian theory of space from the Trendelenburg-Fischer controversy to the early neo-Kantian responses to Helmholtz's objections to Kant. The second part of the book is devoted to the later phase of the debate, after the group-theoretical treatments of geometry by Klein and by Poincaré. In particular, I reconsider Cassirer's position in this debate. Cassirer was not the only one to notice that group theory provides a plausible interpretation of Helmholtz's analysis of the concept of space. The current literature has developed this idea in such a way that the connection with Einstein's space-time theories may be more apparent than in Cassirer's work. However, it seems to me that Cassirer was particularly clear about the empiricist aspect of Helmholtz's approach that can be brought in agreement with a neo-Kantian perspective: the identification of space with a mathematical structure is justified only insofar as this provides us with the most general laws of order of appearance. The hierarchical order of formal relations is only part of the argument for the applicability of mathematics, because the level of generality depends, on the other hand, on the problems concerning measurement within the context of specific theories. Cassirer's view clearly implies that the highest principles have some constitutive role only relative to specific theories and may be subject to revision for the solution of scientific problems. However, he also made it clear that a philosophical account of a priori knowledge should meet the tasks of a general theory of experience and account for some continuity across theory change, even in those cases in which continuity depends not on relations among mathematical structures, but on the mathematical method. In this sense, I suggest that the revision of the Kantian theory of space in Marburg neo-Kantianism opened the door to a very subtle, although today unusual, way to look at the relation between mathematical and natural concepts.

Chapter 1
Helmholtz's Relationship to Kant

1.1 Introduction

Hermann von Helmholtz developed epistemological views in connection with his contributions to various branches of science, including physics, physiology, and the inquiries into the foundations of mathematics. He studied medicine at the Friedrich-Wilhelms Institut in Berlin from 1838 until 1842, where he attended courses in physiology taught by Johannes Müller. Helmholtz was Professor of Physiology from 1849 until 1871 at the universities of Königsberg, Bonn, and Heidelberg, publishing contributions both in physics and physiology. He gave lectures and published papers on the foundations of geometry from 1868 to 1878. In 1871, he became Professor of Physics at the University of Berlin. He was a professor there until 1888, and was founding president of the Physicalisch-Technische Reichsanstalt in Berlin from 1887 until his death in 1894.[1]

Helmholtz's reception of Kant goes back to his earliest epistemological considerations, and further developments are found in Helmholtz's main epistemological writings. Helmholtz's relationship to Kant was much discussed at the time and in more recent studies. Neo-Kantians, such as Hermann Cohen and Alois Riehl, and, more recently, Kant-oriented scholars, such as Michael Friedman, Thomas Ryckman, Robert DiSalle, Timothy Lenoir, and David Hyder, emphasize the connection because, especially if one considers Helmholtz's geometrical papers, Helmholtz formulated both compelling objections to Kant and rejoinders to those objections within the framework of Kant's transcendental philosophy. It might seem that Helmholtz may help to address the question regarding which aspects of Kant's philosophy are in agreement with later scientific developments, including non-Euclidean geometry, and which ones ought to be refurbished or even rejected by someone who is willing to take such developments into consideration.

---

[1] For detailed biographical information on Helmholtz, see Königsberger (1902–1903).

© Springer International Publishing Switzerland 2016
F. Biagioli, *Space, Number, and Geometry from Helmholtz to Cassirer*,
Archimedes 46, DOI 10.1007/978-3-319-31779-3_1

This chapter provides an overview of Helmholtz's remarks on Kant regarding the law of causality, the theory of spatial perception, and the conception of space and time. A more detailed account of Helmholtz's geometrical papers and of his theory of measurement is given in Chaps. 3 and 4, respectively. Since the following overview is far from complete, references to more detailed studies on related issues are cited. In particular, this chapter does not provide a thorough analysis of the development of Helmholtz's relationship to Kant.[2] The proposed topics will enable us to follow what is approximately a chronological order. I believe, however, that the thematic order may be more appropriate to make the points of disagreement with Kant clear, and thereby lead one to appreciate those aspects of Helmholtz's epistemology that indeed admit a Kantian interpretation. The following overview will concentrate specifically on those aspects that were influential in their reception by neo-Kantians, such as Alois Riehl, Hermann Cohen, and Ernst Cassirer.

1.2   The Law of Causality and the Comprehensibility of Nature

One of the guiding principles of Helmholtz's epistemology was the law of causality, which he characterized as an a priori principle in Kant's sense, that is, a principle which is independent of actual experience, because it provides us with a precondition for a possible experience in general. The causal relation was one of Kant's classical examples of an a priori judgment in the introduction to the *Critique of Pure Knowledge*:

> Now it is easy to show that in human cognition there actually are such necessary and in the strictest sense universal, thus pure *a priori* judgments. If one wants an example from the science, one need only look at all the propositions of mathematics; if one would have one from the commonest use of the understanding, the proposition that every alteration must have a cause will do; indeed in the latter the very concept of a cause so obviously contains the concept of a necessity of connection with an effect and a strict universality of rule that it would be entirely lost if one sought, as Hume did, to derive it from a frequent association of that which happens with that which precedes and a habit (thus a merely subjective necessity) of connecting representation arising from that association. (Kant 1787, pp.4–5)

The law of causality is independent of experience because necessity and universal validity are defining characteristics of the causal relation. Kant used the above examples to prove that in both scientific and common uses of the understanding we distinguish this kind of a priori necessity from the subjective necessity of a frequent association of events. He went on to say that: "Even without requiring such examples for the proof of the reality of pure *a priori* principles in our cognition, one could establish their indispensability for the possibility of experience itself, thus

---

[2] For an accurate account of the development of Helmholtz's epistemological views, see Hatfield (1990, Ch.5). More recently, important turning points in Helmholtz's relationship to Kant have been emphasized by Hyder (2009).

establish it *a priori*. For where would experience itself get its certainty if all rules in accordance with which it proceeds were themselves in turn always empirical, thus contingent?; hence one could hardly allow these to count as first principles" (p.5).

Helmholtz famously referred to Kant in his lecture of 1855 "On Human Vision." After enunciating the principle that there can be no effect without a cause, Helmholtz said: "We already need this principle before we have some knowledge of the things of the external world; we already need it in order to obtain knowledge of objects in space around us and of their possibly being in cause-effect relationships to one another" (Helmholtz 1855, p.116). In the physiology of vision, the law of causality is required for inferring the external causes of sensations from nerve stimulation.

On that occasion, Helmholtz was delivering the Kant Memorial Lecture at the University of Königsberg. Helmholtz's agreement with Kant might have been occasioned, at least in part, by such a circumstance (see Königsberger 1902, Vol. 1, pp.242–244). The success of this lecture, along with Helmholtz's later epistemological lectures and writings on related subjects, went beyond even what he might have expected. The neo-Kantians, especially Alois Riehl, appealed to Helmholtz's authority to support their project of a scientific philosophy. In this regard, Riehl called Helmholtz "the founder of a new philosophical era" (Riehl 1922, pp.223–224). Accordingly, the philosophical reception of Helmholtz particularly emphasized his relationship to Kant and did not always pay enough attention to other aspects of Helmholtz's epistemology. Nevertheless, there is evidence that Helmholtz's own considerations in 1855 were not due solely to the occasion. Not only did Helmholtz refer to Kant's law of causality in his *Handbook of Physiological Optics* (1867), but the same law played an important role in Helmholtz's contributions to mechanics. Helmholtz's connection with Kant goes back to his essay *On the Conservation of Force* of 1847. In this essay, Helmholtz gave a proof of an equivalent formulation for the conservation of force, namely, the proposition that all forces can be calculated as functions of distances between pairs of points. Once the direction has been specified, all forces are supposed to be central forces of the kind of Newton's gravitational force between two point-masses, which is directly proportional to the product of the masses and inversely proportional to the square of their distance.

Helmholtz's reference to Kant is implicit in the introduction to his essay. There, Helmholtz distinguished between the experimental part of physics and the theoretical one in a way that is clearly reminiscent of Kant's characterization of natural science in the Preface to the *Metaphysical Foundations of Natural Science*: "A rational doctrine of nature […] deserves the name of a natural science, only in case the fundamental natural laws therein are cognized *a priori*, and are not mere laws of experience. One calls a cognition of nature of the first kind *pure*, but that of the second kind is called *applied* rational cognition" (Kant 1786, p.468). Similarly, according to Helmholtz, the experimental part of physics requires us to infer general rules from single natural processes; the theoretical part is to infer the unobservable causes of such processes from their observable effects. The law of causality is required for making inferences of the second kind. In this connection, the law coincides with the demand that all natural phenomena be comprehensible. Helmholtz wrote:

The final goal of theoretical natural science is to discover the ultimate invariable causes of natural phenomena. Whether all processes may actually be traced back to such causes, in which case nature is completely comprehensible, or whether on the contrary there are changes which lie outside the law of necessary causality and thus fall within the region of spontaneity or freedom, will not be considered here. In any case it is clear that science, the goal of which is the comprehension of nature, must begin with the presupposition of its comprehensibility and proceed in accordance with this assumption until, perhaps, it is forced by irrefutable facts to recognize limits beyond which it may not go. (Helmholtz 1847, p.4)

Helmholtz used his Kantian conception of causality to advocate a mechanistic conception of nature. He isolated matter and force as the fundamental concepts of theoretical physics and pointed out their inseparability: since force and matter are abstracted causes of the same phenomena, none of them can cause observable effects independently of the other.[3]

Helmholtz revised this picture in 1854 in order to reply to the objections raised by the German physicist Rudolf Clausius. Now, Helmholtz recognized that his proof for the centrality of forces presupposed empirically given points and relative coordinate systems. In the case of systems with two points, the positional dependence of the force can be specified as its dependence on distance, and the specification of the direction of the force entails its centrality. Since empirical conditions for the determination of distance and of direction must be specified, the centrality of forces apparently ceases to be deduced a priori. In other words, according to the above distinction between the different parts of physical theory, there seems to be a shift of the centrality of forces from the theoretical to the empirical part of the theory.

Helmholtz explicitly distanced himself from Kant in 1881. By that time, Helmholtz admitted that the centrality of forces is not necessary and can only be assumed as an empirical generalization. Furthermore, he noticed that such an assumption can be called into question if one considers more recent electromagnetic theories. Therefore, Helmholtz declared in an appendix to the second edition of his essay *On the Conservation of Force*, that, in 1847, he adhered to a Kantian conception of causality he was no longer willing to defend. He realized only later that what he called the law of causality is better understood as the requirement that the phenomena be related to one another in a lawful way. One need not assume that force and matter are abstracted causes of the same facts in order to account for their inseparability. It suffices to notice that the inseparability of force and matter is necessary for objectively valid laws to be formulated (Helmholtz 1882, p.13).

Helmholtz's revision suggests that the guiding principle for empirical science is not so much causality, as the demand for the comprehensibility of nature. The comprehensibility of nature is not completely different from the law of causality as originally formulated by Helmholtz: they are both conditions for conceptualizing some natural processes and coherently extending conceptualization to all natural processes. However, in 1881, Helmholtz made it clear that the demand for the comprehensibility of nature differs from causality, because it does not necessarily entail a reduction of all of physics to the mechanistic explanation of nature. Although this

---

[3] On the analogy between Kant's *Metaphysical Foundations of Natural Science* and Helmholtz's introduction to the *Conservation of Force*, see Heimann (1974).

view apparently contradicts the view of natural science advocated by Kant in 1786, Helmholtz seems to rely on Kant's characterization of the empirical part of natural science as guided by methodological or regulative principles. We already noticed that a priori principles in Kant's sense delimit the domain of a possible experience in general. Therefore, Kant called these principles constitutive of the objects of experience. In addition, in the section of the *Critique of Pure Reason* which is the devoted to the Antinomies of Pure Reason, Kant introduced the notion of regulative principles as applying to the world-whole one can only have in concept. As an example, Kant mentioned the principle that there can be no experience of an absolute boundary in the empirical regress of the series of appearances. He called this proposition a regulative principle of reason, because nothing can be said about the whole object of experience, but only something about the rule according to which, experience, suitable to its object, is to be instituted and continued (Kant 1787, pp.548). More generally, Kant called "regulative" the hypothetical use of reason, which he characterized as follows:

> The hypothetical use of reason, on the basis of ideas as problematic concepts, is not properly constitutive, that is, not such that if one judges in all strictness the truth of the universal rule assumed as hypothesis thereby follows; for how is one to know all possible consequences, which would prove the universality of the assumed principle if they followed from it? Rather, this use of reason is only regulative, bringing unity into particular cognitions as far as possible and thereby approximating the rule to universality.
>
> The hypothetical use of reason is therefore directed at the systematic unity of the understanding's cognitions, which, however, is the touchstone of truth for its rules. Conversely, systematic unity (as mere idea) is only a projected unity, which one must regard not as given in itself, but only as a problem; this unity, however, helps to find a principle for the manifold and particular uses of the understanding, thereby guiding it even in those cases that are not given and making it coherently connected. (Kant 1787, p.675)

These sections of the *Critique of Pure Reason* strongly suggest that Kant introduced regulative principles to account for scientific generalization, for which constitutive principles provided a necessary but not sufficient condition. This is confirmed by the fact that in the *Critique of Judgment* Kant supplemented his earlier view of natural science with the Critique of the Teleological Judgment.

Considering the epistemic value of Kant's regulative principles, one may say that Helmholtz tended to assume the comprehensibility of nature as a principle of experience in Kant's sense, though not so much a constitutive principle as a regulative one (see Hyder 2006, pp.4–11). Such an interpretation was predominant in the philosophical reception of Helmholtz, beginning with early neo-Kantianism. In the second edition of *Kant's Theory of Experience* (1885), Hermann Cohen, the founder of the Marburg School of neo-Kantianism, noticed a point of agreement between Kant and Helmholtz on the conception of the law of causality: since such a law is purely logical (i.e., independent of experience), it affects not so much actual experience, as its comprehension. In this connection, Helmholtz's law of causality can be considered a condition of possible experience, though not "in the sense of synthetic possibility strictly speaking" (Cohen 1885, p.452). This is because Helmholtz attributed some kind of transcendental function to the law of causality, as a condition for the conceptualization of experience, but not necessity and universality, which are the characteristics of a priori knowledge in Kant's sense. The empiricist aspect of Helmholtz's approach lies in the fact that such a function depends on the empirical sciences and their advancement. This is the aspect that neo-Kantians, such as Cohen and Cassirer, considered compatible with or even essential for a transcendental inquiry into the conditions of knowledge after radical changes in mathematics and physics.

Schiemann (2009) argues that the connection between Helmholtz and Kant is mainly due to the philosophical reception of Helmholtz. Helmholtz's epistemological views should be reconsidered by clarifying the different meanings of the concept of causality in his writings. Schiemann calls the idea of causality applied to objects of natural research "phenomenal" causality. This is the demand that the phenomena be unequivocally determined by preceding causes. Besides this notion of causality, which is the one Helmholtz presupposes in the Introduction to his essay *On the Conservation of Force*, there is a notion Schiemann calls "noumenal" causality. This should be used to legitimize Helmholtz's realism: namely, the argument that the assumption of an external world is evident. Such an assumption would precede "any causal relations that (merely possibly) exist among phenomena" (Schiemann 2009, p.126). However, I do not see why the fact that the assumption of an external world appears to us to be evident should be more fundamental than causal realism itself. The assumption of ultimate, unknowable causes is particularly problematic in Helmholtz's view, insofar as he distanced himself from the assumption of absolute limits of knowledge. The quote above suggests that limits that cannot be excluded are only those that may be encountered in empirical research. I think that the constitutive/regulative distinction is more appropriate to do justice to Helmholtz's belief in the possibility of a progressive extension of the laws or nature. It is true that this way to understand his conception of science is due, at least in part, to the philosophical reception of his views. However, I do not take this as a mystification of some opposing view: the reception of Helmholtz in neo-Kantianism is a good example of the possibility of a fruitful interaction between philosophy and the sciences.

Michael Friedman and Thomas Ryckman make a similar point without referring to the reception of Helmholtz in neo-Kantianism. Although Helmholtz tended to express himself in terms of a causal realism in his earlier writings, his later insight into the experimental method and its applications in the physiology of vision and in the theory of measurement led him to ground objective knowledge in the lawful connection of appearances. In particular, Friedman (1997) draws attention to the following quote from Helmholtz's 1878 paper "The Facts in Perception":

> I need not explain to you that it is a *contradictio in adjecto* to want to represent the real, or Kant's "thing in itself," in positive terms but without absorbing it into the form of our manner of representation. This is often discussed. What we can attain, however, is an acquaintance with the lawlike order in the realm of the actual, admittedly only as portrayed in the sign system of our sense impressions. (Helmholtz 1878, pp.140–41)

Similar to the interpretation of experiments in physics, the connection of sense impressions depends on a learning process based on the principle of causality or the lawlikeness of nature. Since interpretation is required for our system of signs to have a meaning, Friedman (1997, p.33) considers such a process constitutive of the objects of experience in Kant's sense.

More recently, Hyder (2006) pointed out Helmholtz's connection with Kant for the answer to the question: How are the constitutive principles of experience related to regulative principles? Hyder observed that Kant's regulative principles can be seen from two points of view: "[A]s illegitimate statements concerning the totality of the natural world, or as methodological, meta-theoretical principles concerning the organisation of theories. Only in the latter sense can they be taken to be valid rules for thought" (Hyder 2006, p.17). In Helmholtz's epistemology, the different functions of regulative and constitutive principles correspond to the dual direction of Helmholtz's determinacy requirements. "Upward" determinacy is required to justify the claim that all forces observed in nature must be seen as determinations of a set of basic forces that characterize the various species of matter. At the same time, Helmholtz's principle of positional determinacy introduces the "downward" requirement that the ultimate spatial referents of motive concepts be determined (Hyder 2006, pp.19–20). Helmholtz's remarks on causality do not suffice to attribute to him a Kantian architectonic of knowledge. Nevertheless, I agree with Hyder that the dual direction of Helmholtz's requirements admits an interpretation in terms of a transcendental argument for the determinacy of physical theory. I rely on the neo-Kantian reception of Helmholtz because I believe that it was Cassirer who clarified the consequences of Helmholtz's approach to the theory of measurement for the Kantian system of principles. In the following chapters, I argue that the interpretation of Kant in the Marburg School of neo-Kantianism, on the one hand, and the reception of epistemological writings by scientists and mathematicians, such as Helmholtz, Felix Klein, and Henri Poincaré, on the other, led Cassirer to the view that, despite the fact that the content of Kant's distinctions can be reformulated, the core idea that there are increasingly higher levels of generality in the conditions of experience was confirmed by more recent developments in the history of science.[4]

---

[4] It has been objected that the interpretation of Kant in the Marburg School of neo-Kantianism tends to blur the difference between constitutive and regulative principles (see especially Friedman 2000a, p.117). Since the discussion of this objection will require us to consider Cohen's and Cassirer's arguments for the synthetic a priori character of mathematics in some detail, further references to this debate are given in Chap. 2. For now, it is worth noting that for Cassirer – as well as for Helmholtz – the scope of synthetic a priori knowledge – and, therefore, the validity of constitutive principles – cannot be established once and for all, because it depends on the advancement of science. "The validity of [the] statements [of critical philosophy] is not guaranteed once and for all, but it must justify itself anew according to the changes in scientific convictions and concepts. Here, there are no self-justified dogmas, which could be assumed for their 'immediate evidence' and fixed for all time: the only stable thing is the task of the continually renewed examination of scientific fundamental concepts, which for the critique becomes at the same time a rigorous self-examination" (Cassirer 1907, p.1).

1.3 The Physiology of Vision and the Theory of Spatial Perception

Notwithstanding the developments in Helmholtz's conception of causality and his 1854 revision of his proof of the centrality of forces, I have already mentioned that only one year later in the Königsberg lecture, he endorsed the apriority of the principle of causality in the physiology of vision. His starting point was a general consideration regarding the relationship between philosophy and the sciences in the nineteenth century. Despite the importance of this relationship, nineteenth-century scientists were often skeptical about philosophy. In Helmholtz's opinion, skepticism was a reasonable reaction to the philosophy of nature of Hegel and Schelling and their attempt to predict empirical results by means of pure thought. In order to bridge the gap between philosophy and the sciences, Helmholtz's suggestion was to reconsider Kant's philosophy. Firstly, as Kant's commitment to Newton's mechanics shows, such a gap did not exist at that time. Secondly, and more importantly, the issue of philosophy for Kant was to study the sources of knowledge and the conditions of its validity. According to Helmholtz, every historical period should be confronted with the issue so formulated (Helmholtz 1855, p.89).

To begin with, Helmholtz focused on the theory of perception, because he believed this subject to be one of the most appropriate to explore interactions between philosophy and the sciences. Helmholtz maintained that Johannes Müller's theory of specific sense energies confirmed Kant's theory of representation: Kant pointed out that there are subjective factors of representation – which is confirmed by Müller's proof that sensuous qualities depend not so much on the perceived object as on our nerves (Helmholtz 1855, p.98). Since optical nerves can be stimulated in different ways, visual sensations do not necessarily depend on light. They might be caused, for example, by an electric current or by a blow to the eye. On the other hand, light does not necessarily cause visual sensations; for example, ultraviolet rays cause only chemical reactions. These examples show that sensation is necessary, but not sufficient for an object to be perceived. This is because perception also entails some inference from the subjective factors of perception (i.e., nerve stimuli) to existing objects. The law of causality provides the basis of the validity of such inferences.

Helmholtz developed his view in 1867 in the third part of his *Handbook of Physiological Optics*, the opening section of which is devoted to the theory of perception. He introduced the psychological part of the physiology of vision by noticing that the perception of external objects presupposes a psychic activity that could not yet be reduced to physical concepts. Therefore, physical explanations ought to be avoided, and those psychic activities that enable us to localize some objects are to be better understood as a kind of inference. Unlike inferences properly speaking, however, the corresponding associations are unconscious (Helmholtz 1867, p.430). Helmholtz's view, which became known as the theory of unconscious inferences, offered an empiricist and non-reductionist perspective on human vision.

According to Helmholtz's theory, spatial representations have empirical origins and deserve a causal explanation. Therefore, causal reasoning is required as in Helmholtz's original view. At the same time, Helmholtz's emphasis in 1867 lies in the fact that particular associations are not necessary and can be accomplished in many ways. As we shall see in the following chapters, this is the point of departure from the Kantian theory of space. On the one hand, Helmholtz adopted what he called a Kantian formulation of the problem of knowledge: once the empirical content of knowledge is distinguished from the subjective forms of intuitions, there is the problem of justifying the inference from subjective forms to some objective meaning (Helmholtz 1867, p.455). On the other hand, he pointed out the risk of extending the subjectivity of the general forms of intuition to particular intuitions. This would lead to a nativist theory of vision and to an aprioristic conception of geometrical axioms as propositions given in our spatial intuition. By contrast, empirical explanations presuppose the only data available, namely, the qualities of sensations. Helmholtz advocated a "sign" theory, according to which sensations symbolize their stimuli, but do not bear any resemblance to real entities. Nevertheless, signs must be chosen so that they can stand for relations between such entities. In fact, human beings learn to use signs for practical purposes. The epistemic value of such use lies in its providing us with guiding rules for our actions: signs help us to anticipate the course of events and to produce the sensations we expect (Helmholtz 1867, pp.442–443).

In Helmholtz's view, spatial intuitions ought to be explained as the results of series of sensations that can be associated in various ways. Once we have learned to localize particular objects, the learning process is usually forgotten. Therefore, it might seem as if the laws of spatial intuition were innate in us. According to Helmholtz, nativist theories of vision precluded an explanation of how such laws can be obtained and were not able to provide a justification of the application of such laws to concrete reality. Applicability was assumed as a kind of pre-established harmony between thought and reality. Helmholtz endorsed the view that all spatial intuitions are psychical products of learning, and classified and rejected as nativist those views that presuppose innate, anatomical connections to account for the singularity of vision.[5]

---

[5]This is the fact that we have two eyes, but perceive only one world. Helmholtz maintained that the two retinas produce two sets of sensations that we have to learn to refer to a single object. Therefore, he opposed the "identity hypothesis." Note that Helmholtz opposed nativism with regard to two separate questions. The first question concerns the two-dimensionality of vision. At the time Helmholtz was writing, the dominant view endorsed, among others, by Helmholtz's teacher Johannes Müller, was that a two-dimensional spatial representation is primitively given in vision and only the perception of depth and distance (i.e., the kind of perception that presupposes three-dimensionality) has to be learned. Nevertheless, even before Helmholtz's challenge, such a view had been called into question both by physiologists, such as Steinbuch, Nagel, Classen, and Wundt, and by philosophers such as Cornelius and Waitz, who followed Herbart in deriving all spatial representations from nonspatial sensations through the "fusion" or association of such sensations (see Waitz 1849, p.167). The other question is the explanation of single vision. In this regard, Helmholtz called Müller, Ewald Hering, and all those who supposed the two retinas to be

Since nativist views apparently include Müller's view, Helmholtz seems to call into question his original argument in favor of Kant's theory of representation as well. Does Helmholtz's refutation of nativism entail a refutation of the Kantian theory of space? Or are there points of agreement between Kantianism and Helmholtz's empiricism? Does Kant's philosophy provide a justification for the application of a priori laws to concrete reality other than pre-established harmony? To answer these questions, it will be necessary to take Helmholtz's philosophy of mathematics into account. I shall say in advance that points of agreement between Kantianism and empiricism cannot be set aside, provided that Helmholtz's physiological interpretation of the a priori is rejected. This was, for example, the opinion of Alois Riehl. In his paper of 1904, "Helmholtz's Relationship to Kant," Riehl wrote:

> The critical inquiry into knowledge, the proof of the conditions and limits of its objective validity, is converted [by Helmholtz] into a nativist theory of the origins of our representation, namely, J. Müller's theory of space. The more Helmholtz himself tended to the opposite side, the more he coherently and exclusively took an empiricist direction, and the more he believed, just for this, that he should distance himself from Kant. His relationship to Kant had a development that went hand in hand with his refutation of nativism. (Riehl 1904, p.263)

Arguably, Helmholtz believed that he should distance himself from Kant insofar as he took an empiricist direction. However, Riehl's opinion was that Helmholtz's empiricist insights can be interpreted as a development of Kant's ideas (see also Riehl 1922, p.230).

By contrast, Moritz Schlick, in his comments on the centenary edition of Helmholtz's *Epistemological Writings* (1921), connected Helmholtz's theory of signs with his own project of a scientific empiricism, which, unlike neo-Kantianism, on the one hand, and positivism, on the other, would provide us with a philosophical interpretation of Einstein's general theory of relativity. By that time, Schlick was known to be one of the first philosophers to appreciate the revolutionary import of Einstein's work, and Schlick's book on general relativity, *Space and Time in Contemporary Physics* (1917), was about to appear in a fourth, revised edition (1922). In 1921, Schlick especially appreciated Helmholtz's general perspective on knowledge. On the one hand, Helmholtz clearly distinguished signs from images: "For from an image one requires some kind of similarity with the object of which it is an image" (Helmholtz 1878, p.122). On the other hand, he accounted for our belief in the capability of our system of signs to refer to an external reality as follows:

> To popular opinion, which accepts in good faith that the images which our senses give us of things are wholly true, this residue of similarity acknowledged by us may seem very trivial. In fact it is not trivial. For with it one can still achieve something of the very greatest importance, namely forming an image of lawfulness in the processes of the actual world. (Helmholtz 1878, p.122)

---

anatomically connected with each other nativists (Helmholtz 1867, p.456). On Helmholtz's position in the nativism/empiricism debate, see Hatfield (1990), pp.180–188.

In commenting on this quote, Schlick referred to the first part of his *General Theory of Knowledge* (1918) for the attempt "to show that forming such an image of what is lawlike in the actual, with the help of a sign system, altogether constitutes the essence of all knowledge, and that therefore our cognitive process can only in this way fulfil its task and needs no other method for doing so" (Schlick in Helmholtz 1921, p.166, note 15).

Friedman (1997) calls into question Schlick's connection for two main reasons. Firstly, Schlick bore in mind the development of the axiomatic method in David Hilbert's *Foundations of Geometry* (1899). This method led to a new way to define geometrical concepts by using axioms. A more detailed account of Schlick's considerations is given in Chap. 7. For now, it is noteworthy that Schlick used this method to sharply distinguish geometries as axiomatic systems from spatial perception. Since the formal approach enables us to deal with geometry independently of spatial intuitions, the problem of the interpretation of mathematical structures deserves completely different considerations. In order to establish a univocal coordination between geometrical concepts and the empirical manifold of spatiotemporal events, Schlick introduced another method based on the calculation of spatiotemporal coincidences in Einstein's theory of relativity. This method differs from Helmholtz's, not only because it presupposed a very different mathematical and physical context, but also because it obscured Helmholtz's analogy between the experimental method and the theory of signs. We have already noticed that this analogy led Helmholtz to distance himself from causal realism. Helmholtz's definition of objectivity differs from Schlick's because it does not depend on the reference to a mind-independent reality, but on the lawlikeness of the ordering of appearance. Although Schlick agreed with Helmholtz on the limits of naïve realism, Schlick restricted objective knowledge to spatiotemporal coincidences, because he believed that causal realism holds true for quantitative knowledge.

Secondly, Schlick's views led him to emphasize the contrast between Helmholtz's theory of spatial perception and the Kantian theory of space. More precisely, Schlick referred to Kant's characterization of space (and time) as pure intuitions, namely, as mediating terms between general and empirical concepts. Schlick maintained that Helmholtz made such a mediating term superfluous by pointing out the empirical origin of spatial intuitions. However, Schlick's reading of Helmholtz entails that the concept of space under consideration is a psychological concept, which for Schlick has nothing to do with the physico-geometrical concept of space. "The latter is a non-qualitative, formal conceptual construction: the former, as something intuitively given, is in Helmholtz' words imbued with the qualities of the sensations, and as purely subjective as these are" (Schlick in Helmholtz 1921, p.167, note 20).

Friedman reconsiders the Kantian aspect of Helmholtz's theory for the following reason. The core idea of Helmholtz's geometrical papers relates to his previous studies in the physiology of vision, because he believed that the distinction between voluntary and external movement, and the capacity to reproduce external changes by moving our own body or the objects around us, lies at the foundation of geometrical knowledge. In particular, Helmholtz pointed out the empirical origin of the notion of a rigid body: solid bodies or even parts of our own body work as standards of measurement according to the observed fact that such bodies do not undergo any remarkable changes in shape and size during displacements. Even though Helmholtz's argument contradicts the aprioricity of geometry, it retains the structure of a transcendental argument for the applicability of mathematics in the definition of physical concepts: not only does the definition of rigid bodies depend on the principle of their free mobility, but the same principle presupposes a specific mathematical structure of physical space. According to Friedman, it is in this sense that Helmholtz (1878, p.124) identified such a structure with a form of intuition in Kant's sense: "The same regularities in our sensations, on the basis of which we acquire the ability to localize objects in space, also give rise to the representation of space itself" (Friedman 1997, p.33).

Following Friedman's interpretation, Ryckman (2005, pp.67–75) maintains that the connection with Kant provides us with a consistent reading of Helmholtz's definition of rigid bodies as opposed to Schlick's. Given Schlick's account of geometrical knowledge, such a definition can only be consistent if it tacitly presupposes a conventional definition of rigidity. In this regard, Schlick distanced himself from Helmholtz's empiricism and defended a conventionalist approach to the foundations of geometry. In order to highlight the contrast between Kantianism, empiricism, and conventionalism, which occupies us on several occasions in the rest of the book, the following section provides an introduction to Helmholtz's considerations about the Kantian theory of space.

1.4   Space, Time, and Motion

The most problematic subject in Helmholtz's relationship to Kant is the theory of space and time. I have already mentioned that the goal of Helmholtz's empiricism in the physiology of vision was to show how spatial intuitions can be derived from nonspatial sensations. This way of proceeding apparently called into doubt Kant's analysis of the notions of space and time as forms of appearance.

Kant defined such a form by abstracting from any particular content (i.e., sensation): "Since that within which the sensations alone can be ordered and placed in a certain form cannot itself be in turn sensation, the matter of all appearance is only given to us *a posteriori*, but its form must all lie ready for it in the mind *a priori*, and can therefore be considered separately from all sensation" (Kant 1787, p.34). This gives us the concept of a pure intuition, "which occurs *a priori*, even without an actual object of the senses or sensation, as a mere form of sensibility in the mind" (p.35). On the one hand, space and time as pure intuitions are distinguished from empirical ones. On the other hand, Kant distinguished these notions from general concepts because of their singularity: whereas general concepts – according to the syllogistic logic of Kant's time – are obtained by subsuming a variety of cases under one common element, space (and time) are thought of as essentially single and the manifold in them is obtained by dividing one single space. It followed that any spatial concept presupposes the intuition of space. Therefore, Kant maintained that "all geometrical principles, e.g., that in a triangle two sides together are always greater than the third, are never derived from general concepts of line and triangle, but rather are derived from intuition and indeed derived *a priori* with apodictic certainty" (p.41). But are there such notions as pure intuitions and, if so, how can they be known?

This question is essential to Kant's philosophical project, as his answer to this question offered a basis for his characterization of the judgments of mathematics as synthetic a priori. We have already noticed that these were among Kant's classical example of a priori cognition. More precisely, the a priori cognition that is the object of the transcendental inquiry is any cognition which extends the scope of our knowledge a priori. Therefore, Kant (1787, p.10) famously distinguished those judgments in which the predicate belongs to the subject as something that is (covertly) contained in the latter concepts, which he called analytic judgments or judgments of clarification, from those judgments in which the predicate lies entirely outside the subject, although it stands in connection with it. He called the latter kind of judgments synthetic or judgments of amplification. Kant's examples for analytic judgments are such laws of general logic as the principle of non-contradiction and such propositions as "all bodies are extended." By contrast, all the propositions of mathematics, including numerical formulas and the principles of geometry, are synthetic according to Kant.

Kant's theory of space and time gives us a more specific reason why mathematical judgments amplify our cognition: they do so in virtue of a connection of general concepts and a priori intuitions. The above argument for the synthetic a priori character of the principles of geometry enables Kant to argue for the view that the same principles apply with apodictic certainty to the manifold of experience, insofar as this is determined by the form of intuition.

One of the aspects of Kant's foundation of mathematics which appeared to be problematic in the nineteenth century was that he excluded the concept of motion, which entails empirical factors, from his analysis of space and time. By contrast, motion plays a fundamental role in Helmholtz's considerations: the formation of the concept of space presupposes associations that can be experienced only by moving beings. Not only are visual sensations always given in conjunction with tactile ones, but also the voluntary movement of our own body is necessary for changes of place to be noticed and distinguished from other kinds of changes. The same distinction lies at the basis of Helmholtz's definition of geometrical notions according to the free mobility of rigid bodies.

Kant also admitted a kind of motion that corresponds to the construction of geometrical objects in pure intuition. In a note added by Kant to the second edition of the *Critique of Pure Reason*, he distinguished the motion of objects in space from the description of a space: the former presupposes empirical factors, whereas the latter requires an act of what Kant calls "productive" imagination. He distinguished the productive imagination from the empirical, reproductive one because of its generating power (Kant 1787, p.155, and note). He identified the second kind of motion with that of a mathematical point in the *Metaphysical Foundations of Natural Science*. The description of a space provides us with the a priori part of the theory of motion (Kant 1786, p.489). In other words, the synthesis of the productive imagination, unlike association grounded in the reproductive imagination, is supposed to contribute to the possibility of objective knowledge.

Helmholtz's explanation of how objective knowledge can be obtained differs from Kant's because of its psychological character. The point of disagreement with Kant was made explicit by Helmholtz in Helmholtz (1870). It specifically concerned the foundations of geometry: space, as analyzed by Kant in the Transcendental Aesthetic, is supposed to provide foundations a priori, whereas for Helmholtz geometry is grounded in empirical facts (e.g., in the observation that parts of our own body and solid bodies around us can be displaced without remarkable changes in shape and size). The argument against Kant is that geometrical axioms, owing to their empirical origin, are not necessary. Not only is the logical necessity of geometrical axioms called into question by the possibility of a consistent development of non-Euclidean geometry, but Helmholtz's foundation of geometry provides us with a physical interpretation of this geometry. This clearly calls into question the role of intuition as a source of apodictic certainty when it comes to choosing among different geometries for the representation of physical space.

Helmholtz's objections to Kant are discussed in Chap. 3. For now, it is noteworthy that Helmholtz raised the question whether necessity is an essential characteristic of a priori knowledge. Kant clearly ruled out logical necessity by considering both sensible and intellectual conditions of knowledge. Arguably, non-Euclidean geometry is consistent with Kant's view that pure intuition alone can provide mathematical concepts with objective reality. Kant's argument was directed against Leibniz's and Wolff's attempts to infer the existence of mathematical objects from the lack of contradiction in mathematical concepts. Although Kant could not consider the possibility of non-Euclidean geometry, his approach suggests that, while infinitely many geometries can be considered as logical possibilities, only Euclidean geometry is a real possibility according to the form of outer intuition. Kant acknowledged, for example, the logical possibility of such objects as a two-sided plane figure, since there is no contradiction in the concept of such a figure. "The impossibility arises not from the concept itself, but in connection with its construction in space, that is, from the conditions of space and of its determination" (Kant 1787, p.268).

In drawing attention to the quote above, Friedman considered decisive for the synthetic character of mathematics that "there are logical possibilities, such as the two-sided plane figure, that are nonetheless mathematically impossible: their impossibility consists precisely in their failure to conform to the conditions of pure intuition" (Friedman 1992, p.100). Friedman (1992, Ch.1) argued that the Kantian theory of space is contradicted by the possibility of Euclidean models of non-Euclidean geometries only under the supposition that pure intuition provides us with a model for Euclidean geometry. However, this interpretation overlooks the fact that Kant's model or realization of the idea of space depends on the transcendental proof that the fundamental concepts of the understanding necessarily apply to the manifold of intuition. In order to clarify the role of pure intuition in Kant's definition of mathematics as synthetic a priori knowledge, Friedman rather adopted what he characterized as a "logical" approach. In this view, the focus lies not so much on the question concerning the origin and the justification of geometrical axioms, as on the mathematical reasoning. The possibility of an indefinite iteration of intuitive constructions overcame the difficulties of the syllogistic logic of Kant's time, which was monadic and did not suffice for the representation of an infinite object. Friedman gave a series of examples of geometrical and arithmetical reasoning in which what Kant called "construction in pure intuition" corresponds to what we would represent today as an existential instantiation. I focus on the example of the number series in Chap. 4. For now, it is worth noting that such an approach provides a consistent reading of Kant's claim that all mathematical judgments (including numerical formulas) are synthetic. The construction of infinite domains both in geometry and in arithmetic depends on the successive character of the productive imagination. Therefore, Friedman maintains that the description of a space in Kant's sense presupposes a kinematical conception of geometry: although the objects of geometry are not themselves necessarily temporal, geometrical construction is, nonetheless, a temporal activity (Friedman 1992, p.119).

In a later article from 2000, Friedman considers Helmholtz's philosophy of geometry a plausible development of Kant's kinematical conception. Such a conception emerges from Helmholtz's foundation of geometry on the facts observed concerning our experiences with mobile rigid bodies. He used the same facts to identify the form of spatial intuition as the structure of a manifold of constant curvature. Not only did Helmholtz consider the free mobility of rigid bodies an empirical generalization, but the structure thus characterized includes both Euclidean and non-Euclidean geometries as special cases. Therefore, Helmholtz contrasted this picture of spatial intuition with the older view that spatial intuition is a simple and immediate psychological act and provides us with evident truths. Although Helmholtz seemed to attribute such a view to Kant or sometimes, more specifically, to the "Kantians of strict observance," Friedman's reading suggests that at least "the germ of Helmholtz's kinematical conception is already present in Kant himself" (Friedman 2000b, p.201). In this connection, Friedman reconsiders his original stance by admitting that such a conception would enable us to overcome the limits of the logical approach to the Kantian notion of intuition. Even though existential instantiation satisfies the demand of singularity, it can be objected that Kant's pure intuitions are characterized by immediacy as well. The competing view, which focuses on this second characteristic, can be traced back to Parsons (1969), who also more recently maintained that immediacy for Kant is "direct, phenomenological presence to the mind, as in perception" (Parsons 1992, p.66). Therefore, Friedman called this approach "phenomenological." He presented Helmholtz as a suitable candidate for mediating between these approaches because, on the one hand, geometry for Helmholtz is grounded in the imaginary changes of perspective of a perceiving subject based on actual experiences. On the other hand, Helmholtz ruled out the view of geometrical axioms as evident truths: in Helmholtz's view, the meaning of geometrical axioms depends on operations with rigid bodies.

According to this reading, Helmholtz's main disagreement with Kant depends on how the form of intuition is specified in different ways. Helmholtz foreshadows a relativized conception of the a` priori, insofar as the line between the a priori and the empirical part of the theory of motion is not abolished, but drawn somewhere else:

some of the assumptions earlier considered a priori (i.e., the specific geometrical properties of space) are now ascribed to the empirical part of physical theory and may be subject to revision. However, Helmholtz also acknowledged the possibility of generalizing the a priori assumptions concerning the form of spatial intuition, so as to include all the possible combinations of sense impressions to be found in experiments. Ryckman expresses the same idea by saying that: "Helmholtz argued against the Kantian philosophy of geometry while retaining an inherently Kantian theory of space" (Ryckman 2005, pp.73–74). In Ryckman's view, appreciating this aspect of Helmholtz's argument would enable us to clearly distinguish Helmholtz's definition of rigid bodies from a stipulation by identifying it with a constraint imposed by the a priori form of spatiality itself. Ryckman deems this view inherently Kantian, because the a priori form of spatial intuition provides us with a condition of the possibility of geometrical measurement.

Although I largely agree with this reading, it seems to me that it does not do justice to Helmholtz's emphasis on the empiricist aspect of his approach. Helmholtz argued against the Kantian philosophy of geometry because he did not admit a meaningful use of "pure intuition" as distinguished from psychological intuition. For the same reason, Helmholtz's kinematics has its roots in his psychology of spatial perception, which is at odds with Kant's description of a space as a successive synthesis of the productive imagination. DiSalle (2006), nevertheless, maintains that Helmholtz's account of spatial intuition provides us with a philosophical analysis of the assumptions upon which Kant's "productive imagination" implicitly relies. In order to support this interpretation, DiSalle points out that Helmholtz's "facts" underlying geometry are better understood as rules governing idealized operations with solid bodies. DiSalle refers to Helmholtz's characterization of spatial changes as those that we can bring about by our own willful action and combine arbitrarily. By using the later group-theoretical analysis of space by Henri Poincaré, the same characteristics can be identified with the features of spatial displacements that enable us to treat them as forming a group, and rigidity can be defined as one of the properties of solid bodies left unchanged by the Euclidean group. However, it seems to me that DiSalle can hardly avoid Poincaré's conclusion that such a definition of rigid bodies – contrary to Helmholtz's opinion – would be conventional. According to DiSalle, "Poincaré's group-theoretical account of space (Poincaré 1902, pp.76–91) is only a psychologically more detailed, and mathematically more precise, articulation of Helmholtz's brief analysis" (DiSalle 2006, pp.77–78).[6]

---

[6]Cf. Lenoir (2006). Lenoir reconsiders the empirical aspect of Helmholtz's theory of spatial perception by connecting it with Helmholtz's works on such other empirical manifolds as tone sensations and the color system. Similarly, Kant's definition of space as the form of intuition can be understood as the abstract space of n-dimensional manifolds emptied of all content (Lenoir 2006, p.205). This reading certainly reflects the importance of Helmholtz's psychological standpoint for his analysis of the concept of space. However, it seems to me that Lenoir's reading would rather lead to Schlick's conclusion that Helmholtz actually ruled out the Kantian theory of space by reducing the form of spatial intuition to the purely qualitative and subjective factors of spatial perception.

Another problem of this reading is that Helmholtz's approach led him to a new formulation of the problems concerning measurement. In this regard, there is a deeper disagreement between Helmholtz and Kant. The disagreement can be traced back to a manuscript Helmholtz probably wrote before the publication of his essay *On the Conservation of Force*. The manuscript, which has been made available by Königsberger (1902–1903, pp.126–138), includes a characterization of space and time as general, natural concepts. The most peculiar characteristic Helmholtz attributed to these notions is their being divisible into homogeneous parts, namely, into parts that can be proved to be equal in some respect. Measurement specifically requires divisibility into equal parts. Equality here entails arithmetical equality of numerical values to be assigned to a set of parts, once a single part has been chosen as a unit. At times, it seems that Helmholtz bore in mind Kant's conception of motion as construction in pure intuition. Helmholtz wrote:

> Motion must belong to matter quite aside from its special forces; but then the only remaining characteristic of a determinate piece of matter is the space in which it is enclosed; but since it is robbed of this characteristic as well by motion, we can only speak of its identity if we can intuit the transition from the one space to the other, i.e. motion must be continuous in space. (Königsberger 1902–1903, p.135; Eng. trans. in Hyder 2006, p.35)

It seems that the condition for establishing the equality of spatial magnitudes here is the same as that assumed by Kant in the *Metaphysical Foundations of Natural Science*. Kant wrote: "Complete similarity and equality, insofar as it can be cognized only in intuition, is *congruence*" (Kant 1786, p.493). Note, however, that cognition in intuition does not suffice for Helmholtz's analysis of measurement. This presupposes a metrical notion of equality. Helmholtz's point is that if physical objects are to be related to one another, they must be considered as quantities. Therefore, physics presupposes arithmetic, which is the science of quantitative relations (Königsberger 1902–1903, p.128).

In this regard, Helmholtz's analysis is completely different from Kant's. Consider the following quote from the *Critique of Pure Reason*:

> On this successive synthesis of the productive imagination, in the generation of shapes, is grounded the mathematics of extension (geometry) with its axioms, which express the conditions of sensible intuition *a priori*, under which alone the schema of a pure concept of outer appearance can come about; e.g., between two points only one straight line is possible; two straight lines do not enclose a space, etc. These are the axioms that properly concern only magnitudes (*quanta*) as such.
>
> But concerning magnitude (*quantitas*), i.e., the answer to the question "How big is something?", although various of these propositions are synthetic and immediately certain (*indemonstrabilia*), there are nevertheless no axioms in the proper sense. (Kant 1787, pp.204–205)

Kant maintained that geometry and arithmetic differ both in their methods and in their objects. Therefore, the axioms that for Kant are grounded in the synthesis of the productive imagination concern only magnitudes in general. However, Kant's cognition in intuition, unlike Helmholtz's, does not presuppose the specification of the magnitude of a quantity. This is the issue of arithmetic. The answer to the question how big something is requires not so much construction in pure intuition, as calculation. The synthetic judgments of arithmetic (i.e., numerical formulas) differ from geometrical axioms because they are infinite in number and there is only one way to accomplish the corresponding calculation. By contrast, one and the same geometric construction can be realized in infinitely many ways in principle – for example, by varying the length of the lines and the size of the angles in a given figure while leaving some given proportions unvaried.[7]

This point of disagreement is reflected in Helmholtz's conception of space: spatiality does not follow from the general concept of space as the property of those objects that are in space. In fact, spatial determinations presuppose arithmetic and analytic geometry. In 1870, Helmholtz maintained that analytic geometry provided a general standpoint for a classification of hypotheses concerning space. The aprioricity of the axioms of (Euclidean) geometry is ruled out by the possibility of obtaining a more general system of hypotheses by denying supposedly necessary constraints in the form of outer intuition.

Although I believe that Helmholtz's account of spatial intuition can be made compatible with a relativized conception of the a priori, my suggestion is to reconsider the importance of the philosophical debate about the foundations of geometry for the actual development of such a conception. Helmholtz himself did not seem to provide a conclusive answer to the question about the status of a priori knowledge after theory change. In 1870, he argued against the a priori origin of geometrical axioms because the possibility of formulating different hypotheses contradicts the kind of necessity Kant attributes to a priori knowledge. This is necessity that should result from the forms of intuition in conjunction with the concepts of the understanding. According to Kant, the axioms of (Euclidean) geometry express the conditions under which any outer appearance can be measured. Nevertheless, in 1878, Helmholtz admitted that Kant's conception of space as the form of outer intuition can be generalized so as to include all possible hypotheses. Helmholtz did not change his views about the empirical origin of geometrical knowledge. What he arguably took from Kant is the conviction that the notions of space and time, as forms of intuition, provide us with foundations of mathematics and of the mathematical science of nature. However, he only gradually accepted the variability of the form of outer intuition as a consistent development of the Kantian theory of space. In fact, this was the core of his objections to Kant in 1870. Helmholtz's considerations in 1878 in this regard were also problematic, because he did not discuss the consequences of such a development for the Kantian theory of the a priori. It is quite revealing that Helmholtz did not hesitate to endorse the Kantian theory of the forms

---

[7] On this point of disagreement between Kant and Helmholtz, see Darrigol (2003, pp.548–549) and (Hyder 2006, pp.34–36). Hyder emphasizes that Helmholtz's commitment to physics enables him to see an aspect of measurement overlooked by Kant, namely, the fact that if two points are to determine a single spatial magnitude, this must be congruent with another magnitude determined by those points at a second point of time. The problem with change in place is that this calls into question the identity of the system after the motion with the system before the motion. It does not suffice to appeal to intuitive continuity. The invariance of a system in physics can only be established once magnitudes have been assigned numerical values and compared according to the laws of arithmetic.

of intuition in the case of time. In that case, only one possibility is given, namely, unidirectional, linear time.

As we will see in the next chapter, the philosophical discussion of Kant's Transcendental Aesthetic began long before Helmholtz's considerations, with Herbart's objections to Kant and within the so-called Trendelenburg-Fischer controversy. The neo-Kantians, especially Hermann Cohen, focused, first of all, on Kant's theory of the a priori. These discussions prepared the ground for a very interesting way to deal with the problems raised by Helmholtz. Not only did Cohen admit that there are infinitely many geometrical hypotheses, but his conception of geometrical axioms was perfectly aligned with the relativized conception of the a priori that followed from his interpretation of Kant's transcendental philosophy.

I argue that the philosophical roots of this debate were essential to the development of such a conception and for its extension to the principles of geometry. This will require us to focus on Cassirer. Cassirer's reception of Helmholtz foreshadows the reading discussed above in many ways. At the same time, I think that Cassirer gave more specific reasons for adopting later classifications of geometries by using group theory to express Helmholtz's ideas. Firstly, Cassirer clearly distanced himself from Helmholtz's psychological interpretation of Kant's form of spatial intuition, which can be hardly identified as such abstract concept as the concept of group. Secondly, similar to Cohen before him, Cassirer recognized that the development of analytic methods in nineteenth-century geometry made the assumption of a pure intuition superfluous by clarifying the conceptual nature of mathematical constructions. Therefore, Cassirer redefined the synthetic character of mathematics in terms of a conceptual synthesis able to generate univocally determined objects. This corresponds to the idea that geometrical properties can be defined as relative invariants of a transformation group. Although Cassirer's account of mathematical reasoning foreshadows a logical approach to Kant's notion of intuition, I think that it clearly differs from both logical and phenomenological approaches, because of Cassirer's broader understanding of mathematical method as a paradigm of the symbolic and conceptual reasoning which is required for the definition of physical objects. Cassirer's approach enabled him to make it clear that whereas there can be no agreement between neo-Kantianism and empiricism regarding the origin of mathematical concepts, there are important points of agreement with such empiricists as Helmholtz in the approach to measurement. The use of the abstract concept of group for the representation of motions is justified only insofar as this reflects a double direction of the inquiry into the foundations of geometry: from the mathematical structures to their specifications and from the problems concerning measurements to the development of conceptual tools for their solution. Furthermore, neo-Kantianism and Helmholtz's empiricism agree regarding the view that, owing to the complementarity of these two directions, any system of the conditions of experience must be left open to further generalizations in the course of the history of the sciences.

References

Cassirer, Ernst. 1907. Kant und die moderne Mathematik. *Kant-Studien* 12: 1–49.
Cohen, Hermann. 1885. *Kants Theorie der Erfahrung*, 2nd ed. Berlin: Dümmler.
Darrigol, Olivier. 2003. Number and measure: Hermann von Helmholtz at the crossroads of mathematics, physics, and psychology. *Studies in History and Philosophy of Science* 34: 515–573.
DiSalle, Robert. 2006. Kant, Helmholtz, and the meaning of empiricism. In *The Kantian legacy in nineteenth-century science*, ed. Michael Friedman and Alfred Nordmann, 123–139. Cambridge, MA: The MIT Press.
Friedman, Michael. 1992. *Kant and the exact sciences*. Cambridge, MA: Harvard University Press.
Friedman, Michael. 1997. Helmholtz's *Zeichentheorie* and Schlick's *Allgemeine Erkenntnislehre*: Early logical empiricism and its nineteenth-century background. *Philosophical Topics* 25: 19–50.
Friedman, Michael. 2000a. *A parting of the ways: Carnap, Cassirer, and Heidegger*. Chicago: Open Court.
Friedman, Michael. 2000b. Geometry, construction and intuition in Kant and his successors. In *Between logic and intuition: Essays in honor of Charles Parsons*, ed. Gila Sher and Richard Tieszen, 186–218. Cambridge: Cambridge University Press.
Hatfield, Gary. 1990. *The natural and the normative: Theories of spatial perception from Kant to Helmholtz*. Cambridge, MA: The MIT Press.
Heimann, Peter M. 1974. Helmholtz and Kant: The metaphysical foundations of *Über die Erhaltung der Kraft*. *Studies in History and Philosophy of Science* 5: 205–238.
Helmholtz, Hermann von. 1847. *Über die Erhaltung der Kraft*. Berlin: Reimer. Reprinted with additional notes in Helmholtz (1882):12–75. English edition in Helmholtz (1971): 3–55.
Helmholtz, Hermann von. 1855. Über das Sehen des Menschen. In Helmholtz (1903), 1:85–118.
Helmholtz, Hermann von. 1867. *Handbuch der physiologischen Optik*, vol. 3. Leipzig: Voss.
Helmholtz, Hermann von. 1870. Über den Ursprung und die Bedeutung der geometrischen Axiome. In Helmholtz (1903), 2:1–31.
Helmholtz, Hermann von. 1878. Die Tatsachen in der Wahrnehmung. In Helmholtz (1921):109–152.
Helmholtz, Hermann von. 1882. *Wissenschaftliche Abhandlungen*. Vol. 1. Leipzig: Barth.
Helmholtz, Hermann von. 1903. *Vorträge und Reden*. 2 vols. 5th ed. Braunschweig: Vieweg.
Helmholtz, Hermann von. 1921. *Schriften zur Erkenntnistheorie*, ed. Paul Hertz and Moritz Schlick. Berlin: Springer. English edition: Helmholtz, Hermann von. 1977. *Epistemological writings*. (trans: Lowe, Malcom F., ed. Robert S. Cohen and Yehuda Elkana). Dordrecht: Reidel.
Helmholtz, Hermann von. 1971. *Selected writings*, ed. Russell Kahl. Middletown: Wesleyan University Press.
Hilbert, David. 1899. Grundlagen der Geometrie. In *Festschrift zur Feier der Enthüllung des Gauss-Weber-Denkmals in Göttingen*, 1–92. Leipzig: Teubner.
Hyder, David. 2006. Kant, Helmholtz and the determinacy of physical theory. In *Interactions: Mathematics, physics and philosophy, 1860–1930*, ed. Vincent F. Hendricks, Klaus Frovin Jørgensen, Jesper Lützen, and Stig Andur Pedersen, 1–44. Dordrecht: Springer.
Hyder, David. 2009. *The determinate world: Kant and Helmholtz on the physical meaning of geometry*. Berlin: De Gruyter.
Kant, Immanuel. 1786. *Metaphysische Anfangsgründe der Naturwissenschaft*. Riga: Hartknoch. Repr. in *Akademie-Ausgabe*. Berlin: Reimer, 4: 465–565. English edition: Kant, Immanuel. 2004. *Metaphysical Foundations of Natural Science* (trans: Friedman, Michael). Cambridge: Cambridge University Press.
Kant, Immanuel. 1787. *Critik der reinen Vernunft*. 2nd ed. Riga: Hartknoch. Repr. in *Akademie-Ausgabe*. Berlin: Reimer, 3. English edition: Kant, Immanuel. 1998. *Critique of Pure Reason* (trans: Guyer, Paul and Wood, Allen W.). Cambridge: Cambridge University Press.
Königsberger, Leo. 1902–1903. *Hermann von Helmholtz*. 3 Vols. Braunschweig: Vieweg.

Lenoir, Timothy. 2006. Operationalizing Kant: Manifolds, models, and mathematics in Helmholtz's theories of perception. In *The Kantian legacy in nineteenth-century science*, ed. Michael Friedman and Alfred Nordmann, 141–210. Cambridge, MA: The MIT Press.

Parsons, Charles. 1969. Kant's philosophy of arithmetic. Repr. in Parsons, Charles. 1983. *Mathematics in philosophy: Selected essays*, 110–149. Ithaca: Cornell University Press.

Parsons, Charles. 1992. The transcendental aesthetic. In *The Cambridge companion to Kant*, ed. Paul Guyer, 62–100. Cambridge: Cambridge University Press.

Poincaré, Henri. 1902. *La science et l'hypothèse*. Paris: Flammarion.

Riehl, Alois. 1904. Helmholtz in seinem Verhältnis zu Kant. *Kant-Studien* 9: 260–285.

Riehl, Alois. 1922. *Führende Denker und Forscher*. Leipzig: Quelle & Meyer.

Ryckman, Thomas A. 2005. *The reign of relativity: Philosophy in physics 1915–1925*. New York: Oxford University Press.

Schlick, Moritz. 1917. *Raum und Zeit in der gegenwärtigen Physik: Zur Einführung in das Verständnis der allgemeinen Relativitätstheorie*. Berlin: Springer.

Schlick, Moritz. 1918. *Allgemeine Erkenntnislehre*. Berlin: Springer.

Schiemann, Gregor. 2009. *Hermann von Helmholtz's mechanism: The loss of certainty. A study on the transition from classical to modern philosophy of nature*. Trans. Cynthia Klohr. Dordrecht: Springer.

Waitz, Theodor. 1849. *Lehrbuch der Psychologie als Naturwissenschaft*. Braunschweig: Vieweg.

Chapter 2
The Discussion of Kant's Transcendental Aesthetic

2.1 Introduction

Helmholtz's objections to Kant concerning the origin and meaning of geometrical axioms were influential in the later philosophical debate on the relationship between space and geometry. However, the discussion of Kant's Transcendental Aesthetic in neo-Kantianism was rooted in earlier objections formulated by such philosophers as Kant's successor at the University of Königsberg, Johann Friedrich Herbart, and the neo-Aristotelian Adolf Friedrich Trendelenburg. Herbart was the first to call into question the necessity of the representation of space assumed by Kant. Kant's argument was as follows:

> Space is a necessary representation, *a priori*, which is the ground of all outer intuitions. One can never represent that there is no space, although one can very well think that there are no objects to be encountered in it. It is therefore to be regarded as the condition of the possibility of appearances, not as a determination dependent on them, and is an *a priori* representation that necessarily grounds outer appearances. (Kant 1787, pp.38–39)

Kant's assumption that space is a necessary representation follows from the impossibility of abstracting from the concept of space in the cognition of extended objects. Herbart's objection is that impossibility is meant thereby in a psychological sense. Therefore, Kant's claim does not suffice to infer the necessity of space as a condition of the possibility of appearances. His psychological argument rather suggests that the representation of space is empirically necessary, namely, necessary under the given circumstances (Herbart 1825/1850, p.225).

Trendelenburg's objection to Kant is that the representation of space is not merely subjective, because it presupposes movement. Both Herbart's and Trendelenburg's criticisms of Kant motivated the founder of the Marburg School of neo-Kantianism, Hermann Cohen, to reconsider the role of the Transcendental Aesthetic in the *Critique of Pure Reason*. He concluded that the apriority of some knowledge does not depend on our mental faculties, but rather on its role in the constitution of the objects of experience. Neither intuitive nor intellectual conditions

© Springer International Publishing Switzerland 2016
F. Biagioli, *Space, Number, and Geometry from Helmholtz to Cassirer*,
Archimedes 46, DOI 10.1007/978-3-319-31779-3_2

of experience can be identified as some kind of psychic processes. Since experience, in Kant's sense, entails objective knowledge, the conditions of experience ought to be studied in connection with the history of science.

Cohen's major work on Kant, *Kant's Theory of Experience* (1871a), was the background for later discussions of geometrical empiricism in neo-Kantianism. The different strategies proposed in neo-Kantianism for defending the aprioricity of geometrical axioms are discussed in Chap. 3. The present chapter deals with the philosophical discussion of Kant's Transcendental Aesthetic in which Cohen was involved. The following section provides a general introduction to Kant's characterization of space and time as a priori intuitions, with a special focus on those aspects that were called into question in the debate under consideration. Section 2.3 deals with the Trendelenburg-Fischer controversy about the status of space and time. The development of Cohen's theory of the a priori is analyzed in Sect. 2.4. To conclude, Sect. 2.5 is devoted to the genesis of Kant's Transcendental Aesthetic according to Cohen's student, Ernst Cassirer. In particular, I discuss Cassirer's considerations in support of Cohen's reading of Kant and sketch the connection with Cassirer's interpretation of the logic of knowledge in the history of science.

2.2   Preliminary Remarks on Kant's Metaphysical Exposition of the Concept of Space

Kant's characterization of space and time as forms of intuition goes back to his Inaugural Dissertation of 1770 on "The Form and Principles of the Sensible and Intelligible World." Kant introduced the distinction between the form and the matter of appearance in Kant (1770, Sect. 4) and formulated it in the same terms in the Transcendental Aesthetic:

> I call that in the appearance which corresponds to sensation its matter, but that which allows the manifold of appearance to be intuited as ordered in certain relations I call the form of appearance. Since that within which the sensations can alone be ordered and placed in a certain form cannot itself be in turn sensation, the matter of all appearance is only given to us *a posteriori*, but its form must all lie ready for it in the mind *a priori*, and can therefore be considered separately from all sensation. (Kant 1787, p.34)

This is the section of the *Critique of Pure Reason* which contains Kant's theory of sensibility. Therefore, it was particularly important to him to clarify the notion of intuition by distinguishing it from sensation. Kant deemed sensibility to be the capacity to acquire representations through the way in which we are affected by objects. Understanding is the capacity for producing such representations. Kant made it clear that sensibility alone affords us intuitions, intuition being the only manner in which our knowledge immediately relates to the objects. He called empirical those intuitions that are caused by the objects or related to them through sensation. Kant distinguished these intuitions from representations in which nothing is encountered that belongs to sensation. The notions of space and time analyzed in the Transcendental Aesthetic correspond to the forms of outer and inner intuition, respectively. Therefore, Kant called his exposition of these notions metaphysical. In other words, the forms of intuition ought to be considered by themselves, independently of the matter of the appearances and of empirical motions. The same forms in connection with the theory of the basic concepts of the understanding provide the premises for Kant's proof of the possibility of a priori cognition in the Transcendental Logic.

I have already mentioned that Kant's distinction between intuition and sensation gives us the concept of a pure intuition, "which occurs *a priori*, even without an actual object of the senses or sensation, as a mere form of sensibility in the mind" (Kant 1787, p.35). In addition, the metaphysical exposition of the notion of space includes the following claims:

1. Outer experience presupposes the representation of space.
2. Space is a necessary representation.
3. Space is essentially single, and the manifold in it depends merely on limitations.
4. Space is represented as an infinite given magnitude.

Corresponding claims about time are found in Kant (1787, pp.46–48). The first two claims about space show the characteristics of a priori notions, namely, universality and necessity independent of actual experiences. Kant's first remark is that the representation of space cannot be induced from outer experiences, because spatially ordered experiences already presuppose this representation. His argument for 2 is that one cannot perform abstraction from the representation of space in the cognition of extended objects. Contrary to Herbart's interpretation, performing abstraction here should not be understood in an actual or psychological sense. Space and time are not experienced objects: they are supposed to provide us with ideal conditions for the cognition of any such object.

As noticed in Chap. 1, Kant maintained that for 3, all concepts of space, along with geometrical principles, are derived from a priori intuition. Claims 3 and 4 distinguish intuitions from the concepts of the understanding. The fact that infinite divisibility of space always produces parts of space follows from a construction that can only be indefinitely repeated in pure intuition. Owing to the recursive character of such an operation, a definition in the sense of the formal logic of Kant's time is excluded: parts of space cannot be subsumed under a general concept (Kant 1787, pp.38–40).

Kant's claims have been a source of lively discussion both in themselves and in connection with other parts of the *Critique*. Kant himself emphasized such a connection. The metaphysical exposition of the concepts of space and time is followed by a transcendental exposition of the same concepts. This section includes Kant's considerations on the place of pure intuitions in his philosophical inquiry. Here, Kant made it clear that space and time do not suffice to provide us with knowledge. The sensible conditions of knowledge have been analyzed separately in order to describe knowledge as the application of the concepts of the understanding analyzed later, in the Transcendental Logic, to the manifold of pure intuition. The goal of the transcendental inquiry is to prove that such an application is necessary for knowledge. One of the premises for the transcendental proof is the ideal nature of space and time. In order to play a mediating role between thought and experience, the manifold of pure intuition must share intuitive nature with sensibility, on the one hand, and ideal nature with the products of the understanding, on the other.

Another section of the *Critique* that ought to be taken into consideration is the part of the Transcendental Dialectic that is devoted to the antinomies of pure reason. In particular, the first cosmological antinomy includes Kant's considerations on the infinity of space. This section of the *Critique* indirectly confirms that space has an ideal nature. Suppose that space is an object. It then should have a magnitude. Under this premise, the question whether space is an infinite magnitude cannot be decided on rational grounds. This argument sheds some light on Kant's previous characterization of space as an infinite given magnitude. It should be clear now that the magnitude considered in 4 cannot be of the same kind as those of physical objects: that would contradict the ideal nature of space. "Infinitely given" can only refer to the possibility of indefinitely repeating ideal operations (e.g., division). Immediateness here indicates that constructions in pure intuition can be accomplished in principle.

The discussion of Kant's claims in the Transcendental Aesthetic emerged from a more general discussion of transcendental idealism about the forms of intuition: namely, the view that these forms are not perceivable themselves, because they provide us with general schemas for the ordering of any perceivable phenomena. This view was Kant's premise for the claim that a priori knowledge is independent of experience, because the former is a condition of the possibility of experience in general.

In the following, I focus on the related question whether the characterization of space and time as pure intuitions is essential to the view that mathematics provides us with one of the clearest examples of synthetic a priori knowledge. Kant's argument for the synthetic a priori status of mathematics depends on the singularity and immediacy of the representations of space and time, namely, on the defining characteristics of pure intuitions as opposed to concepts. Most commentators and Kant scholars agree that these characteristics do not compel us to attribute some mathematical properties to the forms of intuition, because in the claims above, Kant deals rather with the more fundamental form of mathematical reasoning, which he calls "construction in pure intuition." This plays a role in all definitions and proofs that presuppose the indefinite repetition of some operation. The controversial aspect of Kant's argument regards the intuitive character of this way of proceeding. Cohen was one of the first to point out that Kant's contraposition between concepts and intuitions depended partially on his reliance on the syllogistic logic of his times. Given the limited expressive power of the general concepts of syllogistic logic, it seemed that the formation of mathematical concepts deserved a different explanation in terms of the transcendental logic. Although Cohen agreed with Kant on this point, he argued for a broader understanding of intellectual knowledge, including arithmetical operations and the idealized constructions of geometry, and called into question the intuitive character of the recursive reasoning in mathematics. Therefore,

Cohen came to the conclusion that the assumption of pure intuitions was irrelevant or even an obstacle to the foundation of mathematics. However, it was Cassirer especially who looked at nineteenth-century mathematics and mathematical logic as sources of ideas for substituting Kant's construction in pure intuition with the formulation of conceptual and symbolic systems of relations. His goal was to defend the view that mathematics is synthetic a priori knowledge against such logicists as Louis Couturat and Bertrand Russell, who denied the theory of pure intuitions because of the analyticity of mathematics. As we will see in Sect. 2.5 in more detail, Cassirer used Cohen's broader understanding of intellectual or conceptual synthesis to reformulate the argument that mathematics is synthetic in terms of the emerging structuralism of nineteenth-century mathematics. It followed that Couturat's, Russell's, and others' insights into the relational character of logic, far from leading to the conclusion that mathematics must be analytic, confirmed the view that mathematics is synthetic in Cassirer's sense.

More recently, a similar approach was adopted by Jaakko Hintikka, who referred to Cassirer for the shift in emphasis from sensation to constructive thinking in Kant's philosophy of mathematics (Hintikka 1974, p.134, note 24). Hintikka's point is that a Kantian perspective on the mathematical method can be made compatible with subsequent developments in mathematics and in logic – which have shown that all mathematical arguments can, in principle, be represented in forms of strictly logical reasoning – if one considers that the distinctive character of Kant's intuitions lies not so much in some immediate, nonlogical evidence, as in singularity. Hintikka observed that construction in pure intuition provides a justification for the use of existential assumptions in mathematics. Such constructions appeared to be the only means available to justify existential assumptions regarding an infinite domain (e.g., the series of natural numbers and the points on a line). If it is so, Hintikka goes on,

> [...] then Kant's problem of the justification of constructions in mathematics is not made obsolete by the formalization of geometry and other branches of mathematics. The distinction between intuitive and nonintuitive methods of argument then reappears in the formalization of mathematical reasoning as a distinction between two different means of logical proof. (Hintikka 1974, p.176)

Hintikka's suggestion for a reformulation of this distinction in logical terms is to identify synthetic arguments in mathematics as the use of rules of quantification (e.g., of existential instantiation) for the introduction of new individuals and analytic arguments as those arguments that can be expressed in monadic logic (see Hintikka 1973, pp.174–199).

Charles Parsons's argument for a phenomenological approach to Kant's pure intuitions is that Hintikka, among others, focuses exclusively on the singularity of space and time. However, Kant also states the immediacy of these representations in the metaphysical exposition of space and time, and the immediacy of space seems to provide a fundamental premise for Claim 4 (see Parsons 1992, p.70). The main disagreement regards not so much the interpretation of Kant's text, as the question whether pure intuition is essential to the view that mathematics is synthetic: whereas the phenomenological approach entails a defense of the Transcendental Aesthetic in this respect, the former line of argument is not committed to the assumption of pure intuitions or to Claim 4.

Michael Friedman answers the same question affirmatively because of the temporal aspect of the constructions considered by Kant: the form of successive progression is common to all iterative procedures. What characterizes geometric constructions in Euclid's geometry is that all geometric objects can be obtained from some given operations (i.e., (i) drawing a line segment connecting any two given points, (ii) extending a line segment by any given line segment, and (iii) drawing a circle with any given point as center and any given line segment as radius) in a finite number of steps. By contrast, no such initial operations are given in algebra and arithmetic. What distinguishes this way of proceeding from existential instantiation is that the instances desired have to be actually constructed, and this requirement seems to be a defining characteristic of mathematical objects in Kant's sense (see Friedman 1992, pp.118–119).[1]

However, it seems to me that a literal interpretation of "successive" progression contradicts another aspect which seems to be no less essential to Kant's characterization of mathematical method. Kant describes the emergence of the mathematical method in the Preface to the second edition of the *Critique* as follows:

> A new light broke upon the first person who demonstrated the isosceles triangle (whether he was called "Thales" or had some other name). For he found that what he had to do was not to trace what he saw in this figure, or even trace its mere concept, and read off, as it were, from the properties of the figure; but rather that he had to produce the latter from what he himself thought into the object and presented (through construction) according to a priori concepts, and that in order to know something securely a priori he had to ascribe to the thing nothing except what followed necessarily from what he himself had put into it in accordance with its concept. (Kant 1787, p.XII)

The decisive step in the introduction of mathematical method lies in the substitution of actual construction with construction according to a priori concepts. Kant made it clear that the temporal order in which some premises are thought does not affect the validity of the hypothetical inferences of mathematics and of a priori knowledge in general.[2]

Cohen clarified this aspect of the notion of the a priori in Kant's work by comparing the Dissertation of 1770 and the first edition of the *Critique of Pure Reason* with the second edition and with the *Prolegomena to Any Future Metaphysics That Will Be Able to Come Forward as a Science* (1783). The view of mathematical method

---

[1] Friedman contrasts his reconstruction with Hintikka's in Friedman (1992, p.65, note).

[2] The quote above is followed by other examples from natural science: "When Galileo rolled balls of a weight chosen by himself down an inclined plane, or when Torricelli made the air bear a weight that he had previously thought to be equal to that of a known column of water, or when in a later time Stahl changed metals into calx and then changed the latter back into metal by first removing something and then putting it back again, a light dawned on all those who study nature. They comprehended that reason has insight only into what it itself produces according to its own design; that it must take the lead with principles for its judgments according to constant laws and compel nature to answer its questions" (Kant 1787, pp.XII–XIII).

that emerged from Cohen's study enabled him and Cassirer to present their shift in emphasis from sensation to constructive thought as a consistent development of the Kantian philosophy of mathematics and, at the same time, to engage in the nineteenth-century debate about the foundations of geometry from a new perspective.

2.3 The Trendelenburg-Fischer Controversy

The neo-Aristotelian Friedrich Adolf Trendelenburg discussed Kant's metaphysical exposition of the concepts of space and time in *Logical Investigations* (1840) as follows. Kant's claims entail that space and time are subjective factors of knowledge. This is because space and time in the Transcendental Aesthetic are considered independently of motion. In particular, Kant, similar to Descartes before him, deemed space not so much a product of motion, as its condition. By contrast, Trendelenburg's view was that spatial notions are derived from movement.

In order to argue for the latter view, Trendelenburg pointed out two other possible conceptions of space and time, namely, as objective factors of knowledge and as both objective and subjective factors. Trendelenburg ruled out the purely objective character of space and time because there is, in fact, a kind of movement that only presupposes ideal operations with mathematical points (i.e., what Kant called "description of a space through productive imagination"). Trendelenburg argued for the subjective/objective character option, because he believed that such a movement corresponds, simultaneously, to a possible realization in physical space. Trendelenburg's objection to Kant was that Kant assumed the purely subjective character of space and time without considering the subjective/objective option. Therefore, none of the claims he made in the metaphysical exposition suffice to prove the subjectivity of space and time (see Trendelenburg 1840, pp.123–133). Claim 4, in particular, already presupposes motion. Infinity here depends on the possibility of indefinitely repeating some operation (p.132).

The Kantian theory of space and time was defended against Trendelenburg by the Hegelian Kuno Fischer in the second edition of *System of Logic and Metaphysics or the Science of Knowledge* (1865). Trendelenburg replied to Fischer in his *Historical Contributions to Philosophy* (1867). The controversy continued until the end of the 1860s, when Trendelenburg published an essay entitled *Kuno Fischer and His Kant*. Trendelenburg's essay was followed by Fischer's *Anti-Trendelenburg* (1870).[3]

Fischer maintained that Kant distanced himself from Descartes precisely because Kant ruled one of the options mentioned by Trendelenburg, namely, that space and time are purely objective. These concepts must be purely subjective for Kant if mathematics is to be possible. The synthetic a priori judgments of mathematics are

---

[3] For further references and discussion of the Trendelenburg-Fischer controversy, see Köhnke (1986, pp.257–268), Beiser (2014, pp.212–215).

grounded in pure intuition. Were space and time real objects, the judgments of mathematics would not be a priori (i.e., universal and necessary). Judgments about such objects would not be distinguished from empirical judgments, which are approximate and revisable.

For Fischer, the remaining, subjective/objective character option, which is the one defended by Trendelenburg, presupposes both the notion of a subjective space grounded in intuition and that of a real space outside us. The two kinds of space should then be correlated to each other by Trendelenburg's theory of movement. Fischer's objection is that this option is contradicted by the singularity of intuitive space stated by Kant in Claim 3. Owing to its singularity, space is a primitive concept: not only does it provide us with foundations of geometry, but the same concept is presupposed for the cognition of extended objects. To sum up, Fischer maintained that a priori concepts, including space and time, can be proved to be subjective or independent of experience, provided that they are determined a priori, as in Kant's metaphysical exposition. At the same time, a priori concepts provide us with conditions of knowledge. Therefore, Fischer rejected Trendelenburg's distinction between ideal constructions and reality, along with his correlation problem, which for Fischer is an unsolvable one (see Fischer 1865, pp.174–182).

2.4   Cohen's Theory of the A Priori

Hermann Cohen studied philosophy and psychology at the universities of Breslau and Berlin, where he attended classes taught by Trendelenburg. During his studies at the University of Berlin, Cohen collaborated, in particular, with Moritz Lazarus and Heymann Steinthal, who applied Herbart's psychological method in the social sciences. It was in the *Zeitschrift für Völkerpsychologie und Sprachwissenschaft*, founded by Lazarus and Steinthal in 1859, that Cohen published "On the Controversy between Trendelenburg and Kuno Fischer" (1871b). Cohen's intervention in the Trendelenburg-Fischer controversy was followed by the first edition of Cohen's major work on Kant, *Kant's Theory of Experience*, which appeared the same year. Both writings foreshadow Cohen's theory of the a priori, whose most detailed presentation is found in the second, substantially revised version of *Kant's Theory of Experience* (1885).

This section considers the development of Cohen's thought during that period, with a special focus on those aspects that were influential in the neo-Kantian discussions on the status of geometrical axioms and the relationship between space and geometry. In particular, I point out that Cohen was one of the first to relativize the notion of a priori. This suggests that the geometry of space cannot be determined independently of empirical science. Geometrical notions can be deemed a priori relative to scientific theories, insofar as geometrical knowledge is required for the interpretation of observations and experiments.

2.4.1 Cohen's Remarks on the Trendelenburg-Fischer Controversy

In his article "On the Controversy between Trendelenburg and Kuno Fischer," Cohen defended the Kantian theory of space and time against Trendeleburg's objections. At the same time, his interpretation of Kant differed considerably from Fischer's.[4] In particular, Cohen called into question Fischer's claim that the aprioricity of space and time should be proved a priori. Fischer overlooked the fact that Kant's standpoint in the *Critique of Pure Reason* differs from that of the Dissertation of 1770 precisely because critical philosophy rules out a priori knowledge by means of concepts alone. This is what distinguishes critical philosophy from dogmatism, which is the belief that knowledge by means of concepts alone is possible. Now, knowledge about a priori knowledge requires the kind of cognition Kant called transcendental and defined as "all cognition that is occupied not so much with objects but rather with our mode of cognition of objects insofar as this is to be possible *a priori*" (Kant 1787, p.25). The goal of the transcendental inquiry is to prove that the concepts of the understanding apply to the manifold of intuition. The domain of a possible experience in general thus delimited coincides then with the domain of a priori knowledge.

Fischer may have been confused, because, in 1770, Kant drew the distinction between concepts and intuitions in the same terms as in the Transcendental Aesthetic, namely, by making the claims mentioned above, especially 3 (see Kant 1770/1912, Sect. 15). Owing to the structure of the *Critique*, however, Cohen maintained that these claims do not suffice to prove that space provides us with a priori knowledge. The metaphysical characterization of such knowledge as necessary and universal must be completed by specifying which knowledge is supposed to be a priori. This requires both sensibility and understanding. Therefore, the metaphysical exposition of the concept of space is followed by a transcendental exposition of the same concept, which Kant defined as "the explanation of a concept as a principle from which insight into the possibility of other synthetic *a priori* cognitions can be gained" (Kant 1787, p.40). However, Kant only introduced geometrical knowledge later in the Transcendental Logic. Furthermore, the first cosmological antinomy confirms indirectly that the intuitive space analyzed in the Transcendental Aesthetic cannot be thought of as a magnitude: it would be impossible to know if space itself is a finite magnitude or an infinite one. Space only provides ideal rules for ordering all possible appearances, which is a precondition for acquiring knowledge about magnitudes (Cohen 1871b, pp.162–163).

In Cohen's interpretation, such a structure sheds light on the peculiar meaning of aprioricity in Kant's critical philosophy. Aprioricity does not entail subjectivity but objectivity: the conditions of knowledge first make experience possible (Cohen 1871b, pp.255–256). Therefore, despite the fact that the distinction between form

---

[4] For a detailed comparison of Cohen's view with Trendelenburg's and Fischer's, see Beiser (2014, pp.478–481).

and matter of the appearance goes back to Kant (1770), Kant in the Transcendental Aesthetic, omitted his previous definition of form as the result of a certain law implanted in the mind (cf. Kant 1770/1912, Sect. 4). It is only in connection with the understanding that the forms of intuition provide us with principles of knowledge. This is what distinguishes the transcendental inquiry from the psychological assumption of innate laws. According to Cohen, Trendelenburg's argument tacitly implies the new meaning of "objectivity" attributed by Kant to a priori knowledge, insofar as subjectivity and objectivity for Trendelenburg, do not exclude each other. Trendelenburg mentioned, for example, the fact that: "When we produce the representation of space through the inner motion of the imagination (subjectively), nothing prevents the space that is produced by the corresponding external movement to be objective" (Trendelenburg 1867, p.222). The inner motion of the imagination in the quote above plays the same role as the successive synthesis of the productive imagination in Kant's sense. We have already noticed that Kant sharply distinguished the productive imagination from the reproductive, psychological notion of imagination, because only the former notion is the middle term between sensibility and understanding, which makes it possible to apply the concept of the understanding to the manifold of intuition. Cognition in pure intuition thereby provides us with the necessary conditions for the cognition of the objects of a possible experience in general.

To summarize, Cohen rejected Fischer's interpretation. At the same time, Trendelenburg's objection to Kant was neutralized: Kant did not need to prove the subjectivity of space and time; in fact they are both subjective and objective or simply objective in the sense of the critical philosophy. Arguably, Cohen's conception of critical philosophy was influenced by Trendelenburg in some respects. Cohen agreed fundamentally with Trendelenburg's thesis that space and time are both subjective and objective, insofar as the ideal rules for the formation of these concepts can be related to empirical reality (see Trendelenburg 1869, pp.2–3).[5] Furthermore, in later writings, Cohen came to the conclusion that the forms of intuition cannot be determined independently of the mathematical treatment of movement in natural science.[6] Trendelenburg's attempt to define spatial notions by means of movement might have played a role in the development of Cohen's thought, although Cohen's

---

[5] For further evidence of Trendelenburg's influence on Cohen, see Köhnke (1986, pp.260, 270–272); Ferrari (1988, pp.29–31) and Gigliotti (1992, p.56). In contrast with this line of interpretation, Beiser emphasizes the difference between Cohen's critical concept of objectivity and Trendelenburg's concept, which, according to Beiser, "is essentially that of transcendental realism, according to which we know an object when our representations correspond to the thing-in-itself, which is given in experience" (Beiser 2014, p.480). A discussion of this assessment would require a closer examination of Trendelenburg's view of objectivity. I limit myself to point out Trendelenburg's influence on Cohen's view that the space, and a priori concepts in general, depend on an interaction between subjectivity and objectivity in the critical sense. However, it is true that Cohen's agreement with Trendelenburg emerges more explicitly in the second edition of *Kant's Theory of Experience*. Further evidence of this is given in the following section.

[6] The importance of the Trendelenburg-Fischer controversy for the development of Cohen's thought is reconsidered in Patton (2005).

motivation is found above all in his reading of Kant's *Critique of Pure Reason*, along with the *Prolegomena*. Cohen's studies on Kant during the 1870s gradually led him to the view that experience is given in knowledge and, therefore, the history of science provides the starting point for the transcendental inquiry into the conditions of experience.

2.4.2 Experience as Scientific Knowledge and the A Priori

After his article on the Trendelenburg-Fischer controversy, Cohen published *Kant's Theory of Experience* (1871a), which was followed by *Kant's Foundation of Ethics* (1877), and *Kant's Foundation of Aesthetics* (1889). During that period, Cohen made his career at the University of Marburg, where he was supported by Friedrich Albert Lange. Cohen was appointed a lecturer there in 1873, and became Lange's successor as professor in 1876.[7]

A second edition of *Kant's Theory of Experience* appeared with substantial revisions in 1885. In the meantime, Cohen had sharpened his distinction between transcendental philosophy and psychology. In the 1871 edition, he endorsed Herbart's conception of the mind as a functional mechanism: this assumption is required for the explanation of psychic processes and, accordingly, for a transcendental proof of the possibility of experience. Despite the fact that Kant called the forms of experience faculties of the mind, these ought to be understood as relational functions to be analyzed both separately and in connection to one another. It is only in such a connection that the forms analyzed by Kant can be proved to make experience possible. In this sense, Cohen compared Kant's transcendental apperception, which is the highest principle of the system of experience, to Herbart's functionalist conception of the mind. Such a conception can be traced back to Kant himself, for example, when he attributes a "function" to the concepts of the understanding. "By a function," Kant writes, "I understand the unity of the action of ordering different representations under a common one" (Kant 1787, p.93). Having a function or spontaneity is what distinguishes the concepts of the understanding from sensibility, which is receptive according to Kant. In emphasizing the spontaneity of knowledge, Cohen goes so far as to say that Kant's theory of experience presupposes not so much faculty psychology, as a "sane" psychology, such as Herbart's (Cohen 1871a, p.164). This was omitted in the 1885 edition, because, in the meantime, Cohen came to the conclusion that the transcendental inquiry cannot be compared to any direction in psychological research. Transcendental philosophy is characterized by its own method.

Arguably, Cohen used the expression "transcendental method," which is not found in Kant's work, to emphasize that Kant came only gradually to characterize the transcendental inquiry and to distinguish it clearly from psychology. Kant's attempt in the *Critique of Pure Reason* was to prove the possibility of a priori

---

[7] For biographical information about Cohen, see Ollig (1979, pp.29–35).

knowledge by analyzing its fundamental elements and the conditions for their synthesis. Therefore, in the *Prolegomena to Any Future Metaphysics That Will Be Able to Come Forward as a Science* (1783), Kant called the method he used in the *Critique* "synthetic" and contrasted it with that of the *Prolegomena*, which he called "analytic." The new method presupposes Kant's former argument for the possibility of knowledge. His goal now is to reconstruct the conditions for knowledge as found in the exact sciences (Kant 1783, Sect. 4).

In Cohen's interpretation, Kant's use of the latter method makes it clear that the transcendental inquiry does not depend on any psychological assumption about the organization of the human mind. Kant studied the conditions of experience, and the experience under consideration was proved to coincide with scientific knowledge. Cohen summarized his interpretation of Kant's theoretical philosophy and made this point explicitly in the first chapter of *Kant's Foundation of Ethics*. As Cohen put it, experience in Kant's sense, is first given in mathematical physics, more precisely, in Newtonian physics (Cohen 1877, pp.24–25).[8] Therefore, in *The Principle of the Infinitesimal Method and its History* (1883), Cohen saw in Kant's transcendental philosophy the culmination of the idealistic tradition of Plato, Descartes, and Leibniz. In order to clearly distinguish these approaches to knowledge, which were based on scientific experience, from the psychological assumptions that were characteristic of the nineteenth-century theory of knowledge,[9] Cohen (1883, p.6) called the inquiry into the logical presuppositions of knowledge a critique of knowledge. He called the advocated kind of idealism scientific or, with reference to Kant, "critical" (Cohen 1885, p.XII).

The connection with the history of science sheds light on the status of a priori knowledge, which is the object of the transcendental inquiry. In fact, Cohen pointed out even in 1871, that universality and necessity do not pertain to a priori knowledge per se. Consider, for example, the metaphysical exposition of space and time. Once it is established by conceptual analysis that some a priori knowledge must be grounded in these representations, the question is, "Which is the knowledge to which universality and necessity pertain?" The metaphysical a priori (i.e., the description of a priori knowledge as perceived by the knowing subject) must be completed by the transcendental a priori, which is the determination of the same as a condition of objective knowledge (Cohen 1871a, pp.10, 34).[10] A priori knowledge

---

[8] For a thorough account of the development of Cohen's thought, see Ferrari (1988). For a clarification of Cohen's notion of experience, see Richardson (2003).

[9] Hans Vaihinger (1876) traced back the origin of the discipline to John Locke's *Essay Concerning Human Understanding* (1690). However, it was only in the nineteenth century that the "theory of knowledge" (*Erkenntnistheorie*) was introduced in contraposition to the "doctrine of knowledge" (*Erkenntnislehre*), which was associated with metaphysics. Regarding the history of the concept of theory of knowledge, also see Köhnke (1981).

[10] This distinction foreshadows Reichenbach's (1920, p.48) distinction between two meanings on the notion of a priori in Kant's work: 1) as valid for all time, and 2) as constitutive of the objects of experience. Reichenbach's argument in *The Theory of Relativity and A Priori Knowledge* (1920) is that although the first meaning of Kant's notion was disproved by Einstein's use of non-Euclidean geometry in general relativity, the second meaning of a priori may be reaffirmed relative to the

can be determined only by specifying the principles for connecting the cognitive functions required to one another. This is the goal of Kant's Analytic of Principles. Cohen's identification of experience as scientific knowledge suggests that the specification of the principles of knowledge cannot be accomplished independently of scientific theories. The transcendental inquiry is the reconstruction of those conditions that are logically presupposed in the sciences. Therefore, the transcendental method proceeds by conceptual analyses and explores purely logical connections. At the same time, the transcendental inquiry must be confronted with the "fact of science" (Cohen 1883, p.5). Cohen acknowledged that a priori principles depend on scientific theories and may be revised as a consequence of theory change. His requirement, in such a case, was that changes extend the field of knowledge, though he did not take into account more specific criteria (e.g., improvement of experimental precision or predictive power). Cohen (1896/1984, p.77) mentioned, for example, Heinrich Hertz's *Principles of Mechanics* (1894). The example suggests that the extension of knowledge and, therefore, of experience in philosophical terms, may be accomplished by a unified physics, which would include, in principle, all known natural phenomena. Specific heuristic principles arguably presuppose this ideal.

Cohen's conception of the a priori required a revision of the Kantian theory of space and time, which was explicit in the 1885 edition of *Kant's Theory of Experience*. Cohen's critical remarks concerned the status of pure intuition. In order to characterize space and time as pure intuitions, Kant distinguished them from general representations, which are the objects of logic, on the one hand, and from motion, which entails empirical factors, on the other hand. Cohen weakened Kant's distinction in both regards. With respect to the distinction between intuitions and concepts, Cohen maintained that Kant's characteristics for intuitions are distinct from those of general representations, but are conceptual nonetheless. Such characteristics concern a special kind of concepts having to do with rules or order. Claims 3 and 4, for example, entail infinite divisibility, which is an operation that can be indefinitely repeated, so that a manifold is constructed. By contrast, general representations are abstracted from given manifolds and cannot include infinite domains (Cohen 1885, pp.125–126).[11]

---

principles of the new theory. I turn back to the idea of a relativized a priori in Chap. 7. In the following chapters, I argue that the same idea has its origin in Cohen's interpretation of Kant and in the application of the notion of a transcendental a priori to the axioms of geometry by both Cohen and Cassirer.

[11] The same problem motivated later logical approaches to Kant's pure intuitions. See Hintikka's essays "Kant's 'New Method of Thought' and His Theory of Mathematics" and "Kant on the Mathematical Method" in Hintikka (1974) and Friedman (1992, Ch.1). The basic idea of these approaches is that Kant's distinction between pure intuitions and the general concepts of syllogistic logic can be clarified by using polyadic logic. According to Friedman, pure intuition as the iterability of intuitive constructions provides a uniform method for instantiating the existential quantifiers we would use today to define such properties as the infinite divisibility or the denseness of a set of points. Such a method is comparable with the use of Skolem functions: instead of deriving new points between two given points from an existential axiom, we construct a bisection function from our basic operations and obtain the new points as the values of this function. Friedman's

On the other hand, Cohen rejected the claim that motion entails irreducibly empirical factors. This is because, in 1883, he called into question Kant's definition of sensation as the matter of the appearances. Cohen pointed out that this definition goes back to Kant's Dissertation of 1770 and reflects his earlier dogmatic standpoint. There is no place for a dualistic opposition between form and matter of appearance in critical philosophy: matter is not so much the effect of a mind-independent reality in our senses, but the result of a conceptual organization of empirical reality within the framework of scientific theories. Kant himself reconsidered his previous definition of sensation in the Analytic of Principles, as he stated that "in all appearances, the real, which is an object of sensation, has intensive magnitude, i.e., a degree" (Kant 1787, p.207). According to Cohen, this principle presupposes Leibniz's infinitesimal method. The so-called matter of sensation can now be redefined by the use of the concept of differential (see Cohen 1883, pp.13–14). In other words, Cohen turned Kant's distinction between intuitions and sensations into a distinction between spatial magnitudes and nonspatial ones. Both kinds of magnitudes must be calculable for the purposes of physics. Empirical contents are not simply given (whether in pure or in empirical intuition), because such calculations presuppose specific mathematical methods. However, as Helmut Holzhey pointed out, Cohen's reception of the metaphysical view of infinitesimals (i.e., as real infinitesimals) led him to distance himself from Kant more significantly:

> Kant's analysis of the relation between sensibility and thought deserved clarification or even revision according to Leibniz's view of infinitesimals; for, in accord with his reading of Leibniz, Cohen defended the view that even geometry determines the objects of knowledge in their reality by using the infinitesimal method. Therefore, geometry as science of space depends not only on intuition, but also on thought. (Holzhey 1986, pp.292–93)

Hozhey made it plausible that the reception of Leibniz played a no less fundamental role in the development of the main ideas of the Marburg School of neo-Kantianism than the reception of Kant, although it is only in the 1890s and in the *Logic of Pure Knowledge* (1902) that Cohen criticized the Kantian theory of space and time as pure intuitions and proposed a sort of "return" to Leibniz (see also Ferrari 1988, pp.186–190). Therefore, Holzhey distinguishes two phases in the reception of Leibniz in Marburg neo-Kantianism: the first goes back to Cohen's *The Principle of the Infinitesimal Method and Its History* (1883), and includes the second edition of *Kant's Theory of Experience* (1885) and Natorp's Habilitation Lecture "Leibniz and Materialism" (1881); the second includes Cohen's *Logic of*

---

conclusion is: "Since the methods involved go far beyond the essentially monadic logic available to Kant, he views the inferences in questions as synthetic rather than analytic" (Friedman 1992, p.65). However, it seems to me that Friedman eliminates the idea of a conceptual construction without discussing it. In the nineteenth century, this idea circulated among both mathematicians and philosophers, and it was the starting point for the argument that mathematics is synthetic in the Marburg School of neo-Kantianism.

Pure Knowledge (1902) and Cassirer's 1902 monograph *The Scientific Foundations of Leibniz's System.*[12]

In relation to the principle of the infinitesimal method, Cohen reconsidered Trendelenburg's remarks about movement in the second revised version of *Kant's Theory of Experience*. Cohen quoted from the following passage of Trendelenburg's *Historical Contributions to Philosophy*: "Objectivity and subjectivity do not exclude one another [...]. For example, if the representation of space is generated subjectively by an inner motion of the imagination, nothing precludes space from being something objective" (Trendelenburg 1867, p.222). For Cohen, this is to say that nothing precludes the space thus generated from providing us with conditions of mathematical physics (Cohen 1885, p.162). Apriority lies in the fact that geometrical constructions, as the results of idealized movements, are necessary presuppositions for the definition of physical concepts.

As we will see in the next chapter, Cohen's distanced himself from Kant more radically in the introduction to the 1896 edition of Lange's *History of Materialism*. There, Cohen argued that Kant's assumption of pure intuition is superfluous, provided that constructions in pure intuitions can be analyzed in terms of ideal operations, as in analytic geometry. During the same period, he envisioned the project of his *Logic of Pure Knowledge* (1902). Cohen identified the logic of pure thought with the logic of scientific knowledge. The logic of pure knowledge entailed a further development of the critique of knowledge. Kant's critique presupposed the distinction between sensibility and understanding. The goal of the transcendental inquiry was to prove the possibility of a synthesis between heterogeneous elements of knowledge. Therefore, Kant introduced the mediating term of pure intuition and defined space and time as forms of outer and inner intuition, respectively. Cohen denied the assumption of pure intuition as an unjustified restriction on the autonomy of thought. He considered space and time as categories of the understanding, and he replaced Kant's synthesis between heterogeneous elements with an internal articulation in pure thought itself (Cohen 1902, pp.160–161).

2.5   Cohen and Cassirer

Ernst Cassirer was born in 1874 in the German city of Breslau (now Wrocław, Poland). During his studies in philosophy at the University of Berlin, he took a course on Kant taught by Georg Simmel. In particular, Simmel recommended *Kant's Theory of Experience* (1871a) by Hermann Cohen to his students. After

---

[12] For a general account of the influence of Leibniz on the Marburg School of neo-Kantianism, see Zeman (1980), who focuses on Cohen's conception of reality and his approach to the infinitesimal method. A more detailed account of the differences between Cohen, Natorp, and Cassirer is found in the aforementioned article by Holzhey and especially in Ferrari (1988). For the distinction between the mathematical and the logical foundations of calculus, which is crucial to understand Cohen's approach to the infinitesimal method correctly, see Schulthess's introduction to Cohen (1883) in Cohen's *Werke*, Vol. 5.

reading Cohen's book, Cassirer decided to move to Marburg to complete his education under the supervision of Cohen and Paul Natorp.[13] Cassirer studied at the University of Marburg from 1896 until 1899. He moved back to Berlin in 1903, and he received his habilitation at the University of Berlin in 1906. Cassirer's early works, *The Scientific Foundations of Leibniz's System* (1902), the first two volumes of *The Problem of Knowledge in Modern Philosophy and Science* (1906, 1907), and *Substance and Function: Investigations of the Fundamental Problems of the Critique of Knowledge* (1910), were deeply influenced by Cohen's interpretation of Kant's transcendental philosophy. Cohen's characterization of the transcendental method as based on the fact of sciene motivated Cassirer to face the problem of whether Kant's assumptions were compatible with later scientific developments, including non-Euclidean geometries, the mathematization of logic, and relativistic physics.[14]

This section offers a discussion of Cassirer's arguments in favor of Cohen's interpretation of Kant's critical philosophy. The first part is devoted to Cassirer's considerations about the conceptions of space and time in the development of Kant's thought. Cassirer's reconstruction was the background for his use of arguments drawn from mathematical logic to support critical idealism. In the second part, I sketch the connection with Cassirer's interpretation of the history of science in terms of the logic of the mathematical concept of function.

2.5.1 Space and Time in the Development of Kant's Thought: A Reconstruction by Ernst Cassirer

In the second volume of *The Problem of Knowledge in Modern Philosophy and Science* (1907a), Cassirer argued in favor of Cohen's interpretation of Kant by emphasizing a possible connection with Leibniz. Cassirer maintained that most interpreters failed to appreciate such a connection because they restricted their attention to Kant's Dissertation of 1770, where Kant first distinguished pure intuitions from concepts. Kant apparently disagreed with Leibniz on the origin of

---

[13] On Cassirer's education and on his relationship to Cohen and Natorp, see Ferrari (1988, Part 2, Ch.1).

[14] In fact Cassirer dealt with a variety of topics in the philosophy of mathematics and the philosophy of science. Of particular interest even to contemporary debates is Cassirer's study on the philosophical aspects of quantum mechanics, *Determinism and Indeterminism in Modern Physics* (1936). On the relevance of Cassirer's views to ongoing debates on structuralist accounts of fundamental physics, see Cei and French (2009) and Ryckman (2015). Although these topics go beyond the scope of the present study, which centers on Cassirer's earlier works, I believe that his overall approach had some of its main motivations in the debate under consideration. The goal of this and the following sections devoted to Cassirer is to reconstruct these motivations in order to shed some light on aspects of his approach that otherwise might be neglected. These are the aspects that relate to Cassirer's neo-Kantian agenda. On Cassirer's use of the transcendental method, see also Ferrari (2010).

mathematical knowledge, which for Leibniz was entirely conceptual, whereas Kant assumed intuitive grounds. Nevertheless, Kant's motivation for the assumption of pure intuitions was the same as Leibniz's: the certainty of mathematical knowledge cannot be grounded in such imprecise representations as empirical intuitions. Cassirer emphasized that both Kant's pure sensibility and Leibniz's conception of space and time as orders of coexistence and of succession presuppose a twofold way to refer to sensation: "Whereas, on the one hand, 'order' per se is something different from the contents to which it refers; on the other hand, these contents can be determined only insofar as they are not confused, but rather present in themselves a general, lawful organization" (Cassirer 1907a, pp.345–346).

Leibniz, for example, distinguished time, which is a simple and uniform continuum, from perceptual sequences, the regularity of which is only approximate. Despite the fact that regular changes (e.g., the earth's rotation) are used for time measurement, the form of time itself must be conceived independently of the assumption that approximate regularities are found in nature: these can only be recognized in comparison with precise laws of motion. Therefore, Leibniz defined time as an "idea of pure understanding," and maintained that it is only because of an idealized and regular representation of motion that the empirical results of the composition of different (possibly irregular) motions can be predicted (Leibniz 1765, p.139).

Similar considerations can be made with regard to space. Whereas Newton's space and time were real and absolute, Leibniz conceived these notions as ideal and relative. According to Cassirer, Kant's conception was Leibnizian in spirit. Kant assumed the ideality of space and time to argue for mathematical certainty. Since he refused to admit that exact knowledge can be abstracted from reality, he located the foundations of mathematics in the forms of pure intuition, which he introduced in opposition to Leibniz's rationalism as a mediating term between thought and reality. Kant tended to assume a relational notion of space and time insofar as, on the other hand, the ordering of the manifold of pure intuition was supposed to provide us with conditions for any possible organization of perception.

Cassirer referred to the collection of Kant *Reflections* published by Benno Erdmann in 1884 to support his interpretation. Cassirer maintained, first of all, that Kant's argument in the Dissertation of 1770 was directed not so much against Leibniz's theory of space and time, as against Christian Wolff's later view that sensibility only produces confused concepts (Cassirer 1907a, p.493, note 29). Given the fact that intuition seemed to have an indispensable role in mathematics (especially in geometry) at that time, Wolff's assumption would compromise mathematical certainty (Kant 1884, no. 414).

Furthermore, Cassirer maintained that the theory of pure sensibility found its motivation in the idea of the system of the principles of knowledge: this idea preceded the thesis that space and time are intuitions, not concepts, in the development of Kant's thought and also worked independently of Kant's later distinction between concepts and pure intuitions. Kant, for example, in a Reflection dated 1768–1770, distinguished between those concepts that are abstracted from sensations and those that are abstracted from the laws according to which the concepts of the first kind are connected. Kant called the former "abstract ideas" and the latter "pure concepts" of the understanding, which is the power of producing conceptual connections or syntheses. Since the concept of space cannot be derived from the perception of extended bodies, Kant deemed it to be a pure concept of the understanding (no. 513). In the Reflection that follows, in Erdmann's edition (no. 514), Kant defined axioms as objective principles of the synthesis that takes place in space and time. He distinguished axioms from logical laws, on the one hand, and from rules (i.e., canons) for a subjective or qualitative kind of synthesis, on the other. These distinctions suggest that space and time correspond to a specific kind of synthesis or function, which for Kant is a capacity of the understanding.[15]

To summarize, it appears that before 1770, Kant had no specific reason to distinguish pure intuitions from concepts. The distinction arguably followed from the claim that space and time are essentially single. However, Cassirer denied that this was a distinctive characteristic of intuition. He drew attention to a Reflection by Kant, also dated 1769, and numbered 274, in the Erdmann edition. Kant here distinguished between empirical and pure knowledge: only the former presupposes sensation. He then divided pure knowledge into singular concepts (e.g., space and time) and pure concepts of reason, which are universals. However, he also called the former "pure intuitions." This suggests that intuitions and concepts do not exclude one another. Something called "intuitive" can be understood as an aspect of something fundamentally conceptual, provided that not all concepts are universals in the sense of traditional, syllogistic logic: some concepts rather indicate univocal ways of ordering (Cassirer 1907a, p.495).

It is apparent that Cassirer endorsed Cohen's interpretation of the Kantian theory of space and time, especially if one considers the second edition of *Kant's Theory of Experience*. At the same time, Cassirer emphasized the importance of Cohen's own stance on the relationship between mathematics and logic. The same year, Cassirer addressed that issue in his paper "Kant and Modern Mathematics" (1907b). In this connection, he distanced himself from Kant more explicitly, insofar as he argued for a logical foundation of mathematics. Cassirer presented his view as a consequence of Cohen's critical logic, on the one hand, and of mathematical logic (i.e., of the so-called "logistic"), on the other. Cassirer wrote:

> The modern critical logic, and the "logistic" as well, distanced itself from the Kantian theory of "pure sensibility." From both standpoints sensibility means in fact an epistemological *problem*, not an autonomous and peculiar *source of certainty*. And the modern critical logic fundamentally agrees with the *tendency* that emerges from the works of Russell and Couturat: it requires a purely logical derivation of the fundamental principles of mathematics,

---

[15] It is noteworthy, however, that for Adickes, Kant might have written Reflection no. 514 (in the Erdmann edition; no. 4370 in the Akademie-Ausgabe of Kant's *Gesammelte Schriften*, Vol. 17, edited by Adickes) between 1776 and 1778 or between 1780 and 1783. Whereas Adickes and Erdmann agree that Reflection no. 513 (no. 3930 in the Akademie-Ausgabe) goes back to 1769. Only no. 513 includes the claim that space and time are concepts of understanding. The definition of axioms as principles of an objective synthesis, in no. 514, is compatible with the theory of pure sensibility proposed by Kant in the *Critique of Pure Reason*. Kant's claim there is that both sensibility and understanding are required for such a kind of synthesis.

by means of which alone we learn to *understand* fully and to rule over conceptually "intuition" itself, namely, space and time. (Cassirer 1907b, pp.31–32)

As pointed out by Ferrari (1988, p.268), Cassirer implicitly distanced himself from Cohen's underestimation of mathematical logic in the *Logic of Pure Knowledge* by considering recent developments in mathematics a "factum" for the transcendental inquiry. According to Cassirer, the validity of the statements of critical philosophy "is not guaranteed once and for all, but it must justify itself anew according to the changes in scientific convictions and concepts" (Cassirer 1907b, p.1). Both Natorp and Cassirer worked at the extension of Cohen's transcendental method to more recent developments in the exact sciences. Natorp's researches go back to the 1890s and culminated in *The Logical Foundations of the Exact Science* (1910), which was an important source of ideas for Cassirer's *Substance and Function* (1910). However, it was especially Cassirer who put these ideas in connection with the woks of such mathematicians as Richard Dedekind, Felix Klein, and David Hilbert, on the one hand, and with Russell's and Couturat's logic of relations, on the other. The main disagreement between Cassirer and Cohen concerned the principle of the infinitesimal method, which Cassirer interpreted not so much in terms of the metaphysical tradition in the foundation of the calculus as in terms of the modern definition of the concept of limit. The structuralist aspects of his philosophy of mathematics enabled him to appreciate the advantage of the logic of relations over traditional logic when it comes to the questions concerning the foundations of mathematical theories. Conversely, Cohen distanced himself from Cassirer in a letter dated August 24, 1910.[16]

Cassirer's reliance on Russell and Couturat seems to be even more puzzling. In the works to which Cassirer referred: *The Principles of Mathematics* (1903) by Bertrand Russell and *The Principles of Mathematics, with a Supplement on Kant's Philosophy of Mathematics* (1905) by Louis Couturat, the authors used more recent achievements in mathematical logic to support the view that the foundations of mathematics deserve a purely logical derivation. Therefore, Couturat contradicted Kant's view that mathematics provides us with a priori synthetic judgments and maintained that all the propositions of mathematics are analytic. For the same reason, mathematicians in the nineteenth century, such as Bernhard Bolzano in Prague and Giuseppe Peano and his school in Italy, reconsidered Leibniz's project of a mathematicized logic. In Germany, the rediscovery of Leibniz was mainly due to the publication of Leibniz's Nachlass – which is an ongoing project initiated by Carl Immanuel Gerhardt – and to Gottlob Frege's connection with Leibniz. It was not by chance that Russell (1900), Couturat (1901), and Cassirer (1902) published three seminal monographs on Leibniz, although their interpretations of Leibniz differed considerably.[17] The discussion which followed prepared the later debate about the

---

[16] A reproduction of the letter is found in in the CD-ROM that accompanies Cassirer (2009).

[17] Whereas Cassirer proposes a comprehensive interpretation of Leibniz's philosophy as a decisive step in the history of scientific idealism, Russell and Couturat focus on Leibniz's logic. However, Russell attributes to Leibniz the attempt to reduce all propositions to the subject–predicate form in line with the Aristotelic-Scholastic tradition. In this reading, Leibniz's reliance on traditional logic

status of mathematical judgments in many ways, and it was Cassirer's starting point for his confrontation with the systematic works of Couturat and of Russell.

As suggested by Heis, Cassirer's twofold relation to Russell in particular can be clarified by distinguishing between different dimensions in Cassirer's argument for his philosophical theory of concepts, which he presented as a theory of the concept of function in contrast with a theory of the concept of substance. On a logical level, Cassirer argued with Russell for an independent logic of relations, which cannot be exhausted by syllogistic. In identifying the latter as the logic of the concepts of substance, Cassirer, similar to Russell, associated the contrast between the old and the new logic with the metaphysical question concerning the status of mathematical objects. On this level, the priority of the concept of function indicates that not every object is an Aristotelian substance – as an object with no essential, irreducibly relational properties. Cassirer's epistemological views, on the other hand, were opposed to Russell's. As Heis puts it: "The fundamental use that Cassirer makes of 'Substanzbegriff' and 'Funktionsbegriff' is [...] not Russellian, but Kantian: it contrasts philosophical views that overlook the epistemic preconditions of various kinds of knowledge with those that recognize the 'functions' that make certain kinds of knowledge possible" (Heis 2014, p.255). Heis goes on to argue that Cassirer's Kantian epistemology led him to distance himself from Russell's theory of the formation of mathematical concepts as falling back into the schema of the concepts of substance. According to this view, the procedures of mathematicians are based on some prior metaphysical assumptions about mathematical objects. By contrast, Cassirer relied on such examples as Dedekind's definition of numbers and the modern concept of energy to argue for the view that all objects (beginning with mathematical objects) are but positions in a relational structure. However, Heis points out that ontological structuralism does not follow directly from Cassirer's epistemic view that knowledge consists of relational structures. The middle term for such an inference is given by Cassirer's methodological considerations about the development of mathematics between the nineteenth and the early twentieth century. From this viewpoint: "Dedekind's view that the natural numbers are positions in a progression [...] simply makes explicit what was implicit in the earlier work by number theorists on these 'extended' number domains, and it allows a unitary coherent conceptual framework in which to talk about all of the various kinds of 'numbers'" (Heis 2014, 263).

In what follows, I rely on Heis (2014) for the clarification of Cassirer's concept of function according to the different levels of the argument summarized above. The emphasis in my own account lies in the fact that the clarification of what the exten-

---

prevented him from accounting for the logical relations that lie at the foundation of mathematical theories and of space-time theories. By contrast, Couturat traces back the basic ideas of the logic of relations to Leibniz's universal characteristic. For a comparison between these readings also in connection with the later debate about Kant and modern mathematics, see Ferrari (1988, Part 2, Ch.4). For a thorough account of Leibniz's ideas and their reception in the nineteenth century (especially in Germany and the United Kingdom) see Peckhaus (1997).

sion of a mathematical structure meant was very much a work in progress at that time and even in the works of mathematicians such as Dedekind and especially Klein it was intimately related to the question concerning the heuristic value of mathematical method. Generalizations in this broader sense included not only the extension of similar domains (e.g., of numerical to larger but likewise numerical domains) but also the discovery of structural similarities between domains across different branches of mathematics (as in Klein's comparative research in geometry) or even between mathematical and physical domains. Cassirer looked at this mathematical tradition for examples of the transposition of the model of the concept of function from mathematical to empirical concepts, and therefore for a confirmation of the synthetic character of mathematics in Cassirer's sense.

In order to highlight this point, my suggestion is to reconsider the transformation of the notion of a conceptual synthesis in the Marburg tradition. Again, Cassirer's article on "Kant and Modern Mathematics" enables us to approach this matter from the standpoint of the nineteenth-century mathematical tradition. Regarding foundational issues in the philosophy of geometry, the problem with Kant's view of mathematics is that later logical analyses of such subjects as the mathematical continuum or geometric space(–s) seem to make the assumption of pure intuitions superfluous. Nineteenth-century geometry led to a completely different understanding of the relationship between space and geometry. As we will see in the following chapters, neo-Kantians such as Alois Riehl defended the Kantian theory of space by pointing out the intuitive origin of spatial concepts. By contrast, Marburg neo-Kantians denied the role of intuition in the construction of mathematical (including geometrical) objects. As for the claim that mathematics is synthetic, Cohen and Cassirer relied not so much on the metaphysical exposition of the concepts of space and time, as on Kant's characterization of mathematical method as construction according to a priori concepts. In the preface to the second edition of the *Critique of Pure Reason*, Kant contrasted actual construction with construction according to a priori concepts by noticing that "in order to know something securely *a priori*," the first to prove that the angles at the base of an isosceles triangle are equal had to "ascribe to the thing nothing except what followed necessarily from what he himself had put into it in accordance with its concept" (Kant 1787, p.XII). Mathematics and natural science provide some of the clearest examples of what Kant also called combination of a manifold in general or synthesis and characterized as follows:

> The manifold of representations can be given in an intuition that is merely sensible, i.e., nothing but receptivity, and the form of this intuition can lie *a priori* in our faculty of representation without being anything other than the way in which the subject is affected. Yet the combination (*conjunctio*) of a manifold in general can never come to us through the senses, and therefore cannot already be contained in the pure form of sensible intuition; for it is an act of the spontaneity of the power of representation, and, since one must call the latter understanding, in distinction from sensibility, all combination, whether we are conscious of it or not, whether it is a combination of the manifold of intuition or of several concepts, and in the first case either of sensible or non-sensible intuition, is an action of the understanding, which we would designate with the general title synthesis in order at the same time to draw attention to the fact that we can represent nothing as combined in the object without having previously combined it ourselves, and that among all representations

combination is the only one that is not given through objects but can be executed only by
the subject itself, since it is an act of its self-activity. (Kant 1787, pp.129–130)

For Kant, knowledge presupposes both the receptivity of the sensibility and the
spontaneity of the understanding. We have already noticed that Kant, therefore,
attributed a function to the understanding alone. At the same time, he made it clear
that an act of the understanding is required in order for any combination in the
object to be conceived. According to Cassirer, Kant's clarification suggests that the
receptive aspect of knowledge depends on its spontaneity in the formulation of
hypotheses. Furthermore, Kant's claim that "we can represent nothing as combined
in the object without having previously combined it ourselves" appeared to be confirmed by the hypothetical status of geometrical assumption in axiomatic systems.

Regarding the notion of synthesis in general, Cassirer believed that critical logic
provides us with a system of the principles of knowledge in Kant's sense:

> Once we will have understood that the same fundamental syntheses, which lie at the foun-
> dation of logic and of mathematics, rule over the scientific articulation of empirical knowl-
> edge, that only these syntheses enable us to establish a lawful order of the appearances, and
> therefore their objective meaning, then, and only then we will obtain a true justification of
> the principles. (Cassirer 1907b, p.45)

This consideration was Cassirer's starting point for the idea of a "logic of objective
knowledge" (p.45). Cassirer's justification of the principles differed from Kant's
because of Cassirer's reliance on the history of science for the characterization of
the fundamental syntheses. In other words, Cassirer distanced himself from Kant's
metaphysical characterization of a priori knowledge and consistently applied the
transcendental method as construed by Cohen to the more recent history of mathematics. It followed that the focus in Cassirer's theory of the concept of function is
not so much on the formation of mathematical concepts as such, but on the aspects
of this formation that admit a transposition from mathematical to empirical research.
This is the main goal of his contrast between the concept of substance and that of
function. The concluding part of this section gives a brief account of the leading
ideas of Cassirer's *Substance and Function* (1910).

2.5.2   Substance and Function

In the preface to *Substance and Function*, Cassirer addressed the following
problem:

> In the course of an attempt to comprehend the fundamental conceptions of mathematics
> from the point of view of logic, it became necessary to analyse more closely the function of
> the concept itself and to trace it back to its presuppositions. Here, however, a peculiar dif-
> ficulty arose: the traditional logic of the concept, in its well-known features, proved inade-
> quate even to characterize the problems to which the theory of the principles of mathematics
> led. It became increasingly evident that exact science had here reached questions for which
> there existed no precise correlate in the traditional language of formal logic. The content of

> mathematical knowledge pointed back to a fundamental form of the concept not clearly defined and recognized within logic itself. (Cassirer 1910, p.III)

The problem lies in the fact that traditional logic was based on general representations and the relation of genus and species. However, this theory of concepts appeared insufficient for the characterization of such mathematical concepts as series and limit. Cassirer relied on mathematical logic in his attempt to overcome this problem by pointing out a different form of concepts, which he identified as the mathematical concept of function. We have already noticed that for Cassirer, the formation of general representations in the traditional sense implies a metaphysical commitment, because it ultimately presupposes the existence of some individuals, from which some common characteristics are obtained by abstracting from any specific quality. Therefore, he called the latter kind of concepts substances. On the contrary, a mathematical function according to Cassirer's definition, which is borrowed from Drobisch (1875, p.22), represents a universal law, which, by virtue of the successive values which the variable can assume, contains within itself all the particular cases for which it holds.

However, Cassirer's consideration was not restricted to mathematics alone. Firstly, such a broad definition as Drobisch's suggests a comparison with natural laws. Secondly, Cassirer's usage clearly implies the Kantian meaning of function as a distinguishing characteristic of the understanding. Therefore, Cassirer's concept of function assumes at least three different meanings when compared to contemporary notions. These meanings include: (i) the logical definition of function as a one-one or many-one relation; (ii) the role or purpose, of particular mathematical or scientific concepts within this or that particular scientific field; (iii) the Kantian meaning of function as a rule-governed activity of the mind (see Heis 2014, p.253). More importantly, as pointed out by Heis, Cassirer advocated the priority of the concept of function (in its different meanings) to account for the structure of scientific theories from a holistic perspective. In this connection, the contrast with the concept of substance corresponds to the fact that even the basic statements concerning measurement presuppose theoretical laws for their interpretation. Insofar as modern physics shed light on this aspect (e.g., by identifying the concept of energy with a constant relation among phenomena associated with a numerical value, instead of looking for an underlying substance), Cassirer pointed out a series of analogies with the history of mathematics. In his view, both the nineteenth-century mathematical tradition and the history of physics and the special sciences show a tendency to shift away from concepts of substance towards concepts of function.

Cassirer's goal in 1910 was to show that the concept of function provided a plausible account of the formation of concepts in the exact sciences and a unitary perspective on knowledge. The same model was supposed to be extendable to arithmetic, geometry, physics, and chemistry. In this sense, the mathematical concept of function assumed both the epistemological and the Kantian meanings mentioned above. More specifically, the role of Cassirer's concept corresponds to the role played by the concept of time in Kant's *Critique of Pure Reason*: the form of time contains the general conditions under which the concepts of the understanding can be applied to any object alone. Kant called the formal conditions of the sensibility, to which the use of the concept of the understanding is restricted, the schema of this concept. He called the procedure of understanding with these schemata "schematism of the pure concepts of the understanding" (Kant 1787, p.179). The reference to Kant's notion of schema is apparent in the following quote from *Substance and Function*:

> In opposition to the logic of the generic concept, which [...] represents the point of view and influence of the concept of substance, there now appears the *logic of the mathematical concept of function*. However, the field of application of this form of logic is not confined to mathematics alone. On the contrary, it extends over into the field of the *knowledge of nature*; for the concept of function constitutes the general schema and model according to which the modern concept of nature has been molded in its progressive historical development. (Cassirer 1910, p.21)

We have already noticed that Kant's concept of function applies to the concepts of the understanding alone. The analogy with time only makes sense in the neo-Kantian interpretation of space and time as pure concepts. Friedman's objection is that the elimination of pure intuition prevented Cassirer from posing the problem of the applicability of mathematics in Kantian terms. The Kantian architectonic of knowledge is called into question: "By rejecting Kant's original account of the transcendental schematism of the understanding with respect to a distinct faculty of sensibility in favor of a teleologically oriented 'genetic' conception of knowledge, Cassirer (and the Marburg School more generally) has thereby replaced Kant's constitutive a priori with a purely regulative ideal" (Friedman 2000, p.117). By contrast, Friedman argues for a relativized conception of the constitutive a priori. However, the quotes above show that Cassirer's goal was to generalize the Kantian notion of schema by identifying it as the mathematical concept of function. This has a constitutive function, insofar as the same model can be transferred from mathematics to physics and provides a necessary condition for the definition of physical objects.

The constitutive aspect of Cassirer's transcendental a priori has been partially reconsidered by Ryckman (1991, p.63): "[A] transcendental-constitutive character is still upheld for the purely formal (but, for Cassirer, synthetic) logical acts underlying concepts."[18] However, Ryckman emphasizes a tension in Cassirer's work: "Cassirer appears to be torn between recognition of a purely abstract 'symbolic' doctrine of scientific concepts congenial with the attenuated and hypothetical realism of Helmholtz (and [...] Schlick), and a desire to avoid any insinuation of a 'metaphysical' form/content dichotomy through a 'dialectical' characterization of the relation of a concept to its content as that of series law to series member" (Ryckman 1991, pp.73–74). It seems to me that this tension only arises under the presupposition of Schlick's understanding of the symbolic doctrine of scientific concepts as a purely abstract doctrine. In the following chapters, I argue that this view of scientific concepts does not do justice to Helmholtz's approach: a dialectical

---

[18] On the constitutive/regulative distinction in Cassirer and on the role of the Marburg School of neo-Kantianism in the prehistory of the relativized a priori, see also Ferrari (2012).

characterization of the relation between formal and empirical factors of knowledge is no less essential to Helmholtz's arguments than to Cassirer's. I consider the works of Helmholtz, Dedekind, and Klein to be fundamental sources of ideas for Cassirer's understanding of symbolic thinking in mathematics as purely formal and synthetic.[19]

To sum up, on the one hand, Cassirer's insights into transformation of mathematics in the nineteenth century led him to adopt a view about space that was closer to Leibniz's than to Kant's. In such a view, space and time are better understood as general forms of the order of coexistence and succession than as pure intuitions. Besides Cassirer's attempt to read Kant's Transcendental Aesthetic in connection with Leibniz's views, Cohen's interpretation of Kant and the idea that the articulation of the principles of knowledge preceded the distinction between pure intuitions and concepts in the development of Kant's thought motivated Cassirer to formulate a generalized version of the Kantian schematism of the pure concepts of the understanding by pointing out the role of the mathematical concept of function in the sciences. Following a line of argument which goes back to Cohen, Cassirer identified the principles of knowledge not so much with immutable constraints imposed upon knowledge by the structure of the mind, but as logical presuppositions for scientific knowledge. On the other hand, both Cohen and Cassirer deemed this a Kantian approach, insofar as Kant's forms of experience, owing to their ideal nature, provide us at the same time with defining conditions of the objects of experience. The idea of *Substance and Function* is reminiscent of the fact that Kant was the first to reduce the metaphysical concept of substance to a category of the understanding. However, a similar consideration holds true for all the forms of experience indicated by Kant, beginning with the forms of sensibility:

> Even space and time, as pure as these concepts are from everything empirical and as certain as it is that they are represented in the mind completely *a priori*, would still be without objective validity and without sense and significance if their necessary use on the objects of experience were not shown; indeed, their representation is a mere schema, which is always related to the reproductive imagination that calls forth the objects of experience, without which they would have no significance; and thus it is with all concepts without distinction. (Kant 1787, p.195)

What is essential to the interpretation of Kant in the Marburg School of neo-Kantianism is the fact that the forms of experience provide us with necessary conditions of knowledge only in their synthesis a priori. This corresponds to Kant's claim that experience rests upon "a synthesis according to the concepts of the object of appearances in general." Therefore, Kant formulated the following principle: "Every object stands under the necessary conditions of the synthetic unity of the manifold of intuition in a possible experience" (p.197). He maintained that "The conditions of the possibility of experience in general are at the same time conditions of the possibility of the objects of experience" (ibid.).

---

[19] I turn back to the discussion about Cassirer and the relativized a priori in Chap. 7, after presenting his argument for the applicability of non-Euclidean geometry.

Cassirer indicated the mathematical concept of function as the clearest expression now available for the view that the cognition of the objects of experience is made possible by a complex of conceptual relations into which empirical evidence is built. The example of non-Euclidean geometry played a key role in the development of this view: Cassirer's interpretation of the logic of knowledge implied a generalization of Kant's form of outer intuition to a system of hypotheses, including both Euclidean and non-Euclidean geometries as special cases. In order to introduce such a view, the following chapters offer a reconstruction of the earlier debate on geometrical axioms initiated by Helmholtz and of the reception of Helmholtz's argument for the possibility of a physical interpretation of non-Euclidean geometry in neo-Kantianism.

References

Beiser, Frederick C. 2014. *The genesis of neo-Kantianism: 1796–1880*. Oxford: Oxford University Press.

Cassirer, Ernst. 1902. *Leibniz' System in seinen wissenschaftlichen Grundlagen*. Marburg: Elwert.

Cassirer, Ernst. 1906. *Das Erkenntnisproblem in der Philosophie und Wissenschaft der neueren Zeit*, vol. 1. Berlin: B. Cassirer.

Cassirer, Ernst. 1907a. *Das Erkenntnisproblem in der Philosophie und Wissenschaft der neueren Zeit*, vol. 2. Berlin: B. Cassirer.

Cassirer, Ernst. 1907b. Kant und die moderne Mathematik. *Kant-Studien* 12: 1–49.

Cassirer, Ernst. 1910. *Substanzbegriff und Funktionsbegriff: Untersuchungen über die Grundfragen der Erkenntniskritik*. Berlin: B. Cassirer. English edition: Cassirer, Ernst. 1923. *Substance and Function and Einstein's Theory of Relativity* (trans: Swabey, Marie Collins and Swabey, William Curtis). Chicago: Open Court.

Cassirer, Ernst. 1936. *Determinismus und Indeterminismus in der modernen Physik*. Göteborg: Göteborgs Högskolas Årsskrift 42.

Cassirer, Ernst. 2009. *Nachgelassene Manuskripte und Texte*. Vol. 18: *Ausgewählter wissenschaftlicher Briefwechsel*, ed. John Michael Krois in collaboration with Marion Lauschke, Claus Rosenkranz and Marcel Simon-Gadhof. Hamburg: Meiner.

Cei, Angelo, and Steven French. 2009. On the transposition of the substantial into the functional: Bringing Cassirer's philosophy of quantum mechanics into the twenty-first century. In *Constituting objectivity*, ed. Michel Bitbol, Pierre Kerszberg, and Jean Petitot, 95–115. Dordrecht: Springer.

Cohen, Hermann. 1871a. *Kants Theorie der Erfahrung*. Berlin: Dümmler.

Cohen, Hermann. 1871b. Zur Controverse zwischen Trendelenburg und Kuno Fischer. *Zeitschrift für Völkerpsychologie und Sprachwissenschaft* 7: 249–296.

Cohen, Hermann. 1877. *Kants Begründung der Ethik*. Berlin: Dümmler.

Cohen, Hermann. 1883. *Das Princip der Infinitesimal-Methode und seine Geschichte: Ein Kapitel zur Grundlegung der Erkenntniskritik*. Berlin: Dümmler.

Cohen, Hermann. 1885. *Kants Theorie der Erfahrung*, 2nd ed. Berlin: Dümmler.

Cohen, Hermann. 1889. *Kants Begründung der Ästhetik*. Berlin: Dümmler.

Cohen, Hermann. 1896/1984. Einleitung mit kritischem Nachtrag zur *Geschichte des Materialismus* von F. A. Lange. Repr. in *Werke*, ed. Helmut Holzhey, 5. Hildesheim: Olms.

Cohen, Hermann. 1902. *Logik der reinen Erkenntnis*. Berlin: B. Cassirer.

Couturat, Louis. 1901. *La logique de Leibniz d'après des documents inédits*. Paris: Alcan.

Couturat, Louis. 1905. *Les principes des mathématiques, avec un appendice sur la philosophie des mathématiques de Kant*. Paris: Alcan.

Drobisch, Moritz Wilhelm. 1875. *Neue Darstellung der Logik nach ihren einfachsten Verhältnissen mit Rücksicht auf Mathematik und Naturwissenschaft.* Leipzig: Voss.

Ferrari, Massimo. 1988. *Il giovane Cassirer e la Scuola di Marburgo.* Milano: Angeli.

Ferrari, Massimo. 2010. Is Cassirer a neo-Kantian methodologically speaking? In *Neo-Kantianism in contemporary philosophy,* ed. Rudolf A. Makkreel and Sebastian Luft, 293–313. Bloomington: Indiana University Press.

Ferrari, Massimo. 2012. Between Cassirer and Kuhn: Some remarks on Friedman's relativized a priori. *Studies in History and Philosophy of Science* 43: 18–26.

Fischer, Kuno. 1865. *System der Logik und Metaphysik oder Wissenschaftslehre,* 2nd ed. Heidelberg: Bassermann.

Fischer, Kuno. 1870. *Anti-Trendelenburg: Eine Gegenschrift.* Jena: Dabis.

Friedman, Michael. 1992. *Kant and the exact sciences.* Cambridge, MA: Harvard University Press.

Friedman, Michael. 2000. *A parting of the ways: Carnap, Cassirer, and Heidegger.* Chicago: Open Court.

Gigliotti, Gianna. 1992. "Apriori" e "trascendentale" nella prima edizione di *Kants Theorie der Erfahrung* di H. Cohen. *Studi Kantiani* 5: 47–69.

Heis, Jeremy. 2014. Ernst Cassirer's *Substanzbegriff und Funktionsbegriff. HOPOS: The Journal of the International Society for the History of Philosophy of Science* 4: 241–270.

Herbart, Johann Friedrich. 1825/1850. *Psychologie als Wissenschaft.* Vol. 2. Repr. in *Herbarts Sämmtliche Werke,* ed. Gustav Hartenstein, 6. Leipzig: Voss.

Hertz, Heinrich. 1894. *Die Prinzipien der Mechanik in neuem Zusammenhang dargestellt.* Leipzig: Barth.

Hintikka, Jaakko. 1973. *Logic, language-games and information: Kantian themes in the philosophy of logic.* Oxford: Clarendon.

Hintikka, Jaakko. 1974. *Knowledge and the known: Historical perspectives in epistemology.* Dordrecht: Reidel.

Holzhey, Helmut. 1986. Die Leibniz-Rezeption im "Neukantianismus" der Marburger Schule. In *Beiträge zur Wirkungs- und Rezeptionsgeschichte von Gottfried Wilhelm Leibniz,* ed. Albert Heinekamp, 289–300. Stuttgart: Steiner.

Kant, Immanuel. 1770/1912. De mundi sensibilis atque intelligibilis forma et principiis. Repr. in *Akademie-Ausgabe,* ed. Erich Adickes 2: 387–419. Berlin: Reimer.

Kant, Immanuel. 1783. *Prolegomena zu einer jeden künftigen Metaphysik die als Wissenschaft wird auftreten können.* Riga: Hartknoch. Repr. in *Akademie-Ausgabe* 4: 253–384. Berlin: Reimer.

Kant, Immanuel. 1787. *Critik der reinen Vernunft.* 2nd ed. Riga: Hartknoch. Repr. in *Akademie-Ausgabe,* 3. Berlin: Reimer. English edition: Kant, Immanuel. 1998. *Critique of Pure Reason* (trans: Guyer, Paul and Wood, Allen W.). Cambridge: Cambridge University Press.

Kant, Immanuel. 1884. *Reflexionen Kants zur kritischen Philosophie.* Vol. 2: *Reflexionen Kants zur Kritik der reinen Vernunft,* ed. Benno Erdmann. Leipzig: Fues.

Köhnke, Klaus Christian. 1981. Über den Ursprung des Wortes Erkenntnistheorie und dessen vermeintlichen Synonyme. *Archiv für Begriffsgeschichte* 25: 185–210.

Köhnke, Klaus Christian. 1986. *Entstehung und Aufstieg des Neukantianismus: Die deutsche Universitätsphilosophie zwischen Idealismus und Positivismus.* Suhrkamp: Frankfurt am Main.

Leibniz, Gottfried Wilhelm. 1765. *Nouveaux Essais sur l'entendement humain.* In *Die philosophische Schriften,* ed. Carl Immanuel Gerhardt, 5. Berlin: Weidmann.

Locke, John. 1690. *An essay concerning human understanding.* London: Printed by Elizabeth Holt for Thomas Basset.

Natorp, Paul. 1881/1985. Leibniz und Materialismis. *Studia Leibnitiana* 17:3–14.

Natorp, Paul. 1910. *Die logischen Grundlagen der exakten Wissenschaften.* Leipzig: Teubner.

Ollig, Hans Ludwig. 1979. *Der Neukantianismus.* Stuttgart: Metzler.

Parsons, Charles. 1992. The Transcendental Aesthetic. In *The Cambridge companion to Kant,* ed. Paul Guyer, 62–100. Cambridge: Cambridge University Press.

Patton, Lydia. 2005. The critical philosophy renewed. *Angelaki* 10: 109–118.

Peckhaus, Volker. 1997. *Logik, Mathesis universalis und allgemeine Wissenschaft: Leibniz und die Wiederentdeckung der formalen Logik im 19. Jahrhundert*. Berlin: Akademie Verlag.

Reichenbach, Hans. 1920. *Relativitätstheorie und Erkenntnis apriori*. Berlin: Springer.

Richardson, Alan. 2003. Conceiving, experiencing, and conceiving experiencing: Neo-Kantianism and the history of the concept of experience. *Topoi* 22: 55–67.

Russell, Bertrand. 1900. *A critical exposition of the philosophy of Leibniz with an appendix of leading passages*. Cambridge: Cambridge University Press.

Russell, Bertrand. 1903. *The principles of mathematics*. Cambridge: Cambridge University Press.

Ryckman, Thomas A. 1991. *Conditio sine qua non?* Zuordnung in the early epistemologies of Cassirer and Schlick. *Synthese* 88: 57–95.

Ryckman, Thomas A. 2015. A retrospective view of *Determinism and indeterminism in modern physics*. In *Ernst Cassirer: A novel assessment*, ed. J Tyler Friedman and Sebastian Luft, 65–102. Berlin: De Gruyter.

Trendelenburg, Friedrich Adolf. 1840. *Logische Untersuchungen*. Berlin: Bethge.

Trendelenburg, Friedrich Adolf. 1867. *Historische Beiträge zur Philosophie*. Berlin: Bethge.

Trendelenburg, Friedrich Adolf. 1869. *Kuno Fischer und sein Kant: Eine Entgegnung*. Leipzig: Hirzel.

Vaihinger, Hans. 1876. Über den Ursprung des Wortes *Erkenntnistheorie*. *Philosophische Monatshefte* 12: 84–90.

Zeman, Vladimir. 1980. Leibniz influence on the Marburg School, in particular on Hermann Cohen's conception of reality and of the "Infinitesimal-Methode." *Studia Leibnitiana Supplement* 21: 145–152.

Chapter 3
Axioms, Hypotheses, and Definitions

3.1 Introduction

The development of non-Euclidean geometry in the nineteenth century led mathe-maticians, scientists, and philosophers to reconsider the foundations of geometry. One of the issues at stake was to redefine the notion of geometrical axiom and to establish criteria of choice among different axiomatic systems in case of equivalent geometries. What are geometrical axioms? What is their origin? How are they related to the concept of space? The possibility of considering a variety of hypoth-eses concerning physical space appeared to contradict Kant's conception of geo-metrical axioms as a priori synthetic judgments. Therefore, Riemann called geometrical axioms hypotheses, and maintained that the geometry of physical space is a matter for empirical investigation. To support such a view, Helmholtz pointed out the empirical origin of geometrical axioms. At the same time, he foreshadowed a conventionalist conception of geometrical axioms as definitions that can be abstracted from our experiences with solid bodies and their free mobility. In order for abstract definitions to apply to the empirical manifold, they require an additional physical interpretation which is not determined a priori.

This chapter is devoted to Helmholtz's objections to Kant and to the early neo-Kantians' strategies for defending the aprioricity of geometrical axioms. This topic requires us to bear in mind the philosophical discussion of Kant's Transcendental Aesthetic and Cohen's theory of the a priori: Cohen was able to readapt his notion of the a priori to the case of geometrical axioms because he fundamentally relativ-ized Kant's notion. For the same reason, Cohen agreed with Helmholtz that the principles of geometry should be determined in connection with those of mechan-ics, and that the methods required in geometry are analytic rather than synthetic or constructive, as in Euclid's geometry.

Another line of argument goes back to Alois Riehl, who argued for the aprioric-ity of Euclid's axioms. Interestingly, Riehl did not deny the empirical origin of three-dimensionality. However, he maintained that the remaining, formal properties

© Springer International Publishing Switzerland 2016
F. Biagioli, *Space, Number, and Geometry from Helmholtz to Cassirer*,
Archimedes 46, DOI 10.1007/978-3-319-31779-3_3

of space suffice to assume that the geometry of space must be Euclidean. This became a standard argument in neo-Kantianism: variants of it are found in Paul Natorp (1901, 1910), Bruno Bauch (1907), and Richard Hönigswald (1909, 1912). Nevertheless, I emphasize the difference between this line of argument and Cohen's in Sects. 3.3 and 3.4. My suggestion is that Cohen developed a more plausible view because he was not committed to any spatial structure independently of empirical science.[1]

3.2 Geometry and Mechanics in Nineteenth-Century Inquiries into the Foundations of Geometry

It was only gradually that the inquiry into the foundations of geometry became a purely mathematical issue. Nineteenth-century mathematicians, such as Carl Friedrich Gauss and Bernhard Riemann, considered the foundations of geometry in connection with mechanics. Such a research program had philosophical as well as mathematical aspects. Gauss called into question the aprioricity of geometry and developed the conviction that the geometry of space should be determined a posteriori. It was such a view that motivated Riemann to reformulate the hypotheses underlying geometry from a more general viewpoint. Riemann explored the mathematical presuppositions for making a choice among hypotheses in physics.

A related question is whether geometrical propositions can be empirically tested. One of the founders of non-Euclidean geometry, Nikolay Lobachevsky, sought to detect whether the sum of the angles in a triangle is equal to or less than 180° by means of astronomical measurements. Sartorius von Waltershausen reported that Gauss had also made such an attempt during his geodetic work (Sartorius von Waltershausen 1856, p.81).

A more refined kind of empiricism was proposed by Helmholtz. He made it clear that geometrical assumptions cannot be tested directly. Such a test must be indirect because of the origin of geometrical axioms. Though Helmholtz maintained that geometrical axioms have empirical origins, he emphasized the role of cognitive functions and inferences in the formation of geometrical notions. Geometrical structures, as idealized constructions, can correspond only approximately to empirical contents presently under consideration. Possibly different (e.g., non-Euclidean) interpretations of the same phenomena cannot be excluded. From Helmholtz's viewpoint, Gauss's objection against the aprioricity of geometry can be reformulated as follows: the form of outer intuition analyzed by Kant a priori does not suffice to account for all possible interpretations of spatial perception. Helmholtz's a posteriori (i.e., psychological) analysis should provide us with a comprehensive classification of hypotheses.

---

[1] Regarding the comparison between Cohen's and Riehl's strategies, see also Biagioli (2014).

3.2.1 Gauss's Considerations about Non-Euclidean Geometry

In the 1820s, János Bolyai and Nikolay Lobachevsky, independently of each other, developed a new geometry that is based upon the denial of Euclid's fifth postulate, namely, the proposition that if a straight line falling on two straight lines makes the interior angles on the same side less than two right angles, then the two straight lines, if produced indefinitely, meet on that side on which the angles are less than two right angles. This postulate is usually called the "parallel postulate," because it is used to prove properties of parallel lines. One of its consequences is the fact that the sum of the interior angles of any triangle equals two right angles. The denial of the parallel postulate leads to the hypothesis that the sum of the interior angles of any triangle is either less or greater than two right angles. Such a development had been anticipated by such mathematicians as Girolamo Saccheri, Johann Heinrich Lambert, and Adrien-Marie Legendre, who sought to prove Euclid's fifth postulate by denying it and obtaining a contradiction from the said hypotheses. Since these and many other attempts to prove Euclid's fifth postulate failed, the theory of parallel lines had lost credibility at the time Bolyai and Lobachevsky wrote. For this reason, their works remained largely unknown at that time.

Carl Friedrich Gauss was one of the first mathematicians to recognize the importance of non-Euclidean geometry. However, he expressed his appreciation of the works of Bolyai and Lobachevsky only in his private correspondence, which was published posthumously in the second half of the nineteenth century. Even before becoming acquainted with these works, Gauss maintained that the necessity of Euclid's geometry cannot be proved. Therefore, in a letter to Olbers dated April 28, 1817, Gauss claimed that "for now geometry must stand, not with arithmetic which is pure a priori, but with mechanics" (Gauss 1900, p.177; Eng. trans. in Gray 2006, p.63). Gauss wrote in a letter to Bessel dated April 9, 1830:

> According to my most sincere conviction the theory of space has an entirely different place in knowledge from that occupied by pure mathematics. There is lacking throughout our knowledge of it the complete persuasion of necessity (also of absolute truth) which is common to the latter; we must add in humility that if number is exclusively the product of our mind, space has a reality outside our mind and we cannot completely prescribe its laws. (Gauss 1900, p.201; Eng. trans. in Kline 1980, p.87)

It is tempting to relate Gauss's opinion to his later claim that, since 1792, he had developed the conviction that a non-Euclidean geometry would be consistent (see Gauss's letter to Schumacher dated November 28, 1846 in Gauss 1900, p.238). However, Gauss's knowledge about non-Euclidean geometry before his reading of the works of Bolyai and Lobachevsky is hard to reconstruct, and the question whether his views about space and geometry presuppose some knowledge of non-Euclidean geometry is controversial. The problem is that Gauss could hardly have possessed the concept of a non-Euclidean three-dimensional space. For the same reason, it might be questioned whether Gauss deliberately undertook an empirical test of Euclid's geometry, as reported by Sartorius: such a test would imply that non-Euclidean geometry is a possible alternative to Euclidean geometry. Furthermore, the measurement Sartorius refers to would not suffice to put Euclidean geometry to the test. Arguably, Gauss might have mentioned that measurement in his inner circle because it incidentally confirmed his conviction that Euclidean geometry is true within the limits of the best observational error of his time.[2]

Nevertheless, Gauss's empiricist insights were influential in nineteenth-century philosophy of geometry. After Gauss's correspondence on these matters was published, it was quite natural to associate Gauss's claims with the survey of geometrical hypotheses presented by Bernhard Riemann in his habilitation lecture of 1854 "On the Hypotheses Which Lie at the Foundation of Geometry." In fact, it was Gauss who chose the topic of the lecture as Riemann's advisor. The lecture was published posthumously in the *Abhandlungen der Königlichen Gesellschaft der Wissenschaften zu Göttingen* in 1867. Riemann's work posed the following problem: since different geometries are logically possible, none of them can be necessary or grounded a priori in our conception of space. How can one choose between equivalent hypotheses? The view that the geometry of space is a matter for empirical investigation became known as the "Riemann-Helmholtz theory of space" (see, e.g., Erdmann 1877). As we will see in the next section, Riemann's views differed considerably from Helmholtz's. Nevertheless, Helmholtz presented his inquiry into the foundations of geometry as a development of Riemann's inquiry. It is true that both Riemann and Helmholtz ruled out the aprioricity of Euclidean geometry. The fact that these views were associated, especially by philosophers, shows that the philosophical reception of Riemann tended to be mediated by Helmholtz.

3.2.2   Riemann and Helmholtz

Both Riemann's and Helmholtz's inquiries into the foundations of geometry played a fundamental role in the discussion on the philosophical consequences of non-Euclidean geometry. It is noteworthy, however, that, in 1854, Riemann might not have known about the works of Bolyai and Lobachevsky. Riemann's survey of geometries includes non-Euclidean geometries. However, this fact was not apparent before Eugenio Beltrami's (1868) proof that Bolyai-Lobachevsky geometry applies to surfaces of constant negative curvature. Furthermore, we know from one of

---

[2] This interpretation of Sartorius's report has been proposed by Ernst Breitenberger (1984). More recently, Jeremy Gray points out that Gauss called into question the necessity of Euclidean geometry because he focused on the problems concerning the definition of the plane and that of parallel lines. However, he did not start with three-dimensional non-Euclidean space as Bolyai and Lobachevsky did (Gray 2006, p.75). On the other hand, Erhard Scholz defends the interpretation of Gauss's measurement as an empirical test of Euclid's geometry in the following sense. Even though Gauss could not have known non-Euclidean geometry at that time, his study of the geometric properties of surfaces enabled him to make heuristic assumptions about physical space. Scholz interprets the limit of approximation in Gauss's experiment as an informal counterpart of the upper limit of a measure of the curvature of space that is compatible with the results of measurement (Scholz 2004, pp.364–365).

Riemann's early fragments about the concept of manifold – which were published by Erhard Scholz in 1982 – that Riemann, unlike the founders of non-Euclidean geometry, was not willing to adopt a purely analytical approach. Regarding the possibility of abstracting from all the axioms that are grounded in intuition (e.g., the claim that two points determine a line) and retaining only those axioms that concern abstract quantities (e.g., the commutative law of addition), he wrote:

> Although it is interesting to acknowledge the possibility of such a treatment of geometry, the implementation of the same would be extremely unproductive, because it would not enable us to find out any new proposition, and because thereby what appears to be simple and clear in spatial representation would become confused and complicated. Therefore, everywhere I have taken the opposite direction, and everywhere in geometry I encountered multidimensional manifolds, as in the doctrine of the definite integrals of the theory of imaginary quantities, I made use of spatial intuition. It is well known that only by doing so one obtains a comprehensive overview on the object under consideration and a clear insight into the essential points. (Riemann XVI, 40^r, in Scholz 1982b, p.229)

In the same note, Riemann proposed adopting an approach based on real affine geometry. In 1854, he replaced his former, global definition of a straight line via linear equations with the locally defined concept of geodesics. Nevertheless, this fragment suggests that one of the guiding ideas of both approaches was that the study of manifolds was a necessary presupposition for the analytic treatment of the foundations of geometry.[3]

At the beginning of his lecture of 1854, Riemann emphasized the originality of his approach by saying that his conception of space differed from most of the conceptions proposed by philosophers and was influenced only by Herbart and Gauss (Riemann 1996, p.653). The idea was to define space as a special kind of magnitude. Therefore, Riemann developed the more general concept of manifold, which he introduced as an n-fold extended magnitude, and tended to conceive, more generally, as a set, along with a class of continuous functions acting on it. He thereby extended Gauss's theory of surfaces to n-dimensional manifolds. Gauss's theory admits such an extension because it enables the study of the intrinsic properties of surfaces, especially curvature, independently of the assumption of a surrounding, three-dimensional space.[4]

Regarding Herbart's influence on Riemann, it is worth noting that "manifold" (*Mannigfaltigkeit*) occurred in philosophical texts to indicate any series of empirical data. In *Psychology as a Science* (1825), Herbart criticized Kant for having analyzed

---

[3] See Scholz (1982b, pp.218–219). According to Scholz, the charge of sterility applies not so much to the new tradition initiated by Gauss, as to the older tradition of Adrien-Marie Legendre (1794), among others. Whereas the older tradition dealt with the foundation of Euclidean geometry, Bolyai and Lobachevsky considered non-Euclidean hypotheses as new propositions. This fact confirms the conjecture that Riemann was not acquainted with the works of these mathematicians, at least when he wrote the note above. However, it is also worth noting that both in the 1854 lecture and in his earlier fragments on the concept of manifold, Riemann's goal was to develop general concepts for a unified approach to the foundations of geometry. Regardless of Riemann's relationship to the aforementioned traditions, his approach differed from any approach that is based exclusively on calculus; cf. Pettoello (1988, p.713).

[4] On Riemann's concept of manifold, see Torretti (1978, pp.85–103); Scholz (1980, Ch.2).

space and time independently of empirical factors and, accordingly, for having assumed a manifold of pure intuition. Herbart maintained that the concepts of space and time are abstracted from empirically given spatial and temporal manifolds. He included spatiality and temporality in the more general concept of a continuous serial form. Other examples of serial forms, according to Herbart, include the linear representation of tones and the color triangle with three primary colors at its corners and the mixing of the colors in the two-dimensional continuum in between (Herbart 1825/1850, vol. 2, Ch.2). These examples suggest that a "spatial" ordering of sense qualities in Herbart's sense does not depend on the nature of the single elements of a manifold – which are nonspatial – but on our construction.

Arguably, Riemann became acquainted with Herbart's philosophy in Göttingen, where Herbart had finished his career in 1841. At the time of Riemann's studies, Herbart's ideas were being lively discussed in the philosophy faculty, and we know from Riemann's Nachlass that he attended classes in philosophy during the same period. Furthermore, Riemann's Nachlass provides evidence of his interests in philosophy and of his commitment to Herbart's epistemology in particular. In a note to his philosophical fragments he declared that the author (i.e., Riemann himself) "is a Herbartian in psychology and in the theory of knowledge (methodology and eidolology), but for the most part he cannot own himself a follower of Herbart's natural philosophy and the metaphysical disciplines related to it (ontology and synechology)" (Riemann 1876, p.476). Herbart's synechology contained his science of the continuum and formed the part of his metaphysics which lies at the foundation of psychology and the philosophy of nature. It is controversial whether and to what extent Herbart might have influenced Riemann. However, arguably Riemann's concept of manifold played a similar role as the concept of a continuous serial form in Herbart's psychology: both Herbart and Riemann were looking for a general concept for a unified treatment of a variety of spaces. This is confirmed by the fact that Riemann, similar to Herbart, mentions color as an example of continuous manifold in Section I.1 of his lecture of 1854.[5]

---

[5] One of the first to emphasize Herbart's influence on Riemann in this respect was Bertrand Russell, who, regarding the quote above from Riemann's Nachlass, wrote: "Herbart's actual views on Geometry, which are to be found chiefly in the first section of his Synechologie, are not of any great value, and have borne no great fruit in the development of the subject. But his psychological theory of space, his construction of extension out of series of points, his comparison of space with the tone and colour-series, his general preference for the discrete above the continuous, and finally his belief in the great importance of classifying space with other forms of series (*Reihenformen*), gave rise to many of Riemann's epoch-making speculations, and encouraged the attempt to explain the nature of space by its analytical and quantitative aspect alone" (Russell 1897, pp.62–63). By contrast, Torretti (1978, p.108) pointed out the difference between Herbart's qualitative continua – which cannot be identified with sets of points – and Riemann's construction of an *n*-dimensional manifold by successive or serial transition from one of its points to the others. Further evidence of Herbart's influence on Riemann was offered by the Riemann Nachlass at Göttingen University library. This material includes Riemann's notes and excerpts from his studies of Herbart. However, the interpretation of Riemann's claims about Herbart remained controversial. Scholz maintains that Herbart influenced Riemann much more in general epistemology than in his particular philosophy of space. According to Scholz, the dominant background of Riemann's conception of

Riemann's issue was to discover the simplest matters of fact from which the metric relations of space can be determined. He called geometrical axioms "hypotheses" because these matters of fact – like all matters of fact – are not necessary: their evidence is only empirical (Riemann 1996, pp.652–653). However, Riemann also gave a more specific reason for the hypothetical character of the said relations. Relations of measure must be distinguished from relations of extension: the former can be varied only continuously, whereas the variation of extensive relations (e.g., the number of dimensions of a manifold) is discrete. It follows that, on the one hand, claims about extensive relations can be either true or false and, on the other hand, claims about metric relations of space can only be more or less probable. The statement, for example, that space is an unbounded, threefold extended manifold is an assumption that is presupposed by every conception of the outer world. Therefore, the unboundedness of space possesses a greater empirical certainty than any external experience. However, its infinite extent, which is a hypothesis concerning metric relations, does not follow from this. Riemann's conclusion is that claims about the infinitely great lack empirical evidence (p.660).

On the other hand, Riemann believed that causal knowledge depends essentially upon the exactness with which we follow phenomena into the infinitely small. The problem here is that the empirical notions on which the metrical determinations of space are founded (e.g., the notion of a solid body and of a ray of light) seem to have no empirical referent. A related problem is this: whereas the ground of the metric relations in a discrete manifold is given in the notion of it, the ground in a continuous manifold must come from the outside. This is because, with the same extensive properties, different metric relations are conceivable. Manifolds of constant curvature, for example, can have an Euclidean or non-Euclidean metric. Riemann's supposition is that space is a continuous, threefold extended manifold which admits an infinite number of possible geometries.

In the concluding remarks of his lecture, Riemann emphasized the relevance of his inquiry to the physical investigation of space. Riemann's starting point was Gauss's view of space as an object rather than a necessary presupposition of research. Therefore, we cannot know from the outset whether space is continuous or discrete. If space is supposed to be a continuous manifold, it follows from the above classification that the ground of the metric relations of space is to be found in the binding forces which act on it. That is to say, the question of the metric relations of space and of the validity of the hypotheses of geometry in the infinitely small depends on natural science. Riemann wrote:

---

manifold is found rather in a tendency of nineteenth-century mathematics to transfer geometric thinking to non-geometric fields, and this tendency was at least partially known to Riemann via Gauss (Scholz 1982a, p.423). Notwithstanding the originality or Riemann's mathematical achievements, Pettoello (1988) reconsiders the influence of Herbart's Leibnizian conception of space as one of several possible serial forms on the general aims of Riemann's 1854 lecture. For another reading opposed to Scholz, also see Banks (2005), who places Riemann in a more direct line from Herbart and his philosophical project of reducing spatial notions to the behavior of inner states in the infinitely small.

The answer to these questions can only be got by starting from the conception of phenomena which has hitherto been justified by experience, and which Newton assumed as a foundation, and by making in this conception the successive changes required by facts which it cannot explain. Researches starting from general notions, like the investigation we have just made, can only be useful in preventing this work from being hampered by too narrow views, and progress in knowledge of the interdependence of things from being checked by traditional prejudices. (Riemann 1996, p.661)

Helmholtz's connection with Riemann goes back to his paper of 1868, "On the Facts Underlying Geometry." The title of Helmholtz's paper is clearly reminiscent of Riemann's title. At the same time, Helmholtz's title announces his view that geometry is grounded not so much in the hypotheses derived from the general theory of manifolds, as in some facts to be induced by observation and experiment. Helmholtz's inquiry is based especially on the free mobility of rigid bodies, which is the observation that some kinds of bodies (i.e., solid bodies) remain unvaried in shape and size during displacements. Helmholtz's interpretation of this fact is the requirement that each point of a system in motion can be brought to the place of another, provided that all points of the system remain fixedly interlinked. According to Helmholtz, the free mobility of rigid bodies and the remaining facts underlying geometry (i.e., n-dimensionality and the monodromy of space) provide us with the necessary and sufficient conditions to obtain a Riemannian metric of constant curvature.

Helmholtz seems to overlook the distinction between the finite level and the infinitesimal one. In fact, the free mobility of rigid bodies does not imply a metric of constant curvature. Helmholtz's proof was corrected by the Norwegian mathematician Sophus Lie, who deduced the same metric from a set of conditions at the infinitesimal level (Lie 1893, pp.437–471).[6]

Furthermore, Riemann's conception of space differs from Helmholtz's conception because Riemann was not committed to the supposition that space is a manifold of constant curvature. Riemann's conjecture was that the curvature of space might be variable at the infinitesimal level, provided that the total curvature for intervals of a certain size equals approximately zero (Riemann 1996, p.661). Since Riemann also considered manifolds of variable curvature, the scope of his inquiry was wider than that of Helmholtz. Nevertheless, Helmholtz's reason for adopting the theory of manifold was the same as Riemann's: they both assumed the concept of manifold was primitive in order to avoid unnecessary restrictions on the conception of space. Since manifolds of constant curvature are continuous, Helmholtz maintained that the geometry of space depends on experience and can be compared to the structure of such empirical manifolds as the color system (Helmholtz 1868, p.40).

In a letter dated April 24, 1869, Eugenio Beltrami made Helmholtz aware of the fact that the pseudospherical circle Beltrami introduced in his "Essay on an Interpretation of Non-Euclidean Geometry" (1868) and "Fundamental Theory of Spaces of Constant Curvature" (1869) satisfied all the properties of space assumed by Helmholtz, and even infinity. Beltrami proved that such a surface provided an interpretation of Bolyai-Lobachevsky geometry. After reading Beltrami's letter

---

[6] For a thorough comparison between Helmholtz and Lie, see Torretti (1978, pp.158–171).

(now available in Boi et al. 1998, pp.204–205), Helmholtz realized that if space is supposed to be a manifold of constant curvature, a choice has to be made between Euclidean and non-Euclidean geometry. Helmholtz accounted for this generalization of the problem in the lecture he gave in Heidelberg in 1870 on "The Origin and Meaning of Geometrical Axioms." On that occasion, he presented a series of thought experiments to prove that the choice of geometry presupposes a series of observations whose laws are not necessarily Euclidean. The next section deals with the philosophical conclusions of Helmholtz's thought experiments, with a special focus on his objections to Kant.

3.2.3 Helmholtz's World in a Convex Mirror and His Objections to Kant

As we saw in the first chapter, Helmholtz's analysis of the concept of space presupposes empirical intuitions. His first remark against Kant in the Heidelberg lecture is that the expression "to represent" or "to be able to think how something happens" can only be understood as the power of imagining the whole series of sensible impressions that would be had in such a case (Helmholtz 1870, p.5). Helmholtz's theory of spatial perception and his thought experiments about the representation of the relations of measure under the hypothesis of a non-Euclidean space were supposed to contradict Kant's assumption of an unchangeable form of intuition underlying any phenomenal changes.[7] Helmholtz's argument was based on Beltrami's interpretation of Bolyai-Lobachevsky geometry on a pseudospherical surface, which is a surface of constant negative curvature. Helmholtz extended Beltrami's interpretation to the three-dimensional case and described what would appear to be the conditions of motion in the imaginary world behind a convex mirror: for every measurement in our world, there would be a corresponding measurement in the mirror. The hypothetical inhabitant of such a world may not be aware of the contractions of the distances she measures, because these would appear to be contracted only when compared with the results of the corresponding measurements outside the mirror. Therefore, she may adopt Euclidean geometry. At the same time, the geometry of her world would appear to us to be non-Euclidean. Helmholtz's conclusion was that both geometries are imaginable. Geometrical axioms cannot be

---

[7] According to DiSalle (2008, p.76), Helmholtz's account of "imagination" is a philosophical analysis of the assumptions upon which the Kantian "productive imagination" relies implicitly. However, there is no mention of the Kantian notion of "productive imagination" in Helmholtz's text. I consider the quote above to be a critical remark against Kant because, as argued below, Helmholtz's argument rules out cognitions other than empirical and intellectual cognitions. In Helmholtz's account of imagination, there is no place for the faculty of pure intuition, which is essential to Kant's account of imagination.

necessary consequences of the form of spatial intuition. In fact, they are not necessary at all, but may be varied under empirical circumstances.[8]

It follows from Helmholtz's considerations that there is an entire class of geometries that may be adopted in physics, namely, the class that corresponds to the manifolds of constant curvature, which can be negative, positive or equal zero. Helmholtz ruled out Kant's form of spatial intuition insofar as this is restricted to the third, Euclidean case. Helmholtz foreshadowed the possibility of generalizing the form of intuition so that non-Euclidean geometries can be included. However, he maintained that, even in such a case, geometrical axioms should not be thought of as synthetic a priori judgments. He concluded his lecture with the following remark:

> [T]he axioms of geometry certainly do not speak of spatial relationships alone, but also, at the same time, of the mechanical behavior of our most fixed bodies during motions. One could admittedly also take the concept of fixed geometrical spatial structure to be a transcendental concept, which is formed independent of actual experiences and to which these need not necessarily correspond, as in fact our natural bodies are already not even in wholly pure and undistorted correspondence to those concepts which we have abstracted from them by way of induction. By adopting such a concept of fixity, conceived only as an ideal, a strict Kantian certainly could then regard the axioms of geometry as propositions given a priori through transcendental intuition, ones which could be neither confirmed nor refuted by any experience, because one would have to decide according to them alone whether any particular natural bodies were to be regarded as fixed bodies. But we would then have to maintain that according to this conception, the axioms of geometry would certainly not be synthetic propositions in Kant's sense. For they would then only assert something which followed analytically from the concept of the fixed geometrical structures necessary for measurement, since only structures satisfying those axioms could be acknowledged to be fixed ones. (Helmholtz 1870, pp.24–25)

Either the axioms of geometry can be derived from experience – as Helmholtz believed – or they express the consequences that are implicit in the definition of rigid body. The Kantian would be left with the conventionalist option of considering geometrical axioms as definitions. We consider the geometrical conventionalism proposed later by Poincaré in Chap. 6. For now, it suffices to notice that Helmholtz himself tended to conceive geometrical axioms as definitions (e.g., of rigidity). He formulated the question concerning the foundations of geometry as follows: "How much of the propositions of geometry has an objectively valid sense? And how much is on the contrary only definition or the consequence of definitions, or depends on the form of description?" (Helmholtz 1868, p.39). Furthermore, Helmholtz explicitly identified axioms as definitions in the case of arithmetic (see,

---

[8]The validity of Helmholtz's conclusion is restricted to the use of geometry in the interpretation of empirical measurements in finite regions of space. It is noteworthy that his thought experiment does not provide a model of non-Euclidean geometry. Not only did Helmholtz present it only as a thought experiment, but Hilbert later proved the impossibility of the pseudospherical model if the entire plan of Bolyai-Lobachevsky geometry is considered. Hilbert's proof rules out, a fortiori, such a model in the three-dimensional case (see Hilbert 1903, pp.162–172). For an interpretation of Helmholtz's thought experiment as an attempt to provide a model of non-Euclidean geometry, cf. Coffa (1991, pp.48–54). Some of the problems of such an interpretation are discussed in Chap. 6.

e.g., Helmholtz 1903, p.27; 1887, p.94). This conventionalist reading of Helmholtz goes back to Schlick, who used Helmholtz's argument for the applicability of non-Euclidean geometry to infer the conventionality of geometry in Poincaré's sense. We turn back to this reading of Helmholtz in Chap. 7.

A similar view has been advocated more recently by Alberto Coffa, who includes Helmholtz in what Coffa called the "semantic" tradition that developed from the nineteenth-century debate about synthetic judgments a priori. From this viewpoint, Helmholtz's views about the origin and meaning of geometrical axioms led to a more general consideration about the status of what Kant called synthetic a priori judgments: "Many fundamental scientific principles are by no means necessarily thought – indeed, it takes great effort to develop the systems of knowledge that embody them; but their denial also seems oddly impossible – they need not be thought, but if they are thought at all, they must be thought as necessary" (Coffa 1991, p.55). A more complex picture emerges if one considers that the reception of Helmholtz's empiricism ramified in at least three branches: 1) empiricism as opposed to Kantianism, 2) conventionalism, and 3) a variant of Kantian transcendentalism according to which the facts underlying geometry provide us, at the same time, with necessary preconditions for the possibility measurement.[9]

Helmholtz's view of geometrical axioms differs from the conventionalist view, because his emphasis is not so much on our freedom in the formulation of definitions, as on the need for a physical interpretation in order for definitions to apply to empirical reality. In Helmholtz's view, such an interpretation should be induced by observation and experiment: the objective meaning of the definitions under consideration presupposes both mathematics and physics. It follows that the principles of geometry may be subject to revision according to mechanical considerations, which could not be the case if these principles were synthetic a priori judgments in Kant's sense or mere definitions. Synthetic a priori judgments cannot be revised, and mere definitions cannot be put to the test empirically, although they can be arbitrarily changed.

Helmholtz's empiricism has been contrasted with conventionalism especially by DiSalle (2006, p.134): what distinguishes Helmholtz from Poincaré is that, in the case of a choice among hypotheses, mechanical considerations are decisive according to Helmholtz, whereas considerations of mathematical simplicity would suffice for Poincaré. According to DiSalle, the limit of the solutions to the problem proposed by Helmholtz and by Poincaré lies in the fact that the idea that space must be homogeneous proved to be an over-simplification when compared to Einstein's

---

[9] This classification was proposed by Torretti (1978, p.163), and a more detailed reconstruction of the reception of Helmholtz in these traditions is found in Carrier (1994). Regarding the transcendental interpretation of Helmholtz, both Torretti and Carrier focus on Hugo Dingler's metrogenic apriorism. Although I agree that the conventionalist reading of Helmholtz overlooked other aspects of his philosophy of geometry, I do not think that the transcendental reading is committed to the aprioricity of Euclidean geometry as advocated by Dingler. Cohen and Cassirer – whose views are not discussed in the aforementioned studies – show that the constitutive function of the preconditions of measurement might as well be compatible with the aprioricity of a system of hypotheses, including non-Euclidean geometries.

general relativity. In addition, even in 1854, Riemann pointed out that some deeper insight into the nature of bodies and their microscopic interactions was required to address the question of the applicability of geometrical concepts to the infinitely small. Nevertheless, DiSalle's reading enables him to relate Helmholtz to the empiricist view that "dynamical principles – principles involving time as well as space – could force revision of the spatial geometry that had been originally assumed in their development. We might say that this view acknowledges the possibility, at least, that space-time is more fundamental as space" (DiSalle 2008, p.91).[10]

Before turning to the reception of Helmholtz in neo-Kantianism, it is worth adding a few remarks about Helmholtz's methodological views. In Chap. 1 (Sect. 1.4), we noticed that Helmholtz's way of explaining the connection between geometry and physics presupposes metrical notions and analytical methods: arithmeticized quantities and calculations are required for physical magnitudes to be measured. By contrast, Kant apparently believed that synthetic or constructive methods are indispensable in geometry. Therefore, he sharply distinguished geometry from arithmetic. Consider Kant's claims about space (see Sect. 2.2). Kant apparently believed that the infinite divisibility of space followed from Claims 3 and 4. It might seem that the homogeneity of space also depends on such claims. Helmholtz's point is that infinite divisibility already presupposes divisibility into equal parts. What makes the Kantian theory of space unclear about this fact is that Kant arguably bore in mind Euclid's method of proof, which rests upon the congruence of lines, angles, and so on. Since the free mobility of rigid bodies is implicit in this way of proceeding, it seems that all metrical notions can be derived from the intuition that is involved in Euclid's proofs; but once the free mobility of rigid bodies is made explicit, the supposition of Euclidean congruence is called into doubt. In the situation imagined by Helmholtz, both measurements in our world and in the mirror may satisfy the free mobility of rigid bodies, so that both Euclidean and non-Euclidean geometries may be adopted. This speaks in favor of Riemann's definition of space as a special kind of manifold. Helmholtz believed that the generality of our classifications presupposed an analytical approach to geometry. He even interpreted Riemann's theory of manifolds as a result of such an approach (Helmholtz 1870, p.12. Cf. Riemann p.XVI, 40^r, already quoted in Sect. 3.2.2).

The disagreement between Kant and Helmholtz regarding the method of geometry has been emphasized by Darrigol (2003, p.549) and by Hyder (2006, pp.34–35). Both Darrigol and Hyder show that Helmholtz's standpoint goes back the

---

[10] However, it seems to me that DiSalle himself relies largely upon the conventionalist reading for the reconstruction of Helmholtz's argument as a conceptual analysis of what Kant called pure intuition (see especially DiSalle 2006). Thus, it might seem that the empirical aspect of Helmholtz's analysis only depends on the objects under consideration, which are physical objects. However, Helmholtz distanced himself from the assumption of pure intuitions, because he believed that even the simplest spatial intuitions presuppose interaction with external reality and deserve an empirical explanation. Therefore, I think that the main issues at stake in his objections to Kant are methodological issues, and cannot be solved by adopting a formalistic account of Kant's spatial intuition. I return to this aspect of Helmholtz's view after considering some of the rejoinders to his objections against Kant in early neo-Kantianism.

manuscript from the 1840s already discussed in Chap. 1. Helmholtz's (1870) meth-odological considerations, after his correspondence with Beltrami, suggest that the use of analytic methods offered a twofold argument against Kant: not only is the metrical aspect of the notion of congruence necessary for geometry to be used in physics, but analytic geometry provides us with a more comprehensive classifica-tion of the hypotheses that can occur in the description of physical space than Euclidean geometry. Regarding the special assumptions to be made, the possibility of imagining the series of impressions that would be had in the case of a non-Euclidean space should confirm Helmholtz's view that geometrical axioms have an empirical origin and the choice between equivalent geometries is to be made on empirical grounds.

Helmholtz made this point clearer in his paper of 1878, "The Facts in Perception." In the second appendix to this paper, Helmholtz called those magnitudes physically equivalent in which under similar conditions and within equal periods of time simi-lar physical processes take place (Helmholtz 1878, p.153). Here, Helmholtz pointed out explicitly the connection between arithmetic, geometry, and measurement fore-shadowed in his manuscript from the 1840s. Two different magnitudes can be com-pared by superposition of a measuring rod. However, this does not suffice for measurement. If the results of measurements with rule and compass are to provide knowledge, magnitudes that have been proved to be equal by a sufficiently exact comparison must manifest equivalence in any further cases. Physical equivalence of two or more magnitudes, as an objective property of the same, requires every com-parison of spatial magnitudes to find a numerical expression and follow the laws of arithmetic. Helmholtz called such a comparison physical geometry, and distin-guished it from the pure geometry that is supposed to be grounded in our spatial intuition.

In 1878, Helmholtz reformulated his objection to Kant as follows. Suppose that spatial intuition and physical space are related to each other as actual (Euclidean) space is related to its (non-Euclidean) image in a convex mirror. In such a case, physical geometry may not necessarily agree with pure geometry regarding the equality of the parts of space. Helmholtz's conclusion was the following:

> If there actually were innate in us an irradicable form of intuition of space which included the axioms, we should not be entitled to apply it in an objective and scientific manner to the empirical world until one had ascertained, by observation and experiment, that the parts of space made equivalent by the presupposed transcendental intuition were also physically equivalent. (Helmholtz 1878, p.158)

This is a realist description of the situation: either pure geometry agrees with physi-cal geometry or the supposedly a priori knowledge founded in spatial intuition is, in fact, an "objectively false semblance" (p.158). Helmholtz maintained that his argu-ment holds true from an idealist viewpoint as well. He distinguished between the "topogenous" factors of localization and the "hylogenous" ones: the former ones specify at what place in space an object appears to us; the latter ones cause our belief that at the same place, we perceive at different times different material things having different properties. Helmholtz then reformulated his argument as follows:

When we observe that physical processes of various kinds can run their course in congruent spaces during the same periods of time, this means that in the domain of the real, similar aggregates and sequences of certain hylogenous factors can come about and run their course in combination with certain specific groups of different topogenous factors, such namely as give us the perception of parts of space physically equivalent. And when experience then instructs us, that any combination or sequence of hylogenous factors which can exist or run its course in combination with one group of topogenous factors, is also possible with any other physically equivalent group of topogenous factors – then this is at any rate a proposition having a real content, and thus topogenous factors undoubtedly influence the course of real processes. (Helmholtz 1878, pp.160–161)

Helmholtz deemed this version of the argument idealistic, because "objectivity" here was used in another sense than in the previous, realistic version. Objective, in the former sense, seemed to depend on the assumption of a mind-independent reality. However, in the idealistic argument, Helmholtz made it clear that the objectivity of scientific measurements does not presuppose that any specific structure is found in the world: all that is presupposed is that some regularity is found in the phenomena. He identified space with the most general structure or group of topogenous factors underlying such regularities. Objectivity, in this sense, depends not so much on the existence of the objects experienced by us in measurement situations presently under consideration, as on the general conditions of measurement.

Is there a place for a Kantian interpretation of objectivity in Helmholtz's epistemological writings? I think so, provided that one can take into account Helmholtz's objections to Kant. This was the goal of the early neo-Kantian discussions of geometrical empiricism.

To conclude this section, it is noteworthy that Helmholtz himself argued for a revision of the Kantian theory of space. After presenting his argument for the objectivity of physical geometry, he wrote:

Kant's doctrine of the a priori given forms of intuition is a very fortunate and clear expression of the state of affairs; but these forms must be devoid of content and free to an extent sufficient for absorbing any content whatsoever that can enter the relevant form of perception. But the axioms of geometry limit the form of intuition of space in such a way that it can no longer absorb every thinkable content, if geometry is at all supposed to be applicable to the actual world. If we drop them, the doctrine of the transcendentality of the forms of intuition of space is without any taint. (Helmholtz 1878, pp.162–163)

Helmholtz's argument contradicts the aprioricity of geometrical axioms in Kant's sense, because it implies that such propositions as those concerning the specific measure of curvature of a manifold have empirical meaning and can be revised according to empirical considerations. Nevertheless, the quote above suggests that Helmholtz identified the more general structure of a manifold of constant curvature as an a priori form of intuition in Kant's sense. Since Helmholtz considered his mathematical description of space the culmination of his previous studies on human vision, he presented his revised version of the Kantian theory of space in psychological terms: "I believe the resolution of the concept of intuition into the elementary processes of thought as the most essential advance in the recent period. This resolution is still absent in Kant, which is something that then also conditions his conception of the axioms of geometry as transcendental propositions" (p.143).

The neo-Kantians seem to have been particularly influenced by this aspect of Helmholtz's approach, because, even regardless of Helmholtz's work in the physiology of vision, the idea of a "resolution" of the concept of intuition into intellectual processes emerged from the discussion of Kant's Transcendental Aesthetic considered in Chap. 2.[11] Similarly, Helmholtz's claim rules out any role of the Kantian notion of pure intuition in the characterization of space. Helmholtz's equivalent for Kant's form of intuition is derived more by abstracting from empirical contents in the mathematical analysis of the concept of space.

Friedman contrasts Helmholtz's and Kant's approaches as follows:

> It is only because there is no room in Kant's own conception of logical, conceptual, or analytic thought for anything corresponding to pure mathematical geometry that there is a place, accordingly, for a wholly nonconceptual faculty of pure spatial intuition. For Helmholtz, by contrast, there is no difficulty at all in formulating pure mathematical geometry conceptually or analytically with no reference to spatial intuition whatsoever (via the Riemmanian conception of metrical manifold), and an appeal to spatial intuition or perception is only then necessary to explain the psychological origin and empirical application of the pure mathematical concept of space. (Friedman 2000, p.202)

However, I do not agree with Friedman that Helmholtz's argument against the assumption of pure intuition was directed not so much against Kant as against his Kantian contemporaries. Not only did Helmholtz explicitly call into question the meaning of such a faculty in the context of nineteenth-century psychology and of nineteenth-century geometry, but the neo-Kantians, among others, agreed with him on precisely the same point.

The next section deals with the application of Cohen's theory of the a priori to the case of geometrical axioms. In the interpretation proposed, this development offers one of the first examples of a relativized conception of the a priori. According to Cohen, what is given a priori is not so much a set of geometrical propositions, as a general classification of hypotheses, where the level of generality of the classifications under consideration depends on the unifying power of mathematics, on the one hand, and on the history of science, on the other. In the case of geometry, this view led Cohen and Cassirer to agree with Helmholtz that Kant's form of outer intuition deserved to be generalized to a wider structure, including both Euclidean and non-Euclidean hypotheses as special cases.[12]

---

[11] The expression "resolution" of the concept of intuition was introduced as an English translation of the German term *Auflösung*. Since the English translation is literal, I do not have any better suggestion. However, it seems to me that in the German original, Helmholtz is clearer about the fact that the advance in the recent period lies in the substitution of those aspects that were formerly regarded as intuitive with conceptual processes.

[12] For a similar reading of Helmholtz's view of the form of intuition, see Ryckman (2005), DiSalle (2006), and Friedman (2000, 2009). I largely agree with these authors that what is characteristic of Helmholtz's view – especially in the 1878 article on "The Facts in Perception" – is that, whereas the specific structure of space depends on empirical considerations, the free mobility of rigid bodies tends to play the role of a minimal, but necessary, precondition of measurement and, therefore, of a possible experience in general in Kant's sense. The same precondition enables Helmholtz to define space as a manifold of constant curvature, including the three classical cases of such manifolds. However, it seems to me that Friedman's and others' talk about a "generalization of the

3.3 Neo-Kantian Strategies for Defending the Aprioricity of Geometrical Axioms

Cohen's theory of the apriori was the background for later neo-Kantian discussions on the origin and meaning of geometrical axioms. Are these necessary, and if so, how can one escape the conclusion that physical geometry provides us with objective knowledge, even in the case that its judgments differ from those that are derived from pure geometry?

This section considers Riehl's arguments for the homogeneity of space and Cohen's conception of the relation between geometry and physics in the second edition of *Kant's Theory of Experience* (1885). Cohen and Riehl proposed two different strategies for defending the aprioricity of geometrical axioms. In Cohen's view, critical philosophy is not committed to the necessity of a set of propositions; its issue is rather to prove that a connection of conditions is necessarily required for knowledge. By contrast, Riehl restricted aprioricity to some fundamental concepts and emphasized necessity and universality as intrinsic properties of Euclidean geometry.[13]

3.3.1 Riehl on Cohen's Theory of the A Priori

Alois Riehl studied at the universities of Vienna, Munich, Innsbruck, and Graz, where he was appointed lecturer in 1873. At that time, Herbart's philosophy was more popular in Austria than in Germany, and Riehl's first philosophical writing, *Elements of Realism* (1870), shows his commitment to Herbart's realism. In 1872, however, Riehl distanced himself from Herbart's conception of reality as a presupposition for our analysis of experience and defended the existence of the things in themselves, that is, of unobservable bearers of the properties that are accessible to us (Riehl 1872a, p.75).[14] Kant's notion of a thing in itself or "noumenon" indicated that which lies outside the boundaries of a possible experience in general. Therefore,

---

Kantian conception of spatial intuition" (see especially Friedman 2009, p.257) is misleading, given the fact that Helmholtz refers, more precisely, to the form of spatial intuition, the form being construed as something inherently conceptual, as emphasized later by Poincaré (see Chap. 6). Regarding Kant's characterization of space and time as pure intuitions, the quotes above show that Helmholtz distanced himself sharply from Kant. Arguably, Friedman bases his claim on his kinematical interpretation of Kant's spatial intuition. Without calling into question the importance of this interpretation, my emphasis in the proposed reading of Helmholtz lies on the fact that Helmholtz's kinematical conception of geometry presupposed the substitution of pure intuition with analytic methods and physical geometry.

[13] In Chap. 7, I argue that the contrast between Cohen and Riehl is reflected in the different strategies that were adopted in later transcendental interpretations of general relativity. Ryckman (2012) calls these strategies rejecting or refurbishing the Transcendental Aesthetic.

[14] On Riehl's transition from Herbart's realism to realism about the things in themselves, see Pettoello (1998, pp.351–357).

in the section of the *Critique of Pure Reason* "On the Ground of the Distinction of All Objects in General into *phenomena* and *noumena*," Kant distinguished the noumenon as a thing in itself from the phenomenon, namely, the thing as it appears to us. Notwithstanding the unknowability of things in themselves by concepts alone, Kant (1787, pp.311–312) emphasized the importance of the concept of noumenon taken in the negative sense, namely, as a boundary concept for the objectivity of knowledge. Riehl's interpretation suggests that objective knowledge depends rather on the positive meaning of noumenon as a thing in itself, and therefore as a necessary presupposition of knowledge.

For similar reasons, Riehl distanced himself from Cohen's critical idealism. Nevertheless, he appreciated Cohen's work on Kant. In 1872, Riehl published one of the first reviews of Cohen (1871) in *Philosophische Monatshefte*. Riehl seems to have especially esteemed Cohen's commitment to Herbart's psychology, on the one hand, and the defense of the Kantian theory of space, on the other. Riehl wrote:

> [Cohen] indicates a sane psychology that can be found everywhere in [Kant's] Critique, and he does not forget to take into account the "empiricist theory" and the meaning of metamathematical inquiries for the theory of the a priori. In this regard the author, owing to his view, which is free from prejudice, would not have struggled to show in detail that the empiricist theory does not contradict [the Kantian] theory, once correctly understood, but rather that metamathematical inquiries confirm it, although they might seem at odds with it. (Riehl 1872b, p.213)[15]

Riehl developed his views in his major work, *Critical Philosophy and Its Meaning for Positive Science*, which was published in three volumes in 1876, 1879, and 1887. A second edition of each volume appeared with revisions and additions in 1908, 1925, and 1926. Riehl (1876) was devoted to Kant's critical philosophy and its prehistory. Riehl's discussion of geometrical empiricism is found in the second volume of his work, *The Sensible and the Logical Foundations of Knowledge*.

Despite Cohen's influence on the development of Riehl's thought, Riehl's mature philosophy differs considerably from Cohen's. In Riehl's view, the origin of critical philosophy lies not so much in the idealistic tradition of Plato, Descartes, and Leibniz, as in the empiricist philosophy of Locke and Hume. Moreover, Riehl defended the existence of the things in themselves: if the possibility of knowledge is to be explained, there must be something real independently of the way we describe it. Therefore, the theory of knowledge has to distinguish between subjective

---

[15] The expression "metamathematics" was introduced in the second half of the nineteenth century in a pejorative sense and originally designated philosophical speculations concerning spaces of more than three dimensions. Mathematicians also used it as synonym of "non-Euclidean geometry." The same expression currently indicates the mathematical study of axiomatic theories in terms of a metalanguage, as in Hilbert's *Beweistheorie* from the 1920s. Since these theories can also be studied from a semantic viewpoint, "metamathematics" can also refer to model theory (see the entry "Metamathematik" in Mittelstraß 1980–1996, p.866). Riehl's reference to "metamathematical inquiries" into the foundations of geometry reflects a transitional phase in the use of the term: "metamathematics" was no longer used in a pejorative sense, but could not have the current meaning. In 1872, Riehl arguably bore in mind the inquiries into the foundations of geometry of Riemann and Helmholtz, as suggested by the association between metamathematics and empiricism.

factors and objective ones. Riehl's example for this way of proceeding was Helmholtz's question concerning the principles of geometry: "How much of the propositions of geometry has an objectively valid sense? And how much is on the contrary only definition or the consequence of definitions, or depends on the form of description?" (Helmholtz 1868, p.39). Riehl's goal was to extend such a question to all principles of knowledge (Riehl 1879, p.4).

Riehl's theory of knowledge differed from Helmholtz's, because Riehl believed that once this question is settled, it should be possible to correctly individuate a priori concepts and prove that these concepts, despite their being subjective, determine objective features of the things we experience (Riehl 1904, p.267). Riehl pointed out that Helmholtz overlooked the objective side of a priori knowledge, because he did not clearly distinguish the notion of "transcendental" from that of "a priori." In the previous quotation, for example, Helmholtz used the expression "transcendental intuition," which is not found in Kant's work. In fact, Kant maintained that:

> [N]ot every *a priori* cognition must be called transcendental, but only that by means of which we cognize that and how certain representations (intuitions or concepts) are applied entirely *a priori*, or are possible (i.e., the possibility of cognition or its use *a priori*). Hence neither space not any geometrical determination of it *a priori* is a transcendental representation, but only the cognition that these representations are not of empirical origin at all and the possibility that they can nevertheless be related *a priori* to objects of experience can be called transcendental. (Kant 1787, pp.80–81)

Kant's forms of intuition are a priori, and the transcendental exposition of the same indicates that the manifold of pure intuition, along with the categories of the understanding, provides us with necessary conditions for empirical knowledge.

Riehl argued against geometrical empiricism because he believed that empirical generalizations, including the free mobility of rigid bodies, depend on a priori principles. In order to face the problem posed by Helmholtz about the structure of space, Riehl maintained that the a priori concepts of geometry suffice to determine the fundamental properties of physical space, especially homogeneity.

3.3.2   Riehl's Arguments for the Homogeneity of Space

In the second volume of *Critical Philosophy,* Riehl discussed Helmholtz's objections as follows. Riehl conceded that three-dimensionality depends on experience and restricted aprioricity to infinity, continuity, and homogeneity. He interpreted the difference between relations of extension and relations of measure as a difference between empirical properties and formal ones, respectively. On the one hand, he endorsed the view that three-dimensionality can only be induced from the regularities we find in motions, which was commonplace in the physiology of the senses. On the other hand, he maintained that a priori properties followed from the definition of the fundamental concepts of point, line, and so on. Since such concepts can be varied (e.g., lines can be mapped into geodesics on surfaces of positive or negative curvature, as in Helmholtz's thought experiments), infinitely many spaces are logically possible.

Riehl admitted that the a priori properties of space are insufficient to single out space among other manifolds: pure mathematics does not determine the geometry of space. However, geometry is not determined by experience either. Otherwise our choice would be contingent. Riehl rejected this consequence as a skeptical one and maintained that homogeneity and the remaining a priori properties of space can be established by specifying the corresponding features of time. The homogeneity of time depends, in turn, on the identity of our consciousness during any succession of impressions we experience. Riehl mentioned, for example, the definition of a straight line as a unidirectional line, pointing out its temporal origin. His consideration regarding the notion of a rigid geometrical figure was the following: magnitudes can be brought to congruent coincidence as a necessary consequence of their being equal. The notion of a rigid geometrical figure is supposed to follow from our knowledge that parts of space are congruent with each other. Obviously the equality of some given magnitudes must be ascertained by measurement; but their being equal can only correspond approximately to the equality of rigid geometrical figures. Therefore, the fact that rigid bodies can be superposed to magnitudes to be measured presupposes the geometrical notion of congruence, not vice versa.

Riehl considered our conception of space an indispensable, subjective factor of measurement. Empirical magnitudes entail additional, objective factors, namely, those factors that find a numerical expression. In order to assign a determinate number to such a magnitude, some measuring standard must be chosen and superposed a certain number of times on the object to be measured. Such a standard is itself a physical body, and the object to be measured is supposed to provide us with the objective ground of our measurement in that case (Riehl 1879, pp.159–165).

Riehl's second argument for the homogeneity of space is derived from classical mechanics: the relativity of motions in classical mechanics presupposes the concept of a body absolutely at rest or immovable space. If material bodies in motion within physical space changed shape and size, one could not regard space itself as the cause of such changes: motion already presupposes immovable space. Therefore, one will hypothesize a physical cause, such as the temperature of matter. Riehl's conclusion is that space, since it is immovable, is also unchangeable and, therefore, homogeneous (Riehl 1879, p.93).

Riehl's motivation for this hierarchy of the sciences is found in the third volume of *Critical Philosophy*, *Theory of Science and Metaphysics* (1887). Such a hierarchy follows from a distinction that has to be made between different meanings in our talk of the world – namely, the logical, the sensible, and the empirical meaning. Riehl's distinction is as follows. As a consequence of the "*I think*," which Kant described as "a representation which must be able to accompany all other representations, and which in all consciousness is one and the same," (Kant 1787, p.132) there is a single space as well as a single time. Owing to the singularity of space and time, every perception belongs to the single world of our senses. However, neither geometric space nor intuitive space provides the empirical world with a unitary structure. We, therefore, need additional, empirical conditions, such as the homogeneity of matter and the conservation of force (Riehl 1887, pp.281–282).

Riehl's distinction apparently entails a hierarchy of meanings and a corresponding hierarchy of the sciences. He maintained that geometric space reflects the invariants of intuitive space, which owing to the singularity of the world of our senses cannot be subject to revision. In the case of irregularities at the level of physical space, Riehl's suggestion is to look for a physical cause.[16]

For these reasons, Riehl defended the priority of pure geometry over physical geometry. Nevertheless, he took into account Helmholtz's considerations in some regards. Firstly, he admitted that Kant's characteristics for intuitive representations (i.e., immediacy and simplicity) do not prevent space and time from being concepts properly speaking. In fact, no concept is independent of some intuitive ground (Riehl 1879, p.107). Riehl did not call into question the convenience of analytic geometry. Nevertheless, he maintained that the existence of mathematical objects depends on intuition. Even if infinitely many spaces are thinkable, the fundamental concepts falling into definitions, according to him, are Euclidean concepts, and the consistency of non-Euclidean geometry can be proved only relatively to Euclidean geometry, which is supposed to be consistent owing to the intuitive content of its concepts. Euclidean models of non-Euclidean geometry provided such a proof. Therefore, Riehl maintained that the priority of Euclidean geometry was confirmed by metamathematical inquiries into the foundations of geometry. At the same time, he pointed out that metamathematics did not provide an answer to the question whether the origin of geometrical axioms is a priori or empirical. In order to vindicate the aprioricity of geometrical axioms, he needed the said arguments for the homogeneity of space.

Secondly, Riehl's description of measurement was influenced by Helmholtz's search for empirical conditions of measurement. However, it seems to me that Riehl misunderstood Helmholtz's conception of objectivity. Analytic methods in geometry are required for empirical generalizations to be possible. As we will see in the next chapter, a suitable method for comparing magnitudes does not itself provide the objective meaning of physical equivalence in Helmholtz's sense. Such a procedure must be generalized so that objects that have proved to be equivalent are also mutually substitutable in any further cases. In order to accomplish such a generalization, it does not suffice to assign numerical values to the magnitudes to be

---

[16] Riehl's differentiation of the semantic levels involved in our notion of space was influential in early logical positivism. We return to Riehl's influence on Schlick in Chap. 7. It is worth noting that a similar approach was adopted by the young Carnap in his Dissertation on *Space* (1922). Carnap distinguished between formal, intuitive, and physical space (Carnap 1922, p.5). Despite the fact that Carnap's analysis of the concepts of space differs considerably from Riehl's, it is likely that Carnap profited from Riehl's distinction of the meanings attached to these concepts (see Heidelberger 2007, p.34). It is apparent from Carnap's references and endnotes that he was familiar with Riehl's work. Additionally, consider that Carnap wrote his Dissertation under the supervision of the neo-Kantian Bruno Bauch, whose paper on "Experience and Geometry in Their Epistemological Relation" (1907) was clearly influenced by Riehl.

measured by repeatedly superposing some measuring standard; the laws of addition must be extended to physical domains.

3.3.3   Cohen's Discussion of Geometrical Empiricism in the Second Edition of Kant's Theory of Experience

In the second edition of *Kant's Theory of Experience* (1885), Cohen criticized geometrical empiricism and proposed a transcendental interpretation of the connection between geometry and experience. The critical part of Cohen's argument followed from the distinction between metaphysical and transcendental a priori he introduced in the first edition of his work and made more precise in *Kant's Foundation of Ethics* (1877). It follows from Cohen's interpretation of the theory of the a priori that necessity and generality (i.e., the characteristics of a priori knowledge) cannot be attributed to any specific assumption independently of its role in scientific knowledge. The a priori knowledge that is the object of the transcendental inquiry (i.e., what Cohen called the "transcendental" a priori) is distinguished from a metaphysical kind of a priori because it coincides with those principles that are implicit in natural science.

In order to reply to Helmholtz's objections to Kant, Cohen considered the structure of the *Critique of Pure Reason*. Kant's argument for the homogeneity of space is found in the Analytic of Principles (Kant 1787, pp.203–205). Since all appearances presuppose the forms of intuition in space and time in general, the representation of specific spaces and times requires what Kant calls "a synthesis of that which is homogeneous." The corresponding principle is that "all intuitions are extensive magnitudes." After enunciating this principle, Kant made it clear that the pure intuition of space as analyzed in the Transcendental Aesthetic does not provide us with geometrical axioms. These are now said to be grounded in the successive synthesis of the productive imagination, which is distinguished from the empirical, receptive one because of its spontaneity. The same synthesis lies at the foundation of the kind of motion that Kant characterized as follows:

> Motion, as action of the subject (not as determination of an object), consequently the synthesis of the manifold in space, if we abstract from this manifold in space and attend solely to the action in accordance with which we determine the form of inner sense, first produces the concept of succession at all. The understanding therefore does not find some sort of combination of the manifold already in inner sense, but produces it, by affecting inner sense. (Kant 1787, p.155)

In a footnote, Kant reformulated the distinction between motion as determination of an object and as action of the subject in the following terms:

> Motion of an object in space does not belong in a pure science, thus also not in geometry; for that something is movable cannot be cognized *a priori* but only through experience. But motion, as description of a space, is a pure act of the successive synthesis of the manifold in outer intuition in general through the productive imagination, and belongs not only to geometry but even to transcendental philosophy. (Kant 1787, p.155, note)

These quotes make it clear that an act of the understanding is required for the sensible manifold to be schematized or unified under the category of quantity. Therefore, Kant said that the understanding produces the combination of the manifold, which is not simply found in inner sense. Correspondingly, motion as description of a space is the product of an act of successive synthesis, whereas the motion of an object in space can be cognized only through experience. Only the first kind of motion is an object of geometry and even of transcendental philosophy, insofar as the successive synthesis of the productive imagination enables Kant to account for the homogeneity of the manifold of outer intuition.

Kant clarified this point in the Analytic of Principles by saying that that which is homogeneous can be specified as a mathematical quantity. As we have noticed in Chap. 1, Kant noticed at this point that only geometry has axioms in the proper sense (i.e., propositions that concern magnitudes as such). Arithmetic, according to Kant, deals with the more specific question: "How big is something?" Unlike geometry, arithmetic has no axioms in Kant's sense, because the answer to this question is based on calculation, rather than construction in pure intuition or the successive synthesis of the imagination.

These sections of the *Critique of Pure Reason* suggest strongly that Kant was not committed to the claim that spatial intuition alone entails Euclidean axioms, even though he might have believed that these are the only possible ones. This consideration enabled Cohen to admit complete freedom of mathematics in the formulation of geometrical axioms (Cohen 1885, p.228). On the one hand, he made it clear that the issue of transcendental philosophy is another one, namely, to ask for the principles that make mathematics applicable: these correspond to the principles of extensive and intensive magnitudes as formulated by Kant in the Analytic of Principles. On the other hand, Cohen sharply distinguished the issue of transcendental philosophy from the psychological issue addressed by Helmholtz. Kant did not ask for a supposedly immediate, subjective way of localizing things. His question concerned the general form of intuition that is supposed to play some role in the development of objective knowledge (Cohen 1885, pp.236–237). The problem with Helmholtz's geometrical empiricism is that the claim that geometrical notions have empirical origins seems to rest upon the naïve realist supposition that magnitudes exist in themselves. So it might seem as if geometrical axioms, as propositions about magnitudes, have to be induced from experience. On the contrary, geometrical principles must be assumed for judgments about magnitudes to be generally valid (p.419). Cohen agreed with the view that geometry is the part of mechanics that lies at the foundation of measurement in line with Newton and with the empiricist tradition. The disagreement with empiricism concerned the notion of experience, which Cohen understood not so much as the origin of knowledge, but as that which is first given in scientific knowledge. Nevertheless, Cohen recognized a kind of transcendental argument in Helmholtz's requirement of free mobility of rigid bodies as a precondition for the possibility of measurement. Owing to its generality, the notion of a rigid geometrical figure can be understood as an ideal. Insofar as the same notion is necessary for measurement, Cohen deemed it a synthetic a priori concept (p.232).

I think that Cohen's interpretation is more accurate than Riehl's, because Cohen took into account the general level of Helmholtz's inquiry when it comes to account for the objectivity of measurement. At the same time, Cohen contradicted Helmholtz's view that, in such an interpretation, geometrical axioms would follow analytically from the definition of a rigid geometrical figure. Cohen maintained that the notion of a rigid geometrical figure is to be better understood as a synthetic concept, insofar as it provides a necessary presupposition for measurement.

Note that Cohen, similar to Riehl, needed some argument for the homogeneity of space to support his view. Since Cohen admitted complete freedom in the formulation of geometrical axioms, however, it is clear that his argument for the homogeneity of space, unlike Riehl's argument, did not have to entail the priority of Euclidean geometry.[17] The concluding section of this chapter offers a discussion of Cohen's argument in comparison with Kant's and Helmholtz's.

3.4 Cohen and Helmholtz on the Use of Analytic Method in Physical Geometry

We have already mentioned that Kant distinguished sharply between geometry and arithmetic. It follows from Kant's distinction that the definition of geometrical notions essentially requires spatial intuition, as in Euclid's way of proceeding. In the second edition of *Kant's Theory of Experience*, Cohen proposed a revision of Kant's argument, which appears to have been influenced by Trendelenburg and by Helmholtz. The argument proceeds as follows. A manifold must be divisible into equal parts in order to be thought of as a magnitude. In this sense, and not simply as infinite divisibility, homogeneity enables the comparison of different parts of space that is required for measurement. Cohen agreed with Helmholtz that measurement requires additive principles for physical magnitudes to be formulated and that such principles cannot be derived from the form of space. On the contrary, the form of space is determined in the synthesis of that which is homogeneous (Cohen 1885, p.419). Cohen thereby emphasized that the principle that all intuitions are extensive magnitudes not only constitutes mathematical magnitudes, but also turns them into objects of possible experience. As Cohen put it, the "transcendental meaning of

---

[17] Cf. Coffa (1991, pp.57–61). Coffa maintains that Riehl, unlike Cohen, tried to read Helmholtz's insights into Kant's writings. In my opinion, this picture ought to be reconsidered. Even though Riehl set no limits to the abstractions that he regarded as merely analytical, he endorsed the priority of Euclidean geometry, whereas Helmholtz ruled out the priority of any specific geometry. According to Coffa, Cohen's reply to Helmholtz is resumed by his "professorial" consideration: "The critics have not understood Kant" (Coffa 1991, p.54). However, I think that Cohen's defense of the Kantian theory of space was anything but conservative. He did not call into question Helmholtz's proof that the free mobility of rigid bodies does not necessarily entail Euclidean geometry. At the same time, Cohen interpreted the notion of a rigid geometrical figure as a synthetic concept a priori: all concepts that are conditions of objective knowledge are synthetic a priori in Cohen's sense.

pure mathematics lies in its connection with natural science" (p.402). This is reflected in Kant's emphasis on the successive character of the synthesis of that which is homogeneous: the same synthesis underlies the a priori part of the general theory of motion.

Owing to his argument for the homogeneity of space, Cohen did not call into question the correctness of physical geometry. His disagreement with Helmholtz, in this regard also with Riehl, rests upon a different formulation of the problem of knowledge. According to Cohen, the issue of critical philosophy is not to distinguish between definitions or mere descriptions, on the one hand, and the objective meaning of mathematical concepts, on the other, but to prove the connection of such concepts with the objects of experience. Therefore, Helmholtz's defense of physical geometry cannot be regarded as an objection to Kant. Recall Helmholtz's comparison between physical geometry and the pure geometry that is supposed to be grounded in spatial intuition. He apparently assumed, firstly, that the form of intuition is endowed with a geometric structure and, secondly, that such a form depends on the makeup of our mind independently of motion. Helmholtz's conclusion then was that spatial intuition cannot provide us with objective knowledge independently of empirical science.

Cohen agreed with Helmholtz's conclusion, while rejecting both of his premises. That Kant's form of intuition is not endowed with a geometric structure follows from his distinction between sensibility and understanding: geometry presupposes a connection of both sources of knowledge. Cohen's revised version of Kant's argument for the homogeneity of space enabled him to reject the second premise as well. The form of intuition is not independent of motion, and the homogeneity of space presupposes both geometry and mechanics (Cohen 1885, pp.233–234). To conclude, Cohen's revision called into question Kant's contrast between arithmetic and geometry. Cohen's axioms are not characteristic of geometry alone. Cohen deemed axioms all principles that makes mathematical theories applicable and identified the axioms of intuitions as laws of addition. It clearly emerges from his considerations that he looked at the example of analytic geometry to explore the connection between pure and applied mathematics.

Cohen made his connection with Helmholtz more explicit in his introduction to the 1896 edition of Friedrich Albert Lange's *History of Materialism*. There, Cohen argued for his understanding of the transcendental a priori as follows: the apriority of mathematical theories cannot be established independently of their use in natural science. A revision of the Kantian theory of space and time as pure intuitions is required, because Kant, in the Transcendental Aesthetic, tended to consider mathematical and physical apriority separately. The proof that the principles of mathematics provides us with a priori knowledge was completed only in the Analytic of Principles. Now, Helmholtz's use of analytic geometry in the representation of motion suggests that such a separation, along with the concept of pure intuition as a source of mathematical certainty, is superfluous. Cohen wrote:

> Newton's preference for the synthetic method of the ancients had a harmful effect on the whole of [Kant's] system insofar as intuition achieved sovereignty beside and before thought. Kant's terminology also caused trouble because the concept of *intuition* collapsed

with that of *sensation*, from which as pure intuition it should have been totally different. But
if such a difference had to be taken seriously, it was not easy to understand why intuition
should be distinguished so sharply from thought. In this regard, modern geometers, such as
Helmholtz, seem to be more Platonist and Leibnizian than Kant was, because they keep the
constructions of geometry in connection with pure thought. (Cohen 1896/1984, p.65)

According to Cohen, what Kant called pure intuition should be understood as a
form of pure thought, because the autonomy of thought is necessary for synthetic
knowledge a priori to take place. Cohen's remark about geometric methods suggests
that pure thought should work as a substitute for pure intuitions for another reason
as well: only thought offers a standpoint general enough to include all possible spatial connections to be encountered in science. We know from the previous section
that Cohen identified geometrical axioms as principles of all outer intuitions and,
therefore, of measurement. He mentioned, for example, the free mobility of rigid
bodies. As for the formulation of the more specific hypotheses under consideration,
Cohen argued for the freedom of mathematics. In other words, he at least implicitly
admitted Helmholtz's view that, in the case of a choice among hypotheses, empirical considerations would be decisive. Furthermore, Cohen agreed with Helmholtz
that a shift from synthetic to analytic methods was required for the purposes of
physical geometry. Geometry, for Cohen, remains synthetic in the philosophical
sense, because the constructions of geometry in connection with pure thought provide us with constitutive principles of an object in general. Recall that conceptual
synthesis for Cohen is the only possible kind of cognition that is possible a priori.
The applicability of mathematics (and of geometry) depends on its synthetic character, as in Kant's original argument. The disagreement with Kant lies in the fact
that Cohen believed analytic methods in geometry to have made the reference to
spatial intuitions superfluous. Cohen pointed out that this development does not
necessarily make geometry analytic in the sense of the philosophical synthetic/ana-
lytic distinction, which depends on the meaning of a priori concepts for the possibility of knowledge. The contraposition between synthetic and analytic methods in
nineteenth-century geometry emerged more specifically from the discussion about
the foundations of projective geometry; we turn back to this topic in Chap. 5.

To sum up, since Cohen defended both the apriericity of geometrical axioms – in
the sense of the transcendental a priori – and the correctness of physical geometry,
his view clearly entails a relativization of the notion of a priori. Geometry is a priori
knowledge because it provides us with the appropriate tools for the mathematical
treatment of physical magnitudes. At the same time, the choice among geometrical
hypotheses depends on their use in physics. Cohen did not emphasize the fact that
such hypotheses may vary. Nevertheless, his approach to the theory of the a priori
goes exactly in this direction: he argued for the conceptual character of the constructions of geometry and, therefore, for mathematical freedom, because the kind
of necessity that he attributed to these constructions (i.e., relative necessity) presupposes conceptual variations. What is required for the apriericity of mathematics is
that all possible cases that occur in experiment can be classified in advance from a
conceptual viewpoint. Therefore, Cohen was able to reinterpret Helmholtz's philosophy of geometry within the framework of a Kantian theory of experience

without being committed to Helmholtz's inference from the hypothetical character of the principles of geometry to their empirical origin.

Friedman, who is one of the main proponents of a relativized conception of the a priori in contemporary philosophy of science, characterizes this view by saying that, instead of global necessary conditions for all human experience in general, we have merely local necessary conditions for the empirical application of a particular mathematical–physical theory at a given time and in a given historical context (Friedman 2009, p.253). He puts a special emphasis on the role of Helmholtz's account of space in the prehistory of the relativized conception of the a priori, because the condition of free mobility of rigid bodies, which represents a natural generalization of Kant's original (Euclidean) conception of geometry, also holds for non-Euclidean geometries of constant curvature. On the one hand, Helmholtz's generalization of the Kantian theory of space is the "minimal" such generalization consistent with the nineteenth-century discovery of non-Euclidean geometries. On the other hand, "it is no longer a 'transcendental' and 'necessary' condition of our spatial intuition, for Helmholtz, that the space constructed from our perception of bodily motion obeys the specific laws of Euclidean geometry" (Friedman 2009, p.257). In this reconstruction, the decisive step towards a relativized conception of the a priori was taken by Einstein and presupposed his engagement in the debate about the origin and meaning of geometrical axioms. We deal with the role of Poincaré in this debate in Chap. 6. For now, it is worth noting that for Friedman, the analogy with Helmholtz lies in the fact that Einstein's approach to measurement seems to require a further generalization of the earlier views to the principle of equivalence:

> Whereas the particular geometry in a given general relativistic space-time is now determined entirely empirically (by the distribution of mass and energy in accordance with Einstein's field equation), the principle of equivalence itself is not empirical in this sense. This principle is instead presupposed – as a transcendentally constitutive condition – for any such geometrical description of space-time to have a genuine empirical meaning in the first place. (Friedman 2009, p.266)

I argued that the idea of a relativized a priori has its roots in the philosophical discussion of the Transcendental Aesthetic and in Cohen's distinction between the metaphysical and the transcendental meanings of the a priori. The transcendental meaning of this notion – which is essential to Cohen's interpretation of the theory of the a priori – is dependent on the history of science and, therefore, it is the scope of the conceptual systems a priori of mathematics. Reconsidering this theory of the a priori sheds light on the fact that Cohen and Cassirer were among the first philosophers to acknowledge the contributions of such scientists as Helmholtz and Einstein to the philosophical discussion about the concept of space. Not only did Cohen deal with Helmholtz's geometrical papers on several occasions, but in the third edition of his *Logic of Pure Knowledge*, which appeared posthumously in 1922, Cohen emphasized the philosophical significance of the correlation between space and time established by Einstein (Cohen 1977, p.198).

The same ideas enabled Cassirer to deepen Cohen's insights into the methodological transformation of nineteenth-century mathematics. Such examples as the analytical reconstruction of the continuum by Richard Dedekind and the group-theoretical treatment of geometry by Felix Klein appeared to confirm the view that the reference to spatial intuitions had been made superfluous. Therefore, Cassirer believed that a broader understanding of mathematical and geometrical constructions as conceptual ones was required for posing the problems correctly concerning measurement.[18]

In order to highlight this point, the next chapter is devoted to the development of Helmholtz's theory of measurement from 1878 to 1887 and its reception by Cohen and Cassirer. Not only does Helmholtz's theory include physical equivalence as a special case, but his contribution to a general theory of measurement was seminal for the neo-Kantian reconstruction of the argument for the applicability of mathematics. Helmholtz sharply distinguished numbers from empirical domains and posed the problem of formulating the conditions for empirical magnitudes to find a numerical expression. Helmholtz's solution to this problem led to an entirely different definition of spatial magnitudes than Kant's definition, which is based on the idea that limitations of space as pure intuition always produce parts of one and the same space. Nevertheless, I argue for Cohen's reading of Helmholtz's solution as retaining the structure of a transcendental argument in Kant's sense: the general concepts of number and of sum must be defined independently of the reference to specific entities, in order for additive principles to work as constitutive principles of the objects of experience .

References

Banks, Erik C. 2005. Kant, Herbart, and Riemann. *Kant-Studien* 96: 208–234.

Bauch, Bruno. 1907. Erfahrung und Geometrie in ihrem erkenntnistheoretischen Verhältnis. *Kant-Studien* 12: 213–235.

Beltrami, Eugenio. 1868. Saggio di interpretazione della geometria non-euclidea. *Opere Matematiche* 1: 374–405. Milano: Hoepli, 1902.

Beltrami, Eugenio. 1869/1902. Teoria fondamentale degli spazi a curvatura costante. *Opere Matematiche* 1: 406–429. Milano: Hoepli, 1902.

Biagioli, Francesca. 2014. Hermann Cohen and Alois Riehl on geometrical empiricism. *HOPOS: The Journal of the International Society for the History of Philosophy of Science* 4: 83–105.

Boi, Luciano, Livia Giancardi, and Rossana Tazzioli (eds.). 1998. *La découverte de la géométrie non euclidienne sur la pseudosphère: Les lettres d'Eugenio Beltrami à Joules Hoüel; (1868–1881)*. Paris: Blanchard.

Breitenberger, Ernst. 1984. Gauss's geodesy and the axiom of parallels. *Archive for History of Exact Sciences* 31: 273–289.

Carnap, Rudolf. 1922. *Der Raum: Ein Beitrag zur Wissenschaftslehre*. Berlin: Reuther & Reichard.

---

[18] I noticed on several occasions that Friedman, among others, made it plausible that the Kantian theory of pure intuitions deserves to be considered in context, namely, before the said transformation. However, it seems to me that Friedman's reconstruction of the prehistory of the relativized a priori fails to appreciate the significance of the neo-Kantian idea of a conceptual synthesis for a reformulation of Kant's arguments in the nineteenth-century context.

Carrier, Martin. 1994. Geometric facts and geometric theory: Helmholtz and 20th-Century philosophy of physical geometry. In *Universalgenie Helmholtz: Rückblick nach 100 Jahren*, ed. Lorenz Krüger, 276–291. Berlin: Akademie Verlag.

Coffa, Alberto J. 1991. *The semantic tradition from Kant to Carnap: To the Vienna station.* Cambridge: Cambridge University Press.

Cohen, Hermann. 1871. *Kants Theorie der Erfahrung.* Berlin: Dümmler.

Cohen, Hermann. 1877. *Kants Begründung der Ethik.* Berlin: Dümmler.

Cohen, Hermann. 1885. *Kants Theorie der Erfahrung,* 2nd ed. Berlin: Dümmler.

Cohen, Hermann. 1896/1984. Einleitung mit kritischem Nachtrag zur *Geschichte des Materialismus* von F. A. Lange. Repr. in *Werke,* vol. 5, ed. Helmut Holzhey. Hildesheim: Olms.

Cohen, Hermann. 1977. *Logik der reinen Erkenntnis.* In *Werke,* vol. 6, ed. Helmut Holzhey. Hildesheim: Olms.

Darrigol, Olivier. 2003. Number and measure: Hermann von Helmholtz at the crossroads of mathematics, physics, and psychology. *Studies in History and Philosophy of Science* 34: 515–573.

DiSalle, Robert. 2006. Kant, Helmholtz, and the meaning of empiricism. In *The Kantian legacy in nineteenth-century science,* ed. Michael Friedman and Alfred Nordmann, 123–139. Cambridge, MA: The MIT Press.

DiSalle, Robert. 2008. *Understanding space-time: The philosophical development of physics from Newton to Einstein.* Cambridge: Cambridge University Press.

Erdmann, Benno. 1877. *Die Axiome der Geometrie: Eine philosophische Untersuchung der Riemann-Helmholtz'schen Raumtheorie.* Leipzig: Voss.

Friedman, Michael. 2000. Geometry, construction and intuition in Kant and his successors. In *Between logic and intuition: Essays in honor of Charles Parsons,* ed. Gila Sher and Richard Tieszen, 186–218. Cambridge: Cambridge University Press.

Friedman, Michael. 2009. Einstein, Kant, and the relativized a priori. In *Constituting objectivity: Transcendental perspectives on modern physics,* ed. Michel Bitbol, Pierre Kerszberg, and Jean Petitot, 253–267. Dordrecht: Springer.

Gauss, Carl Friedrich. 1900. *Werke,* ed. von der Königlichen. Gesellschaft der Wissenschaften zu Göttingen, 8.

Gray, Jeremy J. 2006. Gauss and non-Euclidean geometry. In *Non-Euclidean geometries: János Bolyai memorial volume,* ed. András Prékopa and Emil Molnár, 61–80. New York: Springer.

Heidelberger, Michael. 2007. From neo-Kantianism to critical realism: Space and the mind-body problem in Riehl and Schlick. *Perspectives on Science* 15: 26–47.

Helmholtz, Hermann von. 1868. Über die Tatsachen, die der Geometrie zugrunde liegen. In Helmholtz (1921): 38–55.

Helmholtz, Hermann von. 1870. Über den Ursprung und die Bedeutung der geometrischen Axiome. In *Schriften zur Erkenntnistheorie,* ed. Paul Hertz and Moritz Schlick, 1–24. Berlin: Springer, 1921. English edition: Helmholtz, Hermann von. 1977. Epistemological writings (trans: Lowe, Malcom F., ed. Robert S. Cohen and Yehuda Elkana). Dordrecht: Reidel.

Helmholtz, Hermann von. 1878. Die Tatsachen in der Wahrnehmung. In *Schriften zur Erkenntnistheorie,* ed. Paul Hertz and Moritz Schlick, 109–152. Berlin: Springer, 1921. English edition: Helmholtz, Hermann von. 1977. Epistemological writings (trans: Lowe, Malcom F., ed. Robert S. Cohen and Yehuda Elkana). Dordrecht: Reidel.

von Helmholtz, Hermann. 1887. Zählen und Messen, erkenntnistheoretisch betrachtet. In *Schriften zur Erkenntnistheorie,* ed. Paul Hertz and Moritz Schlick, 70–97. Berlin: Springer, 1921. English edition: Helmholtz, Hermann von. 1977. Epistemological writings (trans: Lowe, Malcom F., ed. Robert S. Cohen and Yehuda Elkana). Dordrecht: Reidel.

Helmholtz, Hermann von. 1903. *Vorlesungen über theoretische Physik. Vol. 1.1: Einleitung zu den Vorlesungen über theoretische Physik,* ed. Arthur König and Carl Runge. Leipzig: Barth.

Herbart, Johann Friedrich. 1825/1850. *Psychologie als Wissenschaft.* Repr. in *Herbarts Sämmtliche Werke,* vol. 6, ed. Gustav Hartenstein. Leipzig: Voss.

Hilbert, David. 1903. *Grundlagen der Geometrie,* 2nd ed. Leipzig: Teubner.

Hönigswald, Richard. 1909. Über den Unterschied und die Beziehungen der logischen und der erkenntnistheoretischen Elemente in dem kritischen Problem der Geometrie. In *Bericht über den III. Internationalen Kongress für Philosophie, 1. bis 5. September 1908*, ed. Theodor Elsenhans, 887–893. Heidelberg: Winter.

Hönigswald, Richard. 1912. *Zum Streit über die Grundlagen der Mathematik.* Heidelberg: Winter.

Hyder, David. 2006. Kant, Helmholtz and the determinacy of physical theory. In *Interactions: Mathematics, physics and philosophy, 1860–1930*, ed. Vincent F. Hendricks, Klaus Frovin Jørgensen, Jesper Lützen, and Stig Andur Pedersen, 1–44. Dordrecht: Springer.

Kant, Immanuel. 1787. *Critik der reinen Vernunft.* 2nd ed. Riga: Hartknoch. Repr. in *Akademie-Ausgabe.* Berlin: Reimer, 3. English edition: Kant, Immanuel. 1998. *Critique of Pure Reason* (trans: Guyer, Paul and Wood, Allen W.). Cambridge: Cambridge University Press.

Kline, Morris. 1980. *Mathematics: The loss of certainty.* Oxford: Oxford University Press.

Legendre, Adrien-Marie. 1794. *Éléments de géométrie.* Paris: Didot.

Lie, Sophus. 1893. *Theorie der Transformationsgruppen*, vol. 3. Leipzig: Teubner.

Mittelstraß, Jürgen (ed.). 1980–1996. *Enzyklopädie Philosophie und Wissenschaftstheorie.* Stuttgart: Metzler.

Natorp, Paul. 1901. Zu den logischen Grundlagen der neueren Mathematik. *Archiv für systematische Philosophie* 7: 177–209, 372–384.

Natorp, Paul. 1910. *Die logischen Grundlagen der exakten Wissenschaften.* Leipzig: Teubner.

Pettoello, Renato. 1988. Dietro la superficie dei fenomeni: Frammenti di filosofia in Bernhard Riemann. *Rivista di storia della filosofia* 4: 697–728.

Pettoello, Renato. 1998. De Herbart à Kant: Quelques considérations sur le réalisme de Alois Riehl. *Revue de Métaphysique et de Morale* 102: 347–366.

Riehl, Alois. 1870. *Realistische Grundzüge.* Graz: Leuschner & Lubensky.

Riehl, Alois. 1872a. *Über Begriff und Form der Philosophie: Eine allgemeine Einleitung in das Studium der Philosophie.* Leipzig: Haacke.

Riehl, Alois. 1872b. Zur Aprioritätslehre. *Philosophische Monatshefte* 8: 212–215.

Riehl, Alois. 1876. *Der Philosophische Kriticismus und seine Bedeutung für die positive Wissenschaft. Vol. 1: Geschichte und Methode des philosophischen Kriticismus.* Leipzig: Engelmann.

Riehl, Alois. 1879. *Der Philosophische Kriticismus. Vol 2: Die sinnlichen und logischen Grundlagen der Erkenntnis.* Leipzig: Engelmann.

Riehl, Alois. 1887. *Der Philosophische Kriticismus. Vol. 3: Zur Wissenschaftstheorie und Metaphysik.* Leipzig: Engelmann.

Riehl, Alois. 1904. Helmholtz in seinem Verhältnis zu Kant. *Kant-Studien* 9: 260–285.

Riemann, Bernhard. 1876. *Gesammelte mathematische Werke und wissenschaftlicher Nachlass*, ed. Richard Dedekind, Heinrich Martin Weber, Max Noether, and Wilhelm Wirtinger. Leipzig: Teubner.

Riemann, Bernhard. 1996. On the hypotheses which lie at the foundation of geometry. In *From Kant to Hilbert: A source book in the foundations of mathematics*, ed. William Bragg Ewald, 652–61. Oxford: Clarendon. Originally published as: Über die Hypothesen, welche der Geometrie zu Grunde liegen. *Abhandlungen der Königlichen Gesellschaft der Wissenschaften zu Göttingen* 13(1867): 133–152.

Russell, Bertrand. 1897. *An essay on the foundations of geometry.* Cambridge: University Press.

Ryckman, Thomas A. 2005. *The reign of relativity: Philosophy in physics 1915–1925.* New York: Oxford University Press.

Ryckman, Thomas A. 2012. Early philosophical interpretations of general relativity. *Stanford Encyclopedia of Philosophy.* http://plato.stanford.edu/entries/genrel-early/. Accessed 13 Nov 2015.

Sartorius von Waltershausen, Wolfgang. 1856. *Gauss zum Gedächtnis.* Stuttgart: Hirzel.

Scholz, Erhard. 1980. *Geschichte des Mannigfaltigkeitsbegriffs von Riemann bis Poincaré.* Boston: Birkhäuser.

Scholz, Erhard. 1982a. Herbart's influence on Riemann. *Historia Mathematica* 9: 413–440.

Scholz, Erhard. 1982b. Riemanns frühe Notizen zum Mannigfaltigkeitsbegriff und zu den Grundlagen der Geometrie. *Archive for History of Exact Sciences* 27: 213–231.

Scholz, Erhard. 2004. C. F. Gauß' Präzisionsmessungen terrestrischer Dreiecke und seine Überlegungen zur empirischen Fundierung der Geometrie in den 1820er Jahren. In *Form, Zahl, Ordnung: Studien zur Wissenschafts- und Technikgeschichte. Ivo Schneider zum 65. Geburtstag*, ed. Menso Folkerts, Ulf Hashagen, and Rudof Seising, 355–380. Stuttgart: Steiner.

Torretti, Roberto. 1978. *Philosophy of geometry from Riemann to Poincaré*. Dordrecht: Reidel.

Chapter 4
Number and Magnitude

4.1 Introduction

Helmholtz was one of the first to address the problem of clarifying the relationship between the concept of number and that of magnitude. Despite the traditional definition of arithmetic as the theory of quantities, numbers cannot be identified as magnitudes.[1] Numbers can only represent magnitudes in measurement situations. In order to justify the use of numbers in modeling measurement situations, specific conditions are required. The study of these conditions is now known as measurement theory and has its origin in the works of Otto Hölder (1901) and Norman Robert Campbell (1920).[2]

As we have seen in the previous chapters, the problems concerning measurement occupied Helmholtz as a physicist and were crucial in his approach to geometry. Furthermore, as a physiologist, he must have been familiar with the discussion about the measurability of sensations which followed Gustav Fechner's and Wilhelm Wundt's attempts to measure psychological processes.[3] Dealing with similar problems in such a variety of contexts arguably led Helmholtz to pose the problem of formulating general conditions of measurement. Helmholtz's conditions are found in "Counting and Measuring from an Epistemological Viewpoint," which appeared in 1887 in a Festschrift dedicated to the German philosopher and historian of ancient

---

[1] Following the current English usage – which goes back to Russell (1903) – I call properties standing in the relation of being greater or less than something "magnitudes." "Quantities" refers to objects possessing magnitudes. These terms are translated in German to "Mass" and "Quantität," respectively. Since the meaning of these and of related concepts (e.g., of "Größe") changed considerably between the second half of the nineteenth century and the beginning of the twentieth century, more details about the transformation of these concepts in the German-speaking world are given below.

[2] For a standard formulation of measurement theory, see Krantz et al. (1971), and on the origins, see Diez (1997).

[3] Contextualizing Helmholtz's approach to measurement, Michael Heidelberger (1993) suggests that Helmholtz must have had psychological measurement in the back of his mind when he formulated his theory of measurement.

© Springer International Publishing Switzerland 2016
F. Biagioli, *Space, Number, and Geometry from Helmholtz to Cassirer*,
Archimedes 46, DOI 10.1007/978-3-319-31779-3_4

Greek philosophy Eduard Zeller. As the title suggests, Helmholtz's article is divided into two parts: the first part is devoted to the concept of number; in the second part, Helmholtz poses the problem of specifying the conditions for cardinal numbers to express magnitudes.

With regard to the general goal of Helmholtz's study, he is usually acknowledged as one of the forerunners of measurement theory.[4] However, the first part of Helmholtz's argument has been much discussed both during his lifetime and in more recent literature on account of Helmholtz's attempt to ground arithmetic in the psychological fact of the time sequence. Joel Michell (1993, pp.195–196) compares Helmholtz's conception of number with Newton's definition of number as the abstracted ratio of any quantity to another quantity of the same kind (Newton 1728, p.2). Although Helmholtz referred to internal rather than external facts, both his conception of number and Newton's seem to imply a classical conception of measurement as the discovery of a matter of fact. By contrast, Michell points out that the modern, representational view of measurement is made necessary by any view of the ontological status of numbers that removes them from the empirical domain. In this view, measurement is characterized as the numerical representation of some empirical domain under specified conditions.[5]

A more historical approach to Helmholtz's theory of measurement has been adopted by Olivier Darrigol (2003). Regardless of the classical/representational dichotomy, Darrigol offers a detailed reconstruction of Helmholtz's sources, of his achievements, and of the reception of his views. As pointed out by Darrigol, Helmholtz's references include Hermann and Robert Grassmann's formalist foundation of arithmetic, Paul Du Bois-Reymond's phenomenological definitions of number and quantity, and Adolf Elsas' Kantian criticism of measurement in psychology. On the latter issue, Helmholtz was probably aware of the debates in which participants included Wilhelm Wundt, Johannes von Kries, and Eduard Zeller. Furthermore, Helmholtz must have been acquainted with James Clerk Maxwell's discussion of temperature measurement. Given the variety of these sources, it is clear that Helmholtz relied upon them only insofar as they enabled him to propose a consistent and original approach to measurement. In order to adequately represent the historical significance of Helmholtz's contribution, Darrigol emphasized the influence of Helmholtz's conception of magnitudes on such mathematicians as Otto Hölder and Henri Poincaré and on such scientists as Ernst Mach and Pierre Duhem.

While relying on Darrigol's seminal work for the general presentation of Helmholtz's theory of measurement, the present chapter focuses especially on the philosophical aspects of Helmholtz's work. This requires us to reconsider the connection between Helmholtz's analysis of measurement and his inquiry into the foundations of geometry. In 1878, Helmholtz indicated the requirement of a numerical representation as a defining characteristic of equality in physical geometry, as we saw in Section 3.2.3. In 1887, he developed the argument for the objectivity of

---

[4] For a reformulation of Helmholtz's conditions of measurement in terms of a more recent version of measurement theory, see Diez (1997, pp.171–175).

[5] On the distinction between the classical and the representational views, see Michell (1993, p.189).

measurements by specifying the conditions for assigning numerical values to physical magnitudes, including the distance between a pair of points. Helmholtz (1887) sheds light on his view about the connection between measurement and the foundations of mathematics, as pointed out especially by DiSalle (1993). Therefore, this paper is crucial for a better understanding of Helmholtz's empiricism in mathematics. Furthermore, Helmholtz (1887) contains one of the clearest expressions of his naturalistic interpretation of Kant's forms of intuition. Helmholtz deemed the time sequence a psychological basis for the development of the theory of numbers. He believed that the psychological origin of arithmetical axioms was required to clearly distinguish numbers from external objects and to correctly formulate problems concerning the use of numbers in measurement.

The psychological part of Helmholtz's inquiry has been much discussed, even at the time of writing, and may not overcome compelling objections formulated by Edmund Husserl (1891) and Gottlob Frege (1893), among others. At the same time, Helmholtz's psychological considerations offer one of the clearest formulations of his argument for the applicability of mathematics. The argument can be summarized as follows: additive principles can be established independently of the entities to be measured, although they are necessary for judgments about quantities to be valid. My suggestion is that this argument retains the structure of a transcendental argument, that is, the argument that some knowledge is independent of experience, because it is a condition for the possibility of experience. Therefore, it is no accident that such neo-Kantians as Cohen and Cassirer paid particular attention to Helmholtz's paper of 1887. Cohen was one of the first philosophers to review Helmholtz's paper. And Cassirer devoted an important section of *Substance and Function* (1910) to it.

The following section of this chapter offers a discussion of Helmholtz's argument as he presented it in 1887. In order to clarify his philosophical assumptions, I emphasize a development in Helmholtz's conception of objectivity during his discussions with Jan Pieter Nicolaas Land and with Albrecht Krause at the end of the 1870s. Arguably, it was in that connection that Helmholtz realized that the conditions of measurement deserved a more comprehensive analysis. Section 4.3 provides a discussion of the main objections against Helmholtz's treatment of the theory of numbers, beginning with Cohen's objections. In Sect. 4.3.3, I argue that a less problematic, alternative version of Helmholtz's argument for the use of numbers in measurement is found in Cassirer (1910).

4.2 Helmholtz's Argument for the Objectivity of Measurement

As we saw in the previous chapter, Helmholtz argued for the objectivity of measurements in physical geometry. In 1878, Helmholtz directed his argument against the unconditional validity of Euclidean geometry, namely, the geometry that was supposed to be grounded in spatial intuition. Is this a Kantian assumption? Though

Kant could not be confronted with a choice between hypotheses concerning physical space, the assumption of the unconditional validity of Euclidean geometry appears to have been commonplace in nineteenth-century attempts to defend the aprioricity of geometry, with the remarkable exception of Cohen. Thus, it might seem that Helmholtz's argument was directed against the Kantian theory of space altogether.

In this section, I argue that this is not necessarily the case. Helmholtz himself emphasized that his argument was mainly directed against a particular interpretation of the Kantian theory advocated by Jan Pieter Nicolaas Land and by Albrecht Krause, among others. The full argument, as Helmholtz presented it in 1887, even retained the structure of a transcendental argument in Kant's sense, namely, of an inquiry into the preconditions for the possibility of measurement. It appears that the discussions with Land and with Krause motivated Helmholtz to clarify his view of objectivity. Whereas, before 1878, Helmholtz tended to identify objectivity with the mind-independent existence of specific objects (i.e., of rigid bodies), objectivity in his later writings depends on general conditions for the validity of empirical judgments. The argument is not unproblematic, because it entails a shift from formal conditions to empirical ones. Nevertheless, it can receive a consistent interpretation in terms of a Kantian argument. I discuss the main objections to Helmholtz (1887) in the next section. This section provides an account of Helmholtz's discussions with Land and with Krause and a reconstruction of Helmholtz's argument for the objectivity of scientific measurement.

4.2.1 Reality and Objectivity in Helmholtz's Discussion with Jan Pieter Nicolaas Land

Helmholtz revised his paper on the origin and meaning of geometrical axioms and translated it into English for the British journal *Mind* in 1876, six years after his public lecture in Heidelberg. In the English version of his paper, Helmholtz summarized the outcome of his inquiry into the foundations of geometry as follows:

1. The axioms of geometry, taken by themselves out of all connection with mechanical propositions, represent no relations of real things. When thus isolated, if we regard them with Kant as forms of intuition transcendentally given, they constitute a form to which any empirical content whatever will fit and which therefore does not in any way limit or determine beforehand the nature of the content. This is true, however, not only of Euclid's axioms, but also of the axioms of spherical and pseudospherical geometry.

2. As soon as certain principles of mechanics are conjoined with the axioms of geometry we obtain a system of propositions which has real import, and which can be verified or overturned by empirical observations, as from experience it can be inferred. If such a system were to be taken as a transcendental form of intuition and thought, there must be assumed a pre-established harmony between form and reality. (Helmholtz 1876, p.321)

In the first sense, the axioms of geometry are not synthetic. Therefore, there seems to be a gap between geometry and empirical reality. In the second sense, the assumption of a transcendental form of intuition presupposes an idealistic argument Helmholtz usually rejects (see especially, Helmholtz 1862, p.164). Since the assumption of a pre-established harmony between form and reality is unjustified, the connection between geometry and reality is problematic. Nevertheless, a few years later, in 1878, Helmholtz solved the puzzle by saying that "space can be transcendental without geometrical axioms being so" (Helmholtz 1878a, p.149). Helmholtz's solution depends on his distinction between the general properties of space (e.g., three-dimensionality and constant curvature), on the one hand, and further specifications, including not only Euclidean axioms, but also the axioms of spherical and pseudospherical geometry (i.e., the other two cases of manifolds of constant curvature), on the other.[6] Helmholtz's claim suggests that whereas the form of outer intuition can be identified as that of a threefold extended manifold of constant curvature, the specific axiomatic system associated with such a manifold depends on the laws of mechanics governing the behavior of rigid bodies. Given Helmholtz's naturalization of the form of spatial intuition, both this form and its specifications depend ultimately on observation and experiment. That might explain why Helmholtz refrained from calling space a priori in Kant's sense and adopted the ambiguous expression "transcendental space." We have already noticed that Helmholtz's use of the notion of transcendental referring to space (and to time) is problematic, because Kant himself denied that space can be transcendental. Nevertheless, Helmholtz's usage reflects the fact that his characterization of spatial intuition by means of the free mobility of rigid bodies is part of his justification of the objectivity of spatial measurements: the form thus derived differs from the assumptions that can be put to the test, because it provides us, at the same time, with a general framework for the interpretation of measurement.[7]

---

[6] Since space is characterized as a threefold extended manifold of constant curvature, more specific properties of space include the three classical cases of such a manifold. For this interpretation of Helmholtz's distinction between general and specific properties of space, see also Friedman (1997, p.33), Ryckman (2005, p.73), Pulte (2006, p.198), and Hyder (2009, pp.190–191). References to opposing interpretations, beginning with Schlick's, are given in Chap. 6.

[7] As it will become clear after discussing Helmholtz's comparison between space and time, my emphasis lies not so much in Helmholtz's naturalized interpretation of the forms of intuition – which I consider problematic – as in the fact that his argument for the objectivity of measurement retains, nonetheless, the structure of a transcendental argument. Cf. DiSalle (2006, p.129) for a different account of Helmholtz's relationship to Kant on this point: "Helmholtz's derivation of the general form of the Pythagorean metric from the axiom of free mobility reaffirms an important part of Kant's view, namely, that the visual perception of space and the geometry of space have a common basis. But if that basis is nothing more than an empirical fact that might have been otherwise, then the postulates of geometry have no claim to necessity." It seems to me that DiSalle here fails to appreciate the significance of Helmholtz's distinction between the general and specific properties of space: although acquired, the general notion of space provides us with necessary preconditions for the possibility of measurement, and, therefore, plays some role in the constitution of the objects of experience. Only the specific properties might have been otherwise and have no claim to necessity.

It was Jan Pieter Nicolaas Land who motivated Helmholtz to clarify his notion of objectivity. In 1877, Land published a paper entitled "Kant's Space and Modern Mathematics" in *Mind*. Land's point was that Helmholtz had overlooked the distinction between objectivity and reality. Whereas common sense regards the phenomena as real things, science regards them as signs for real things. This is because objective knowledge presupposes some interpretation of the data of sense perception. Physics agrees with common sense as far as metrical properties are concerned and we are counting and measuring. However, we cannot attach real import to analytic geometry, which "has but a conventional connection with the data of intuition, and merges into pure arithmetic" (Land 1877, p.41).

Land admitted that the axioms of geometry, taken by themselves out of all connection with mechanical propositions, represent no relation between physical objects. Axioms concerning the parts of space do not affect the bodies that fill such parts at a given moment. In this regard, Land agreed with Helmholtz: Euclidean axioms do not differ from those of spherical or pseudospherical geometry. Nevertheless, Land maintained that the form of spatial intuition which is actually given is that analyzed in Euclidean axioms (Land 1877, p.46). This is because, for Land, analytic geometry presupposes Euclidean intuitions about the fundamental concepts of geometry. Regardless of the fact that curvature is an intrinsic property of surfaces, Land, similar to many of his contemporaries, believed that spherical or pseudospherical surfaces can only be characterized as constructions in three-dimensional Euclidean space (see also Krause 1878, p.40; Riehl 1925, pp.218–219).

Helmholtz's reply appeared in *Mind* in 1878 as the second part of the paper on the origin and meaning of geometrical axioms (Helmholtz 1878b). The German version of the paper appeared the same year as the third appendix to the paper on "The Facts in Perception," which we discussed already in Section 3.2.3. The reply is that the objectivity of measurement can be accounted for in terms of both a realist and an idealist worldview. In particular, we have already mentioned that the idealist argument shows a development in Helmholtz thought. In 1878, he did not exclude the possibility of a transcendental way to bridge the gap between geometry and empirical reality. He identified the form of outer intuition as the group of spatial transformations or as the physically equivalent groups that remain invariant under material changes. Geometry captures a fundamental feature of empirical reality insofar as such a group is required for measurements to be repeatable. The idealist version of Helmholtz's argument differs from the Kantian theory of space because of Helmholtz's emphasis on the possibility of physically equivalent groups, depending on standards of approximation in empirical research. Therefore, the form of intuition can be specified in terms of of different axiomatic systems, including non-Euclidean geometries. Nevertheless, insofar as Helmholtz's form of intuition provides us with preconditions for the possibility of measurement, these play some role in the definition of physical magnitudes and can be compared with constitutive principles in Kant's sense.

4.2.2 Helmholtz's Argument against Albrecht Krause: "Space Can Be Transcendental without the Axioms Being So"

Krause's objection to Helmholtz is found in his essay on *Kant and Helmholtz on the Origin and Meaning of Spatial Intuition and Geometrical Axioms* (1878). Krause addressed the following question: Can one state different properties of space and, consequently, different geometrical axioms? In order to answer this question, Krause considered the relationship between the sense organs and the brain. He maintained that the Kantian theory of space is compatible with the requirement that spatial relations be univocally determined through their connection with the brain, whatever form or size the sense organs may have. Krause's view was that any variation or hypothesis of different spaces is based on one and the same space, whose properties depend on higher cognitive functions. Otherwise the form of our intuition would vary according to our sense organs, whose spatial features are contingent. Therefore, Krause criticized Helmholtz's attempt to draw spatiality out of sensations. In particular, Krause called into question Helmholtz's argument that a comparison between our space and its image in a convex mirror should provide us with intuitions we never had. According to Krause, such intuitions are impossible. He considered plain surfaces, as well as curved ones, as the boundaries of a three-dimensional body. It followed that straight lines cannot be identified as "straightest" lines or geodesics in spherical and pseudospherical surfaces. For the same reason, Krause denied the possibility of extending the concept of curvature to more than two-dimensional manifolds, according to Riemann's theory of manifolds. For Krause, the curvature of space cannot be measured, because anything endowed with direction already lies in space (Krause 1878, p.84).

Krause's further question was: Are the laws of spatial intuition expressed by the axioms certain? His answer was that, since spatial intuition is necessary for the construction of every geometrical object, the certainty of geometrical axioms cannot be called into question. He opposed the so-called "Riemann-Helmholtz theory of space," because this would lead to the skeptical consequence that there are no geometrical axioms properly speaking. According to Krause, geometrical axioms either provide us with immutable truths or cannot provide us with knowledge at all. Regarding the possibility of revising geometrical axioms, Krause's view was that we should not trust our measurements when they contradicted the axioms, because measurements are at least as approximate as natural laws. By contrast, geometry is exact knowledge.

Helmholtz's reply to Krause appeared as the second Appendix to the paper on "The Facts in Perception" under the title "Space Can Be Transcendental without the Axioms Being So" (Helmholtz 1921, pp.149–152). Firstly, Helmholtz made it clear that the empiricist theory of vision did not entail that the spatial features of our sense organs determine the objects in their shape and size.[8] Secondly, and more importantly,

---

[8] Krause's description is an oversimplification of the theory of local signs, which would entail, for instance, that a child sees in smaller way than an adult, for his eyes are smaller. However, this

he pointed out that the Kantian theory of knowledge is not committed to Krause's assumptions, which are derived from a nativist theory of vision. Therefore, Krause's argument can be falsified from a philosophical point of view: once nativist assumptions are rejected, space can be transcendental without the axioms being so.

We have already mentioned that the interpretation of this claim is controversial, not least because of Helmholtz's attribution of "transcendental" to space, which is in open contradiction with Kant (1787, pp.80–81). We return to the debate about the meaning and the consequences of Helmholtz's claim in Chap. 6. For now, it suffices to notice that Helmholtz did not exclude the possibility of a Kantian interpretation of the form of outer intuition, provided that the empiricist theory holds true for geometrical axioms. Kant identified spatial intuition as the form underlying any phenomenal changes. In Helmholtz's view, the possibility of giving a physical interpretation of non-Euclidean geometry showed that such a form can be specified in different ways.

Regarding Krause's objections to nineteenth-century inquiries into the foundations of geometry, Helmholtz replied that the measure of curvature is a well-defined magnitude which generally applies to n-dimensional manifolds. This consideration nullified Krause's attempt to show that three-dimensional Euclidean space is a necessary assumption for the interpretation of non-Euclidean notions. Helmholtz's point is that we must give reasons for our assumptions. Curiously enough, Krause did not take into account the results of scientific measurements because of their limited accuracy. However, he did not need measurements to be convinced of the correctness of those axioms that were supposed to be grounded in spatial intuition. In this case, Krause reassured himself with appraisals by "visual estimation." That is, for Helmholtz (1878a, p.151), "measuring friend and foe by different standards!"

Helmholtz did not say much about the convenience of regarding space as a transcendental concept. On the one hand, he made it clear that if the form of intuition is transcendental, it cannot be given immediately. On the other hand, the assumption of such a form must not contradict the objectivity of scientific measurements. What is the relation between space and geometrical axioms? Does the assumption of a general form of outer intuition provide a premise of Helmholtz's argument for the objectivity of measurement? Or does the claim about the empirical status of geometrical axioms simply depend on a distinction between metrical and extensive properties in Riemann's sense? In fact, Krause overlooked this distinction. Helmholtz's objection to Krause, however, goes deeper: by dismissing such well-defined magnitudes as the measure of curvature, and by mistrusting scientific procedures, Krause fails to account for the possibility of measurement. Furthermore, it is noteworthy that one of the general characteristics of space, according to Helmholtz, is constant curvature, which is also a metrical property. Nevertheless Helmholtz considered it a necessary presupposition of measurement.

Before handling these questions, it may be helpful to notice that Helmholtz's reply was anticipated in many ways by Benno Erdmann in his essay on *The Axioms*

---

assumption is contradicted by the most familiar experiences (Krause 1878, p.39). Not only did Helmholtz rule out such assumptions, but Krause overlooked that Helmholtz's explanation of visual perception was psychological rather than physiological (see Hatfield 1990, p.182).

of Geometry: A Philosophical Inquiry into the Riemann-Helmholtz Theory of Space (1877). Erdmann considered both Helmholtz's and Riemann's epistemologies a formal kind of empiricism, according to which our representations are only partial images of things which coincide with them in every quantitative relation (space, time, and natural laws) while differing from them in every qualitative one. The assumption of a pre-established harmony between sensations and their causes is called into question, because our mental activities are supposed to originate from our interaction with the world.[9] The empirical occasion for the formation of concepts does not provide us with spatial determinations; rather, we form spatial concepts in order to organize our sensations. Therefore, the form of space must be distinguished from its empirical content.

Regarding the philosophical meaning of the inquiries into the foundations of geometry, Erdmann pointed out that both Riemann's 1854 survey of the hypotheses underlying geometry and Helmholtz's thought experiments of 1870 contradicted the rationalist opinion that spatial intuition is independent of experience. If rationalists were right, space could not undergo any changes. By contrast, Riemann and Helmholtz showed that space admits different geometries. However, they neither answered the question of whether our inference from our representations to the existence of things is correct (which is a matter of controversy between idealism and realism), nor did they rule out other kinds of empiricism. In addition to formal empiricism, Erdmann distinguished between sensism, according to which our representations are images of things, and a refined kind of apriorism, which assumes that our representations, even though they are completely different from things, may correspond to them in each and every single part. Erdmann argued for apriorism as follows. He maintained that the concept of space can be specified both geometrically and analytically. On the one hand, the system of metric relations can be derived from spatial intuition, which is supposed to be singular and directly given, and yet capable of an infinite number of variations. On the other hand, Riemann showed that a generalized metric can also be developed analytically, so that the original system becomes a special case. Now, this prompts the question of how the geometrical and analytical interpretations of geometrical concepts are related. In order to answer this question, Erdmann used the whole/part opposition, which is characteristic of his apriorism. He wrote:

> The fact that our spatial intuition is single is not contradicted: we can only conceptualize the general intuition of a pseudospherical or spherical space of a certain measure of curvature. Such uniqueness, however, is not absolute anymore because we can fix homogeneous parts of those spaces intuitively and compare them with the metrical relations between partial representations of space. But the concepts of such spaces show in their development all the clearness and distinction enabled by the discursive nature of conceptual knowledge. Therefore, we may also speak about a concept of space. At the same time, however, we

---

[9]Cf. Krause's misunderstanding of the theory of local signs discussed above. Hatfield points out that Helmholtz considered spiritualist as well as materialist identifications of psychic activities with the material world to be metaphysical views, lacking explanatory power. By contrast, "[Helmholtz's] explanation ascribed the origin of our spatial abilities to the acquisition of rules for generating spatial representations, the acquisition process being guided by causal commerce with external objects" (Hatfield 1990, p.191).

clearly cannot form it directly without a diversion into the concept of magnitude. (Erdmann
1877, p.135)

Erdmann alluded to Helmholtz's thought experiments. Helmholtz's world in the
convex mirror showed that an intuitive comparison between different metrical systems is possible, though only locally: in order to make such a comparison, one
should not start from space itself, but from its parts. This corresponds to the fact that
Helmholtz relied on Riemann, not so much for the distinction between relations of
measure and relations of extension, as for the approach to the definition of space as
a special kind of extended magnitude: the concept of space presupposes that of
magnitude, not vice versa. At the same time, Erdmann advocated the Kantian view
that space as a whole is an intuition, not a concept.

Helmholtz's argument differs from Kant's, because it goes from the parts to the
whole and is not compatible with the conclusion that space is an intuition.
Nevertheless, he appreciated Erdmann's work on the axioms of geometry and considered it a reliable discussion of that subject in philosophical terms (Helmholtz
1878a, p.149). In my opinion, Helmholtz's appreciation is due to the fact that
Erdmann, unlike Land and Krause, sought to explain how the concepts of space and
of magnitude are related. Similarly, in order to construct the concept of space,
Helmholtz began with the most basic relationship between spatial magnitudes –
namely, their congruence. The general properties of space, especially constant curvature, depend on the free mobility of rigid bodies, which is required for spatial
magnitudes to be congruent. Since manifolds of constant curvature admit different
geometries, narrower specifications (e.g., the axioms of congruence) must be distinguished from the general principles of measurement. Erdmann's considerations
shed light on the difference between Helmholtz's conception of extensive magnitude and Kant's definition of extensive magnitudes as parts of one and the same pure
intuition of space: by relying upon his account of congruence for the construction of
the concept of space, Helmholtz makes the reference to pure intuition superfluous.
More recently, a similar consideration has been made by Darrigol: "Although
[Helmholtz's] definition of quantity seems reminiscent of Kant's 'extensive quantity', there are notable differences. Helmholtz does not relate his definition to the
intuition of space and time. He gives a definition of equality (*Gleichheit*) that can be
applied to any physical property. The definition of quantity implies divisibility into
equal parts, whereas for Kant mere divisibility is enough" (Darrigol 2003,
pp.257–258).[10]

To sum up, Helmholtz's replies to Land and to Krause suggest that the objectivity
of measurement depends on general conditions, which include Euclidean axioms as
special cases. Helmholtz did not reject the interpretation of the conditions required
as conditions of experience in Kant's sense. However, such an interpretation remains
problematic. On the one hand, Helmholtz's focus, in that context, is on the points of
disagreement with Kant: Kant's form of intuition imposes unjustified restrictions on

---

[10] For a comparison between Helmholtz and Kant on the concept of magnitude, see also Hyder
(2006).

empirical research unless one is willing to adopt a generalized form of intuition to be determined by the use of physical geometry. On the other hand, Helmholtz's defense of the objectivity of scientific measurements emphasized the lack of a comprehensive study of the conditions of measurement. My conjecture is that the discussion with Krause motivated Helmholtz to formulate the problem from a new viewpoint, which is explicit in 1887. Then, in order to account for the possibility of measurement, Helmholtz addressed the question of what conditions are required for the use of numbers to express physical magnitudes, including the distance between a pair of points.

In 1887, Helmholtz especially emphasized the Kantian aspects of his epistemology. Not only did he restate that space can be transcendental, but he referred "transcendental" to the form of intuition of time as well. He maintained that the axioms of arithmetic are related to the form of inner intuition as geometrical axioms are related to the form of outer intuition. Borrowing Erdmann's expression, one may say that, in both cases, the relation between intuitions and axioms depends on the formal-empiricist view that some metrical relations are common to subjective and objective experiences. Helmholtz's comparison between geometry and arithmetic in their relation to space and time is misleading, for two reasons. Firstly, Helmholtz's reference to Kant in this connection suggests that a similar comparison is found in Kant. However, we know from the previous chapters that Kant put more emphasis on the difference between geometry and arithmetic, because he believed that arithmetic has no axioms. As pointed out by Friedman, there is no evidence that arithmetic for Kant stands to time as geometry does to space. In the transcendental exposition of the concept of space, Kant explained the synthetic a priori knowledge of geometry in terms of the pure intuition of space. However, he did not mention arithmetic in relation to time. Instead, Kant (1787, p.49) identified the a priori science whose possibility is explained by the pure intuition of time as the general doctrine of motion. He called number "a concept of the understanding" (Kant 1787, p.182). This and other passages also quoted by Friedman suggest that the science of number is itself entirely independent of intuition, and that only its application concerns intuitive objects – namely, objects which are to be counted (Friedman 1992, p.106).

Secondly, Helmholtz himself seems to introduce a fundamental difference between geometry and arithmetic in their relation to space and time. As Darrigol put it:

> Both in geometry and in arithmetic, Helmholtz derived a whole system from the basic fact (free mobility of rigid bodies, ordering in time) and some definitions. The parallel ends here. In geometry, several constant-curvature geometries are compatible with the basic fact, so that experience (together with mechanical axioms) is required to decide between these multiple options. In arithmetic, the basic fact is sufficient to induce a single system of arithmetic (as was nearly the case in Helmholtz's geometry before he became aware of Lobachevski's geometry). External experience is no longer needed to decide between different sets of axioms; rather, external experience is needed to determine which physical properties can be measured by numbers. In one case, the application decides the axioms; in the other, the axioms control the applications. (Darrigol 2003, pp.555–556)

In my reading, this disparity between geometry and arithmetic sheds light on the fact that Helmholtz seemed to find the notion of transcendental less controversial when referred to time. Although for him, both forms of intuitions are acquired, geometry also has an empiricist aspect, in the sense that the specific metric of such a form is a matter for empirical investigation. By contrast, there is only one structure that corresponds to the form of the intuition of time, and the same structure is presupposed in all applications to the empirical domain. Given the relative simplicity of this case, my suggestion is to take a closer look at Helmholtz's arguments regarding the relation between time and arithmetic to gain insight into his use of the notion of transcendental. I suggest that the claim that time is transcendental corresponds to the fact that the laws of addition – which control the applications – play the role of constitutive principles of physical magnitudes. In other words, his argument for formal empiricism seems to presuppose a transcendental argument in Kant's sense. Since Helmholtz's premises differ considerably from Kant's, one might say, more precisely, that – with different premises – Helmholtz's argument for the applicability of mathematical concepts to empirical reality retains the structure of a transcendental argument.[11]

4.2.3   The Premises of Helmholtz's Argument: The Psychological Origin of the Number Series and the Ordinal Conception of Number

In the introduction to "Counting and Measuring," Helmholtz summarized his remarks on transcendental intuition as follows. Geometrical axioms cannot be derived from an innate intuition independently of experience. This claim does not rule out the view of space as a transcendental form of intuition, but rather what Helmholtz considered to be an unjustified interpretation of this view by Kant's successors. According to Helmholtz, these turned the Kantian theory of knowledge into the metaphysical endeavor to reduce nature to a system of subjective forms.[12] We have already noticed that a legitimate idealistic interpretation of the transcendental role of space in Helmholtz's sense should take into account an inner/outer opposition, which is reflected by the opposition between physical and pure geometry. In 1887, Helmholtz used his theory of knowledge to account for the origin and meaning of the axioms of arithmetic. He wrote: "[I]f the empiricist theory – which I besides others advocate – regards the axioms of geometry no longer as propositions

---

[11] For a reconstruction of Helmholtz's argument in comparison with alternative formulations of the same argument by Hölder and Cassirer, see also Biagioli (2014).

[12] Recall that Helmholtz had already contrasted Kant's theory of knowledge with the idealist philosophy of nature of Schelling and Hegel in Helmholtz (1855). Helmholtz's conception of the interaction between subjective and objective factors of knowledge had its roots in his interpretation of Kant and in his reception of the philosophy of Fichte (see Köhnke 1986, pp.151–153; Heidelberger 1994, pp.170–175).

unprovable and without need of proof, it must also justify itself regarding the origin of the axioms of arithmetic, which are correspondingly related to the form of intuition of time" (Helmholtz 1887, p.72).

The parallel with space suggests that time can be deemed transcendental in the same sense: a transcendental argument is necessary for the axioms of arithmetic to be valid for the empirical manifold. The axioms are the following propositions:

AI. If two magnitudes are both equal with a third, they are equal amongst themselves.

AII. The associative law of addition: $(a + b) + c = a + (b + c)$.

AIII. The commutative law of addition: $a + b = b + a$.

AIV. If equals are added to equals, their sums are equal.

AV. If equals are added to unequals, their sums are unequal.

In order to introduce the argument, Helmholtz distanced himself from a formalistic view of arithmetic. He wrote:

> I consider arithmetic, or the theory of pure numbers, to be a method constructed upon purely psychological facts, which teaches the logical application of a system of signs (i.e. of the numbers) having unlimited extent and an unlimited possibility of refinement. Arithmetic notably explores which different ways of combining these signs (calculative operations) lead to the same final result. This teaches us, amongst other things, how to substitute simpler calculations even for extraordinarily complicated ones, indeed for ones which could not be completed in any finite time. (Helmholtz 1887, p.75)

Apart from testing the internal logicality of our thought, such a procedure would appear to be a mere game of ingenuity with fictitious objects. By contrast, Helmholtz emphasized that the axioms of arithmetic are, at the same time, laws of addition; and additive principles of the same kind are required for physical magnitudes to be compared. The goal of Helmholtz's inquiry into the foundations of the theory of numbers was to provide a natural basis for our use of symbols and a proof of their applicability. Therefore, he deemed arithmetic "a method constructed upon purely psychological facts."

According to Helmholtz, the clarification of this point required a complete analysis of the concept of number. In a certain sense, it is clear that the "naturalness" of the number series is merely an appearance: the choice of number signs is a matter of stipulation, and the so-called natural numbers are but arbitrarily chosen signs. All the same, their series is impressed on our memory much more firmly than any other series of objects as a consequence of its frequent repetition. Ordinal numbers acquire a paradigmatic role in the recollection of all other sequences. In this sense, the series of numbers reflects the characteristics of inner intuition: "The present representation is thereby contrasted, in an opposition pertaining to the form of intuition of time, as the succeeding one to the preceding ones, a relationship which is irreversible and to which every representation entering our consciousness is necessarily subject. In this sense, orderly insertion in the time sequence is the inescapable form of our inner intuition" (Helmholtz 1887, p.77). This requires us to designate each step in the series without gaps or repetitions, as in the decimal system.

Helmholtz maintained that the complete disjunction thereby obtained is "founded in the essence of the time sequence" (p.77). He expressed this fact as follows:

AVI. If two numbers are different, one of them must be higher than another.

AVI entails that ordinal relations are asymmetric and transitive. From AI, it follows that equality is transitive and symmetric instead. From transitivity (i.e., if $a = b$ and $b = c$ then $a = c$) the validity of AI for the series of the whole numbers follows. A generalized form of the remaining axioms can be derived from Grassmann's axiom:

$$(a + b) + 1 = a + (b + 1).$$

The associative law of addition, for example, is generalized as follows:

$$R + b + c + S = R + (b + c) + S,$$

where capital letters denote the sum of arbitrarily many numbers. More precisely, Helmholtz makes (implicit) use of the principle of mathematical induction whenever he extends a relation between a number and its successor to the entire series with the phrase "and so on without limit" (see DiSalle 1993, p.519; Darrigol 2003, p.551).[13]

Once addition was defined in the terms of Grassmann's axiom, Helmholtz introduced the following axiom:

AVII. If a number c is higher than another one a, then I can portray c as the sum of a and a positive whole number b to be found.

Helmholtz's goal was to extend the laws of addition, especially AVII, to cardinal numbers. He described the method of numbering off for the purpose of addition as correlating an ordered sequence $(n + 1)$, $(n + 2)$… to the series of the whole numbers. He then correlated a first series preserving a certain sequence to a second series having variable sequences. Given two numbers n and $(n + 1)$, on the one side, and two symbols ε and ζ, on the other, there are two possible manners of correlation:

$$(a) n \to \varepsilon, (n + 1) \to \zeta$$
$$\text{or } (b) n \to \zeta, (n + 1) \to \varepsilon.$$

If a) is substituted for b), the second series α, β, γ, etc. can be put into one-to-one correspondence with the series $(n + 1)$, $(n + 2)$, etc. By continued exchanging of

---

[13] Darrigol suggests that Helmholtz was influenced by the Grassmann brothers, Hermann and Robert, who constructed numbers by iterated connection of a single unit or element. They defined operations and derived their properties by mathematical induction. Evidence for this suggestion is Helmholtz's use of Grassmann's axiom, along with the fact that he refers to the Grassmann brothers' way of proceeding in the introductory section of "Counting and Measuring."

neighboring members of a group, one can bring about any possible sequence of its members without gaps or repetitions. According to Helmholtz's theorem: "Attributes of a series of elements which do not alter when arbitrarily neighbouring elements are exchanged in order with each other, are not altered by any possible alteration of the order of the elements" (Helmholtz 1887, p.85).

As a consequence of this theorem, cardinal numbers can be defined as follows: If the complete number series from 1 to n is needed in order to correlate a number to each element of the group, then n is called the cardinal number of the members of the group. The corresponding proposition for AVII is that the total number of the members of two groups that have no member in common equals the sum of the cardinal numbers of the members of the two single groups. Another consequence is that the commutative law of addition can be generalized as follows. Given the associative law of addition, by AIII, and by his theorem, Helmholtz inferred that:

$$R + a + b + S = R + (b + a) + S = R + b + a + S.$$

Summing up, Helmholtz advocated an ordinal conception of number: the laws of addition apply, first of all, to ordinal numbers, but he needed a theorem in order to prove that the same laws apply to cardinal numbers as well. It might seem that this technicality does not provide us with a justification of Helmholtz's empiricist theory. Note, however, that Helmholtz's conception of addition is not, strictly speaking, arithmetical. Arithmetical addition is a paradigmatic case. However, there are different kinds of numbering even within the theory of numbers, and we do not know from the outset whether the laws of addition apply to the specific case of numbering that ascertains the cardinality of a set. If Helmholtz's theory of knowledge is to be justified, the same principles must be extended to empirical domains.

The psychological origin of the number series and the ordinal conception of number were Helmholtz's premises for the following argument:

> The concept of addition described above [...] coincides with the concept of it which proceeds from determining the total cardinal number of several groups of numerable objects, but has the advantage of being obtainable without reference to external experience. One has thereby proved, for the concepts of number and of a sum – taken only from inner intuition – from which we started out, the series of axioms of addition which are necessary for the foundation of arithmetic; and also proved, at the same time, that the outcome of this kind of addition coincides with the kind which can be derived from the numbering of external numerable objects. (Helmholtz 1887, p.87)

As in the case of space, Helmholtz's analysis of the form of time differs from Kant's, because Helmholtz's forms admit a psychological development. Nevertheless, Helmholtz refers to Kant insofar as the forms of intuition are supposed to entail principles for correlating mathematical theories with empirical reality. One of these principles is the free mobility of rigid bodies. The principles required for the use of numbers to express magnitudes in general are the laws of addition. Insofar as the extension of the laws of addition to empirical domains provides us, at the same time, with a definition of the magnitudes under consideration, the same principles play the role of constitutive principles in Kant's sense.

Then why does Helmholtz present this argument as a confirmation of his original empiricist view? As pointed out by DiSalle, this argument – more than the parallel between space and time – sheds light on the empiricist aspect of Helmholtz's philosophy of mathematics as a whole:

> Helmholtz's empiricism did not seek to reduce mathematical truth to empirical truth. Instead, it described arithmetic as the theory of a process, "based on psychological facts," that identifies and systematize lawlike relations in nature, "reducing the varied manifold of things and changes before us to quantitative relations" (Helmholtz 1887, p.[103]). For Helmholtz, the truth of arithmetic remained a matter of internal consistency, but its objective significance lay in the correspondence it provided between subjective operation and objective states of affair. (DiSalle 1993, p.519)

It is because of this characteristic of Helmholtz's empiricism that there seems to be a shift from the formal to the empirical in the formulation of the conditions of measurement. I believe that the dialectical tension between these two aspects of Helmholtz's work can be resolved in different ways, including the standard reading of Helmholtz which goes back to Schlick. My suggestion is to reconsider the constructive aspect of the operations and processes that, for Helmholtz, lie at the foundation of mathematics, because I believe that this led to a more plausible way to vindicate Kant's philosophy of mathematics than those which were based on the theory of pure intuitions. In particular, in Section 4.3, I rely upon a reformulation of Helmholtz's argument by Cassirer. Furthermore, I think that Helmholtz's approach is empiricist in yet another sense: the objective significance of mathematical laws (i.e., their application to empirical domains) depends on the history of science. Therefore, the Marburg School of neo-Kantianism agreed with Helmholtz that the system of the principles of experience cannot be delimited once and for all. In order to highlight this aspect of Helmholtz's approach, the concluding part of this section deals with Helmholtz's use of his argument to develop a general theory of measurement.

4.2.4 The Composition of Physical Magnitudes

Helmholtz defined magnitudes as those objects, or attributes of objects, which allow a distinction into greater, equal or smaller when compared with similar ones.[14] His question was: "What is the objective sense of our expressing relationships between

---

[14] The German term *Größe* was used to translate both Latin terms *magnitudo* and *quantitas*. This translation overlooks that *magnitudo* is a more general concept, whereas *quantitas* entails both extension and the possibility of a numerical representation (see Du Bois-Reymond 1882, p.14, and note). A similar consideration can be made with regard to the difference between continuous quantities and discrete ones, which in Latin were called *quanta* and *quantitas*, respectively. The German translation for both terms was *Größe*, and was introduced by Christian Wolff in his *Mathematisches Lexicon* (1716). Wolff did not explicitly distinguish between discrete and continuous quantities. Nevertheless, such a distinction is implicit in Wolff's use of multiply/diminish (*vermehren/vermindern*) for those quantities whose comparison presupposes the laws of arithme-

real objects as magnitudes by using denominate numbers [i.e., numbers together with a unit]; and under what conditions can we do this" (Helmholtz 1887, p.75)? The answer requires a physical interpretation of AI and AVII. Recall that Helmholtz had already clarified the notion of physical equaivalence between spatial magnitudes in his reply to Land: if two such magnitudes are to be measured, it must be possible to assign a cardinal number to them. Helmholtz completed his argument in 1887. He then made it clear that AI provides us with a definition of equality: as a definition, AI does not have objective meaning in itself. Measurement presupposes the additional requirement that two similar objects, when interacting under suitable circumstances, allow the observation of a particular outcome which does not occur, as a rule, between other pairs of similar objects. Helmholtz called the procedure that enables us to accomplish such observation the method of comparison. The simplest geometrical structure for which a magnitude is specifiable is, for example, the distance between a pair of points. The method of comparison, in this case, is to verify whether pairs of points can be brought into congruent coincidence. The condition for assigning a numerical value to the distance is that the points remain fixedly linked for at least the time of our operation.

Helmholtz's argument, however, would be incomplete without a second step. Once we have found a suitable method of comparison, the particular outcome of the interaction between two similar objects is supposed to remain unaltered if the two objects are interchanged. This procedure must be generalized so that objects that have proven to be equal are also mutually substitutable in any further cases. To clarify this point, let us return to Helmholtz's example. The concept of length presupposes something more than that of distance. Whereas distance entails only a distinction of equal and unequal, length also presupposes an opposition of greater and smaller. If two pairs of points a, b and a, c, of differing distance, coincide at a and are placed in a straight line so that a portion of this line is common to both, then either b falls upon the line ac or c upon the line ab. This fact leads to a more general consideration: once we know whether two magnitudes are equal or unequal, in order to measure them in the second case, the greater must be calculable as the sum of the smaller and their difference. That is to say, if AI is to acquire objective meaning, it must be also given a physical interpretation of AVII or the principle of homogeneity of the sum and the summands. This is because equality can be attributed to two or more objects only if they are compared from some point of view. Their comparison with regard to magnitude requires that equal or unequal magnitudes be homogeneous. This consideration is purely logical and shows that homogeneity (*Gleichartigkeit*) is a more fundamental property than equality (*Gleichheit*). In addition, Helmholtz gives the following physi-

---

tic, and make bigger or smaller (*vergrößern/verkleinern*) for geometric figures (see Cantù 2008). Notwithstanding the ambiguity of the term, Helmholtz addressed the problem of establishing the conditions for a numerical representation of magnitudes. In this sense, he contributed to clarifying a distinction which is explicit in the first axiomatic theory of measurement by the German mathematician Otto Hölder (1901). Hölder formulated axioms of quantity (*Quantität*) and introduced the concept of measure (*Maß*) after his proof that the ratios of quantities of the same kind can be expressed by positive real numbers and can be summed arithmetically (Hölder 1901, Sect. 10). Further references about Helmholtz's influence on Hölder are given later on in this section.

cal interpretation: the issue of whether the result of connection remains the same when parts are exchanged must be decided by the same method of comparison with which we ascertained the equality of the parts to be exchanged (Helmholtz 1887, p.96). In other words, measuring procedures must be repeatable.

Given the homogeneity of the sum and the summands, the remaining laws of addition can be applied to the composition of physical magnitudes. Helmholtz's theorem concerning cardinal numbers corresponds to the following proposition: "A physical method of connecting magnitudes of the same kind can be regarded as addition, if the result of the connection – when compared as a magnitude of the same kind – is not altered either by exchanging individual elements with each other, or by exchanging terms of the connection with equal magnitudes of the same kind" (Helmholtz 1887, p.96).

To sum up, it does not suffice to have a method of comparison. If the results of measurements are to be univocally determined, the composition of physical magnitudes requires the method of addition.[15] The method of comparison must be completed by that of addition, because the objectivity of measurements presupposes both observable results and univocal determinability.

Helmholtz then showed that magnitudes that can be added are also divisible. Every magnitude occurring can be regarded as the sum of a cardinal number of equal parts to be chosen as units. This choice is conventional, and it might happen that the magnitudes under consideration are not expressible without remainder. Even in that case, the unit can, nevertheless, be divided again in the usual manner, so that the measurement can attain any degree of precision. Whereas complete precision is attainable only for rational proportions, the value of irrational proportion can be enclosed between arbitrarily reducible limits. Thereby, one can calculate all continuous differentiable functions of irrational magnitudes. This way of proceeding remains insufficient for the calculation of everywhere continuous, but nowhere differentiable functions, such as those introduced by Weierstrass in the theory of functions. But it suffices for the purposes of the theory of measurement, insofar as such a kind of discontinuity has not yet been encountered in geometry or physics.[16]

---

[15] For Helmholtz's distinction between the method of comparison and that of addition, see also Helmholtz's introduction to his lectures on theoretical physics from 1893. After recalling that AI is a general definition of equality, he wrote: "The principle in its general formulation is clearly false. For example, an object can have the same weight as another, and the latter can have the same color as a third object. It does not follow that the first object equals the third one. But the principle is correct and it is of great importance, insofar as it applies to magnitudes that can be compared by using the same method of observation. We call such magnitudes homogenous relative to the method of observation" (Helmholtz 1903, p.27). Helmholtz clarified the distinction between the two methods as follows: "The method of comparison does not provide us with an answer to the question: Which of the unequal magnitudes is the greatest? [...] Only the method of addition also determines the concepts of smaller and greater" (p.36).

[16] Darrigol (2003, p.553) pointed out that Helmholtz's notion of divisibility apparently entails not only the possibility of regarding a given quantity as the sum of a number of equal quantities, but also the possibility of approximately expressing a given quantity as a multiple of a fixed unit, and of indefinitely improving the approximation by introducing a series of subunits. In other words, Helmholtz implicitly assumed the Archimedean property as well, as this is the property that, along with the existence of a difference, allows the arbitrarily precise approximation of ratios by rational numbers.

In 1901, the German mathematician Otto Hölder showed how to obtain equivalent results by introducing an axiomatic theory of quantity. Given a set of axioms, including Dedekind's axiom of continuity as formulated in Dedekind (1872), Hölder proved theorems equivalent to the assumptions of divisibility and the Archimedean property. He provided a new formulation of the theory of proportions by using a representation theorem. The theorem states that, for every ratio of quantities (i.e., for each two quantities which are given in a determined order), there exists a determined number, which can be either rational or irrational (Hölder 1901, p.23).

Helmholtz's theory of measurement foreshadows this development in two main respects. Firstly, Helmholtz's issue is the same as Hölder's: once numbers have been clearly distinguished from the objects to be counted, a set of conditions is needed to account for the use of numbers in modeling measurement situations. Secondly, Helmholtz's and Hölder's theories of measurement have similar consequences for the distinction between intensive and extensive magnitudes. Even though Helmholtz maintained that irrational proportions lack a numerical expression, he denied the view that continuity is an intrinsic property of some magnitudes. His point was that the extension of additive principles proceeds independently of the supposition that there might be a difference in nature between extensive magnitudes and intensive ones. Therefore, he substituted the classical distinction between extensive and intensive magnitudes – which might be misunderstood as an essential one – with a relative distinction between additive and nonadditive magnitudes. Such a distinction must be made, because the equation expressing a natural law cannot quantitatively determine a nonadditive magnitude (i.e., a coefficient) until all other magnitudes occurring in it have already been determined as additive magnitudes. However, Helmholtz emphasized the provisional character of his distinction, and he did not exclude the case that new discoveries occasionally enable us to turn nonadditive magnitudes into additive ones (Helmholtz 1887, p.93). Arguably, Helmholtz bore in mind the discussion about the measurability of sensations (see Heidelberger 1993; Darrigol 2003). Therefore, he distanced himself from Elsas's *On Psychophysics* (1886): Elsas denied that sensations can be measured by identifying sensations as essentially intensive magnitudes. At the same time, Heidelberger and Darrigol emphasized the significance of this discussion for posing the problem of measurability in general. Therefore, I interpret the previous quote as a general remark about the additive/nonadditive distinction in the history of science.

The same consequence apparently follows from Hölder's axiomatization: owing to its formal nature, the theory of quantity can be applied to various domains and coherently extended to natural phenomena by identifying physical compositions that satisfy the established laws. In this sense, Hölder adopted Helmholtz's approach. He referred to Helmholtz in the introduction to his theory of measurement (see Hölder 1901, p.2, note). A thorough analysis of the consequences of such an approach for the distinction between intensive and extensive magnitudes is found in Hölder's major epistemological work, *Mathematical Method: Logico-Epistemological Inquiries into Mathematics, Mechanics, and Physics* (1924, pp.88–89). The philosophical distinction between intensive and extensive magnitudes should not play any role in the mathematical treatment of the theory of quantity, because such a distinction would impose unjustified restrictions on the interpretation of additive principles.[17]

However, there is an important difference between Helmholtz's study of the conditions of measurement and later axiomatic theories of quantities: Helmholtz's emphasis on the physical interpretation of the laws of addition suggests that the conditions he is looking for are not so much formal, as empirical.[18] This is the aspect of Helmholtz's inquiry that seems to be most problematic. As we will see in the next section, Hermann Cohen was one of the first to call into question the empirical character of Helmholtz's conditions. Another compelling objection against Helmholtz's previous example for a method of comparison was formulated by Henri Poincaré. Helmholtz's condition for the comparison of spatial magnitudes by superposition of one such magnitude was that its points remain fixedly linked for the time of measurement. Helmholtz seemed to believe such a supposition to be derived from our experiences with solid bodies. Indeed, he maintained that the "well-known" method of comparing distances by bringing two pairs of points into congruent coincidence requires time (Helmholtz 1887, p.92). However, the dynamical factor might have been mentioned for another reason as well: measurement requires us to assign numerical coordinates to the points within a system which needs to remain identifiable in motion (see Hyder 2006, pp.34–35). In other words, Helmholtz assumed the free mobility of rigid bodies both as a consequence of the existence of such bodies and as a rule for our choice of measuring rods. Poincaré's point was that Helmholtz defined as rigid those bodies whose properties are invariant under rigid transforma-

---

[17] Hölder wrote: "The mathematical development of the theory of quantity presupposes that there subsist some relations between some objects, and specifically that there subsists a relation of composition between these objects, which satisfy the established laws formally. The empirical interpretation of the objects and of their relations per se does not affect the development of the theory. Therefore, the same theory can be applied to completely different objects. If, for example, we substitute degrees of sensation and their differences for the points and the segments of a line, and assume, firstly, that such differences can have equal or different magnitudes and, secondly, that the same laws apply as those which in the case of segments were formulated as axioms, we obtain the same theory of measurement. For the same reason, the usual distinction between 'extensive' and 'intensive' magnitudes, as found in philosophical texts, is not essential to mathematics" (Hölder 1924, pp.78–79).

[18] On the relationship between Helmholtz and Hölder, see also Darrigol (2003, pp.563–565). Cf. Michell (1993, pp.195–196), who argues that Helmholtz and Hölder – insofar as he was influenced by Helmholtz – defended a classical conception of measurement as the discovery of a matter of fact. I think that reconsidering Helmholtz's philosophical ideas might help us to see why it is hard to find a place for Helmholtz's theory of measurement in the classical/representational dichotomy. On the one hand, Helmholtz removed numbers from external experiences, as in the representational approach: his attempt was to prove that the laws of addition are grounded in inner intuition and can, nevertheless, be extended to empirical manifolds. On the other hand, his view entailed something more than the representational view: in order to account for the use of numbers in measurement, Helmholtz formulated a transcendental argument, which finds an echo in Hölder's argument that magnitudes that can be intuited can also be constructed. Regarding Hölder's argument, see Biagioli (2013). I am grateful to Joel Michell for letting me know that, after reading Darrigol 2003, he changed his view about his earlier classification of Helmholtz and Hölder as exponents of a classical view of measurement.

tions. How then can he define these transformations? If this definition presupposes the existence of rigid bodies, Helmholtz's reasoning is circular (Poincaré 1902, p.60).

Poincaré's solution was that approximately rigid bodies can be distinguished from nonrigid ones, because they can be easily brought back to their initial position after displacement. It does not follow that physical bodies behave as geometric solids. Since the free mobility of rigid bodies cannot be induced from experience, it must be stipulated. In other words, the formation of geometrical concepts presupposes not so much actual experiences with solid bodies, as an idealized construction of the same.

A similar problem arises if one considers Helmholtz's physical interpretation of the homogeneity of the sum and the summands. Helmholtz's condition for adopting the method of addition in physics amounts to the requirement that measuring procedures be repeatable. But this can hardly be considered a purely empirical condition, insofar as the repeatability of measurements presupposes the regularity of nature. What is Helmholtz's justification for such an assumption? On the one hand, the comprehensibility of nature tends to play the role of a regulative principle in Kant's sense. On the other hand, such requirements presuppose Helmholtz's argument for the applicability of mathematics: since inner and outer experiences originally interact in the development of the forms of intuition, it is no accident that the same rules apply to both kinds of experience. However, the extension of the laws of addition to various domains – including empirical ones – proceeds progressively and can only be anticipated, but not accomplished a priori.

As pointed out by Lydia Patton in regard to anther application of Helmholtz's theory of signs in fluid dynamics, there is a key problem for Helmholtz's psychological justification for the a priori postulate of regularity as governing the sign system: Given the fact that scientific theories do not map on to the known properties of the objects considered independently, what is our justification for constructions in the sign system? Whereas Helmholtz believed that the regularities of nature that that system describes nonetheless map on to regularities in nature, Marburg neo-Kantians found this answer unsatisfactory and posed the problem of the validity of a priori postulates from the perspective of a Kantian theory of experience (Patton 2009, p.285).

Is there a place for a Kantian understanding of experience in Helmholtz's theory of measurement? What distinguishes Kantianism from empiricism, on the one hand, and from conventionalism, on the other? In order to handle these questions, we will have to take into account later developments in the debate on the philosophical consequences of non-Euclidean geometry. Poincaré's objections to Helmholtz are discussed in Chap. 6. We now return to the discussion about Helmholtz's theory of number. It was in the context of this debate that Cohen called into question Helmholtz's notion of experience. At the same time, Cohen and Cassirer reformulated Helmholtz's argument for the applicability of mathematics in Kantian terms.[19]

---

[19] For an illuminating comparison between Helmholtz and Marburg neo-Kantians on the theory of signs and the a priori, see also Patton (2009).

4.3 Some Objections to Helmholtz

As we saw in the last section, the ordinal conception of number provides the premises for Helmholtz's argument for the applicability of the laws of arithmetic to physical magnitudes. The argument for the possibility of scientific measurements also holds true for physical geometry. The possibility of a numerical representation is the distinctive characteristic of physical equivalence, including congruent coincidence between pairs of points. In order for such concepts as that of a rigid geometrical figure to apply to particular physical bodies, rigidity must be defined by specifying the algebraic equations that relate the points of a system to each other. It was also mentioned that Helmholtz's argument entails a shift from conventional factors of measurement to empirical ones. Is this shift justified? Since this is a crucial point, it may be helpful to discuss a series of objections against Helmholtz's conception of number in some detail in the first place. Some of these objections followed immediately after the publication of "Counting and Measuring" by Helmholtz and of another contribution to the Zeller Festschrift, "On the Concept of Number" by Leopold Kronecker. Similar to Helmholtz, Kronecker used the empirical genesis of numerical notations (e.g., of the decimal system) to argue for an ordinal understanding of the notion of number in general. It followed from Kronecker's analysis that all calculi can be reduced to operations with whole numbers originally defined as ordinal numbers (Kronecker 1887, p.273).

Cohen, Husserl, and Frege charged Helmholtz's ordinal conception with circularity: despite his definition of numbers as arbitrarily chosen signs, Helmholtz's argument for the possibility of assigning numbers to counted objects presupposes an arithmetical definition of number. In particular, we will consider Dedekind's definition of 1888. A more sympathetic reading of Helmholtz's argument was given in 1910 by Cassirer, who interpreted Helmholtz's foundation of the theory of number as an example of the use of symbols in science. In Cassirer's view, the shift from symbols to empirical facts is justified, insofar as the use of symbols reflects the relational nature of thought, which for Cassirer is constitutive of the objects of experience.

4.3.1 Cohen, Husserl, and Frege

Hermann Cohen reviewed Helmholtz's article in the 24th volume of *Philosophische Monatshefte* (1888). Cohen's starting point was a general remark about Helmholtz's notion of experience. Helmholtz's argument for the applicability of the laws of arithmetic in physics suggests that experience should be identified as the domain of natural science. As we saw in the second chapter, according to Cohen, this is the kind of experience Kant's transcendental inquiry is concerned with, namely, scientific experience. However, Helmholtz's naturalization of Kant's forms of intuition presupposes a different notion of experience, as individual, psychological experience. Cohen's first remark is that the issue at stake is the constitution of the object of experience, and this depends not so much on some psychic process, as on a connection of general conditions for the comparison of physical magnitudes. However, the major problem with the naturalization of the forms of intuition is that it obscures the point of Kant's theory of pure sensibility, which is that there are different levels in the constitution of the objects of experience. The spatiotemporal ordering of the appearances in pure intuition provides us with the appropriate manifold for the use of the concepts of the understanding. As we saw in Chap. 3, Cohen agreed with Helmholtz that nineteenth-century analytic geometry had made pure intuition superfluous. Nevertheless, a corresponding differentiation of levels is required to better understand how the concepts of homogeneity, equality, and magnitude are related to one another. The fundamental concept for Cohen is that of magnitude, which presupposes both space and time, along with the arithmetical concept of number (Cohen 1888, p.268).

The main objection to Helmholtz concerns the definition of number. Numbers, as used in measurement, must be clearly distinguished from counted objects. In posing the problem of the conditions for numbers to represent magnitudes, Helmholtz took a substantial step towards a clarification of concepts. However, his psychological foundation of the number series obscures his distinction. His characterization of counting reflects the ambiguity in his conception of experience. The time sequence, as such, cannot be identified as the scientific procedure of numbering a series of ordered elements, because the law of order already presupposes arithmetical knowledge. For the same reason, Helmholtz's definition of numbers as arbitrarily chosen signs is misleading. The proper use of signs depends on their intended meaning, which is established by the theory of numbers. This point is crucial for the theory of measurement as well. The description of counting as assigning labels to some groups of things overlooks the constitutive function of numbers as tools of knowledge. The laws of addition provide us with additive principles for physical magnitudes, because magnitudes in physics cannot be estimated otherwise than in terms of numbers. Therefore, Cohen distanced himself from Helmholtz's characterization of numbers as signs:

> The sign means at most an image, the faithful image, which the imagination draws. The sign bears the mark of resignation [...]: as if we had to content ourselves with identification marks, without being able to ask about the reason and justification for their being coined. [...] The risk with the term "sign" lies in the superficial view that things in themselves exist in a determinate way, and can be made countable only by accident. Instead, numbers are tools for producing things as scientific objects. Our problem is to account for the role of numbers in the origin of objects. The ambiguous term "sign" opens the door to the mystical view, which affects numbers ever since Pythagoras. However, we have to approach the theory of numbers as apprentices, not as theologians. (Cohen 1888, p.272)

Cohen goes on to say that Helmholtz's definition of numbers as signs reflects his narrower view of the concept of magnitude as extensive only. By contrast, Cohen relies on his principle of intensive magnitudes to account for the epistemic value of numbers as infinitely small. The calculus offered the example of numbers that constitute, rather than simply designate, objective reality. Therefore, Cohen (1888, p.276) maintained that the principle of intensive magnitude enables us to bridge the gap between the subjective qualities of sensation and the objective realm of science. Apparently, Elsas bore in mind Cohen's conception of reality when he defended the view that intensive magnitudes are fundamental and cannot be reduced to extensive magnitudes. However, Cohen's argument differs significantly from Elsas's. Whereas Cohen emphasized the epistemic value of numerical concepts, Elsas denied that intensive magnitudes can find an adequate mathematical expression (Elsas 1886, p.68).[20]

Without addressing the problem of measurement, Edmund Husserl formulated a similar objection in the *Philosophy of Arithmetic: Psychological and Logical Investigations* (1891). Husserl entitled Helmholtz's treatment of the theory of numbers as a kind of nominalism and argued for the naturalness of the number series as follows:

> If it is conceded that the numbers are nothing other than arbitrary signs in an arbitrarily established order, then certainly the designation of this ordering as a "natural" one is misleading. Any other ordering of the signs could just as adequately have been taken to be the number sequence. However, those who speak of a natural ordering in the domain of numbers surely do not mean the ordering of arbitrary signs, but rather of certain concepts designated by means of them. Whichever we consider, whether the "ordinal" numbers or the "cardinals" (both terms taken in the true sense), we always come to the result that the sequential order is one grounded through the nature of these concepts themselves. (Husserl 1891/1970, p.185)

When we talk about natural numbers, we are usually referring to numerical concepts, not to numerals. Once it is made clear that numerals do not provide us with numerical concepts, there is no reason to begin with ordinal numbers rather than cardinal ones: the meaning of these concepts depends on the arithmetical treatment of the theory of numbers, and, from this point of view, cardinal numbers can be derived from ordinal ones, and vice versa. Recall that Helmholtz introduced ordinal numbers firstly because their ordering is supposed to be grounded in the time sequence. He then proved that cardinal numbers can be ordered in the same way, because this fact holds true for variable sequences of signs. Husserl's objection against this way of proceeding is that, in both cases, order depends on the arithmetical meaning of the concepts under consideration.

Husserl's argument was related to the mathematical debate about the concept of number at that time. In *The Foundations of Arithmetic: A Logical-Mathematical Investigation into the Concept of Number* (1884), Frege identified the natural numbers as cardinal numbers, which he defined as classes of equinumerous classes. His remark against Helmholtz was very close to Husserl's. In the second volume of *The Basic Laws of Arithmetic* (1893), Frege charged Helmholtz with circularity: The concept of number and that of sum, as taken in their arithmetical meaning, are tacitly presupposed (Frege 1893/1966, p.139). Frege rejected Helmholtz's view of

---

[20] For a detailed discussion of Cohen (1888), also in relation to the later debate on the concept of number, see Ferrari (2009).

arithmetic as a method based on psychological facts and argued for a purely logical foundation of arithmetic based on his system of logic.

In Dedekind (1888), independently of Frege, Richard Dedekind advocated the view that "arithmetic is part of logic" by proposing a definition of number "entirely independent of the notions or intuitions of space and time" (Dedekind 1901, p.31). Despite the fact that there are both technical and philosophical differences between Frege's and Dedekind's treatment, Dedekind's proof that the characterization of the natural numbers is categorical filled a gap in Frege's treatment and lent plausibility to the idea of a logical foundation of arithmetic.[21] However, Dedekind's foundation was based on the assumption of the concept of set as a fundamental concept. Frege distanced himself from Dedekind because of his own commitment to the intensional point of view: whereas for Frege everything should be stated in terms of concepts, a set could only be the extension of a concept.[22] This disagreement reflects the fact that Frege's and Dedekind's logicist approaches were based on a different view of logic. A comparison would require us to add more details about the prehistory of logicism.[23] For the purposes of the present chapter, I limit myself to pointing out that the current understanding of logic as formal logic led to a misunderstanding of Dedekind's approach as psychological. However, it appears that Dedekind's elaboration of his conception of numbers was motivated by the attempt to defend the ordinal nature of the notion of number in a more plausible way than Helmholtz's and Kronecker's. Dedekind's argument entails a refutation of Helmholtz's psychologism, on the one hand, and of Kronecker's reductionism, on the other. In order to highlight the logical aspect of Dedekind's approach, the next section offers a brief discussion of his definition of numbers as free creations of the human mind.[24]

---

[21] See Potter (2000, pp.83–84): "Dedekind proved the validity of the method of defining functions by means of recursive equations and derived from this a demonstration that his characterization of the natural numbers is categorical; no trace of either result is to be found in Frege until his later *Grundgesetze*, and by the time he wrote that work he had access to Dedekind's monograph. But it is worth noting that once these omissions from Frege's account are made good, there is a clear sense in which the two treatments are equivalent."

[22] See Gray (2008). Further points of disagreement emerged only after the discovery of the antinomies of set theory. "Frege's and Dedekind's views were rather similar, and were taken to be so until the whole question of the relation between a set and the extension of a concept went from being elementary to being very problematic indeed" (Gray 2008, p.168).

[23] For a broader understanding of logicism as to include "parallel" variants of logicism such as Dedekind's and Frege's, see Ferreirós (1999) and Reck (2013).

[24] According to Dummett (1991, p.49), Dedekind's idea, "widely shared by his contemporaries, was that abstract objects are actually created by operations of our mind. This would seem to lead to a solipsistic conception of mathematics; but it is implicit in this conception that each subject is entitled to feel assured that what he creates by means of his own mental operations will coincide, at least in its properties, with what others have created by means of analogous operations. For Frege, such an assurance would be without foundation: for him, the contents of our minds are wholly subjective; since there is no means of comparing them, I cannot know whether my idea is the same as yours." More recent scholarship initiated by Tait (1996) and Ferreirós (1999) reconsidered the logical aspect of Dedekind's abstraction from all the properties that have to do with a particular representation of mathematical domains (including spatial and temporal intuitions) in order to obtain a categorical characterization of mathematical structures. For further evidence that

4.3.2 Dedekind's Definition of Number

Dedekind's definition of natural numbers is found in his essay of 1888, "The Nature and Meaning of Numbers." In the preface to the first edition of his essay, Dedekind claimed that his intervention in the debate on the concept of number was occasioned by the appearance of Helmholtz's and Kronecker's articles. He wrote: "The appearance of these memoirs has induced me to publish my own views, in many respects similar but in foundation essentially different, which I formulated many years ago in absolute independence of the works of others" (Dedekind 1901, p.31, note). Dedekind contrasted Helmholtz's and Kronecker's views with a Kantian conception of counting as a synthesis of the understanding: counting depends on the ability of the mind to relate things with one another and to let a thing correspond to a thing or represent a thing by another. "Without this ability no thinking is possible" (ibid.). Therefore, Dedekind deemed numbers "free creations of the human mind" (Dedekind 1901, pp.34, 68).[25] He did not deny that the acquisition of such an ability depends on frequently repeated experiences and empirical generalizations. However, he called into question the idea that the empirical genesis exempts us from giving a proof of the basic facts of arithmetic. He defended the view that nothing should be accepted without a proof. In particular, in 1888, he proved a theorem of complete induction, which is a generalization of mathematical induction. The theorem justifies both definitions and proofs by induction, namely, those definitions and proofs that presuppose an inference from a number n and its successor in the series of numbers $n + 1$ to the entire system of numbers.

We have already mentioned that Helmholtz assumed the principle of complete induction without giving a proof. Dedekind agreed with Helmholtz that the notion of a cardinal number had been erroneously considered simple in the received view. For Dedekind, however, it does not suffice to assume some psychological facts underlying arithmetic. Every step in the formation of the concept of number must be made explicit and generally understandable. The point of disagreement with

---

Dedekind distanced himself from a literal understanding of the notion of metal creation – and, therefore, anticipated the objection above – cf. Reck (2003, pp.385–394). In what follows, I argue that Dedekind's response to Helmholtz sheds further light on his critique of psychologism. I rely on the aforementioned literature for the interpretation of his definition of number as an expression of a variant of "logicism" or "logical structuralism."

[25] It is controversial whether Dedekind advocated a Kantian view or opposed Kant, also because in this and other quotes he did not explicitly refer to Kant in support of his own characterization of counting. Potter argues that Dedekind distanced himself from the philosophy of mathematics which emerged from the *Critique of Pure Reason*, since Dedekind's view "abandons the passivity of intuition, and ascribes to our intellect the power to represent entities of our own creation directly. This interpretation is unKantian, not merely because it credits us with a capacity Kant took to belong only to God, but because, according to it, we are capable of intuiting objects not lying within any structure of which we have an a priori grasp" (Potter 2000, pp.102–103). However, it seems to me that Dedekind's idea is reminiscent not so much of intellectual intuition, as of the idea of a purely conceptual or symbolic synthesis as advocated by Marburg neo-Kantians, among others. Regarding the dispensability of pure intuitions, Dedekind's view was in line with many of his contemporaries', including the neo-Kantians.

Kronecker, on the other hand, is that Dedekind's logical foundation of the theory of numbers was not committed to reductionism. Dedekind assumed the infinity of the set of the natural numbers from the outset. And his analysis of continuity in 1872 already showed that the characterization of numbers as free creations of the human mind applies to irrational numbers as well. He believed that every numerical domain, without restrictions, depends on the ability of the mind to relate things to one another.

Dedekind called a system infinite if it can be put into one-to-one correspondence with a proper subset of itself (a subset is proper if it is not equal to the set which includes it). The set of natural numbers, for example, can be put into one-to-one correspondence with the set of even numbers and exhibits the structure of a simply infinite set. Notice, however, that the structure identified by Dedekind as a simply infinite set does not presuppose the example of natural number; his categorical characterization of natural numbers begins rather with the proof that an infinite set exists. Therefore, his example is the totality S of all things which can be objects of thought. The proof depends on the fact that if s signifies an element of S, then the thought s' that s can be an object of my thought is itself an element of S. S is infinite because the mapping that takes s to s' is one-to-one and there are elements in S which are different from all such thoughts s'.[26]

Dedekind's characterization of a simply infinite set presupposes, in addition, his definition of a chain as a set with a transformation to itself (nowadays a set A is said to be closed with respect to a relation R if whenever x belongs to A and xRy, it follows that y belongs to A). Given a one-to-one function φ from any element of a system N to its successor and a fundamental element 1 not contained in φN, the series of natural numbers can be generated as the chain of 1. Dedekind proved that all simply infinite systems are similar to the number series and, accordingly, to one another (Dedekind 1901, pp.93–92). In current terminology, Dedekind's proof establishes the categoricity of the second order Peano axioms of arithmetic, namely, the proposition that models of these axioms are isomorphic to one another.

Notice that Dedekind made a specific assumption in Helmholtz's axiomatization unnecessary, namely AVI. This was precisely the assumption that was supposed to be grounded in the essence of the time sequence. The system of natural numbers can

---

[26]This proof became one of the most controversial parts of Dedekind's account of number after the discovery of Russell's antimony. Even regardless of the antinomies of set theory, the same proof cannot provide a semantic consistency proof. Gödel later showed that neither Dedekind's nor others' proofs could overcome the problems concerning the provability of consistency for arithmetic. Furthermore, Dedekind's notion of the totality of the objects of thought has been largely misunderstood as psychological (see, e.g., Potter 2000, p.100). I agree with Reck (2003) that the categorical characterization of numbers as specific kinds of simple infinities shows rather the characteristics of a logical proof in Dedekind's sense. Regarding Dedekind's definition of infinity, it is noteworthy that Dedekind took a decisive step towards a modern conception of mathematics as the study of abstract structures. As pointed out by Ferreirós, in contrast to the traditional definition, Dedekind based the notion of natural number on a general theory of finite and infinite sets. "The familiar and concrete was thus explained through the unknown, abstract, and disputable. [...] He was defining the infinite through a property that Galileo, and even Cauchy, regarded as paradoxical, for it contradicted the Euclidean axiom 'the whole is greater than the part'" (Ferreirós 1999, p.233).

be treated independently of the time sequence, because the structure analyzed by Dedekind is the more general concept of a progression. Therefore, it suffices to introduce < as a transitive and asymmetric relation. Numbers are then generated as those elements that are in the said relation to each other. Dedekind wrote:

> If in the consideration of a simply infinite system N set in order by a transformation φ we entirely neglect the special character of the elements; simply retaining their distinguishability and taking into account only the relations to one another in which they are placed by the order-setting transformation φ, then are these element called *natural numbers* or *ordinal numbers* or simply *numbers*, and the base-element 1 is called the *base-number* of the *number series N*. With reference to this freeing the elements from every other content (abstraction) we are justified in calling numbers a free creation of the human mind. (Dedekind 1901, p.68)

This quote shows that Dedekind called numbers "ordinal" in a completely different sense than Helmholtz. This corresponds to the fact that Dedekind did not account for abstraction as a psychic process. The term abstraction here indicates that all the properties which have nothing to do with numbers qua numbers, including the properties of spatiotemporal objects, have to be disregarded in order for arithmetic properties to find a rigorous and general expression. Abstraction in Dedekind's sense has its counterpart in the idea of a "free creation," because the univocal determinability of numbers as elements of a simply infinite system provides us, at the same time, with the defining characteristics of numbers.[27] Despite the fact that Dedekind also called the elements thereby generated ordinal numbers, one should not forget that simply infinite systems also provide the fundamental structure for the development of the theory of cardinal numbers. Dedekind did not presuppose specific objects (i.e., ordinal or cardinal numbers), because all that is needed for the definition of the series of natural numbers is the basic element, along with the injective function that maps every element into its image in the system. Abstraction thus construed justifies the view of numbers as free creations of the human mind, because there is no difference in principle between numbers and their images: by definition, an image is called successor of the preceding number in the series and it also belongs to the system.

Is the advantage of Dedekind's approach over Helmholtz's conception of number purely technical? And if not, what does Dedekind add to a philosophical understanding of arithmetical objects? Russell's classical objection in the *Principles of Mathematics* (1903) was that ordinals, according to Dedekind, should be "nothing but the terms of such relations as constitute a progression. If they are anything at all, they must be intrinsically something; they must differ from other entities as points

---

[27] For a similar account of the notion of abstraction in Dedekind's work, see Tait (1996). The meaning of Dedekind's "creation" in this connection has been clarified especially by Reck (2003, p.400): "[Simple infinity] is identified as a new system of mathematical objects, one that is neither located in the physical, spatio-temporal world, nor coincides with the previously constructed set-theoretic infinities. [...] what has been done is to determine uniquely a certain 'conceptual possibility', namely a particular simple infinity. Which one again, i.e., what is the system of natural numbers now? It is that simple infinity whose objects only have arithmetic properties, not any of the additional, 'foreign' properties objects in other simple infinities have."

from instants, or colours from sounds" (Russell 1903, p.249). Since Dedekind's characterization of the system of numbers is categorical, Russell seems to refer to numbers taken as singular objects.

The next section deals with Cassirer's defense of Dedekind's characterization of the system of numbers as a necessary and sufficient condition to univocally determinate the elements in it according to their arithmetical properties. In this connection, Cassirer formulated what I call an internal objection to Helmholtz. Unlike Husserl and other critics, Cassirer did not call into question Helmholtz's argument for the fundamental role of order in the formation of numerical concepts. Nevertheless, Cassirer used Dedekind's definition of number to disentangle the notion of order from psychological aspects. With regard to Dedekind's definition, Cassirer advocated a relational view of mathematical objects in general.

4.3.3 An Internal Objection to Helmholtz: Cassirer

Cassirer maintained that Dedekind's way of generating the number series is a more natural development of the Kantian conception of number than Helmholtz's psychological interpretation. Not only did Kant distinguish pure intuitions, which are the objects of the transcendental inquiry, from empirical intuitions, but, owing to the structure of the *Critique of Pure Reason*, the concept of number cannot be derived from the pure intuition of time. Kant identified number as a concept of the understanding in the following passage from the "Schematism of the Pure Concepts of the Understanfing:"

> The pure image of all magnitudes (*quantorum*) for outer sense is space; for all objects of the senses in general, it is time. The pure schema of magnitude (*quantitatis*), however, as a concept of the understanding, is number, which is a representation that summarizes the successive addition of one (homogeneous) unit to another. Thus number is nothing other than the unity of the synthesis of the manifold of a homogeneous intuition in general, because I generate time itself in the apprehension of the intuition. (Kant 1787, p.182)

The same quote lends plausibility to Cohen's thesis that Kant was not committed to the claim that axioms concerning magnitudes are given in intuition. These axioms correspond rather to the principles linking the concept of number to the manifold of a homogeneous intuition in general. Cohen read Kant's argument in connection with the problem of measurement. Cohen's own answer was that because of the conceptual origin or number, the theory of numbers provides us with the appropriate tools for the definition of physical magnitudes.

In "Kant and Modern Mathematics," Cassirer (1907, p.34, note) noticed that Kant's conception of number depends not so much on some psychological assumptions regarding the form of our temporal intuition, as on the conceptual determination of such a form as an ordered one. In this sense, Cassirer adopted Hamilton's (1835) definition of algebra as a science of pure time or order in progression. Cassirer emphasized the purely logical character of the concept of order, as analyzed in mathematical logic. But is this definition in line with Kant's? In order to highlight this point, it might be helpful to make a comparison with Friedman. As noticed above, Friedman, among others, showed that a parallel between geometry and arithmetic in Kant would be inconsistent with Kant's claim that arithmetic has no axioms because it has no objects of intuition as magnitudes.

> [Arithmetic and algebra,] as techniques of calculation, are in turn independent of the specific nature of the objects whose magnitudes are to be calculated. In the theory of magnitude itself we assume absolutely nothing about the nature and existence of the magnitudes to be thereby determined: we merely provide operations (such as addition, subtraction, and also the extraction of roots) and concepts (above all the concept of ratio) for manipulating any magnitudes there may be. What magnitudes there are is settled outside the theory itself – for Kant, by the specific character of our intuition. (Friedman 1992, pp.113–114)

In support of his reading, Friedman points out that, for Kant, calculation is possible even when there are no numbers for the determination of a magnitude. Kant, for example, denied that there are irrational numbers, because he believed that the determination of magnitude in that case is only possible by means of infinite approximation to a number.[28]

We turn back to the status of irrational numbers, after a brief exposition of Dedekind's definition, in Chap. 5. For now, it is noteworthy that Cassirer emphasized the intellectual character of the notion of a progression in general for very different reasons. According to him, abstraction from the nature and existence of magnitudes provides us with a more general standpoint for their univocal determination in the empirical use of mathematical theories. He especially appreciated Dedekind's view of number because of the reversal of the traditional way of defining mathematical objects by referring to magnitudes already given outside the theory. To quote again from Ferreirós (1999, p.233), Dedekind broke with tradition and explained "the familiar" through the "unknown, abstract, and disputable."

More explicitly than in "Kant and Modern Mathematics," Cassirer distanced himself from Kant in *Substance and Function*. He wrote:

> Since in the determination of its concept only the general moment of "progression" is retained, everything that is here said of number is valid with regard to every progression in general; it is thus only the serial form itself that is defined and not what enters into it as material. If the ordinal numbers in general are to exist then they must, so it seems, have some "inner" nature and property; they must be distinguished from other entities by some absolute "mark," in the same way that points are different from instants, or tones from colors. But this objection mistakes the real aim and tendency of Dedekind's determinations. What is here expressed is just this: that there is a system of ideal objects whose whole content is exhausted in their mutual relations. The "essence" of the numbers is completely expressed in their positions. And the concept of position must, first of all, be grasped in its greatest logical universality and scope. The distinctness required of the elements rests upon purely conceptual and not upon perceptual conditions. The intuition of pure time, upon which Kant based the concept of number, is indeed unnecessary here. (Cassirer 1910, pp.39–40)

---

[28] Friedman refers in particular to Kant's exchange with August Rehberg in 1790. For further details about Kant's view of irrational magnitudes, see Friedman (1992, pp.110–112).

It has already been mentioned that Russell called into question Dedekind's definition of number precisely because everything that is said here about number is valid
with regard to every progression in general. Cassirer's reply is that numbers taken
singularly do not have some defining characteristic or inner nature besides their
mutual relations. Therefore, the concept of position, "grasped in its greatest logical
universality," is necessary and sufficient to distinguish any element of the system of
numbers from any other element. For the same reason, Cassirer recognized that
sensible conditions – whether empirical or pure – have been made superfluous: all
aspects that do not have to do with arithmetic are disregarded on account of
Dedekind's abstraction, and numbers are defined or created – as Dedekind put it –
according to their position within a categorical structure.

We know from Sect. 2.5 that, in 1907, Cassirer argued for a connection between
logicism and critical idealism. We can now see that his interpretation of logicism is
tied to Dedekind's approach to the theory of numbers (as pointed out by Heis 2011).
Indeed, Cassirer usually mentioned the example of Dedekind's definition of number
to support the thesis that there tends to be a shift from concepts of substance to
concepts of function in the transformation of scientific concepts (see, e.g., Cassirer
1907, pp.7–21; 1910, Ch.2; 1929, p.407; 1957, Ch.4). The connection with critical
idealism lies in the fact that individuals first present themselves as a result of a conceptual synthesis. As in the case of natural numbers, abstraction from allegedly
existing entities and from their qualities prepares the ground for a more precise and
comprehensive characterization of arithmetical objects: such objects as numbers are
fully determined only as members of a series. At the same time, critical idealism is
committed to the proof that conceptual syntheses of the same kind provide us with
conditions for the cognition of empirical objects. Therefore, Cassirer argued for an
extension of the syntheses of mathematics to mathematical physics. This is the
aspect of Cassirer's approach towards mathematics that corresponds to a transcendental proof of the possibility of knowledge in Kant's sense: mathematical syntheses must be proved to provide the necessary presuppositions for any phenomenal
connection and experience of nature (Cassirer 1907, p.45).

This again is the line of argument we found in Helmholtz (1887). Unlike
Helmholtz, however, Cassirer emphasized that nature, thus construed, is the result of
a complex connection of conceptual and sensible conditions, which is made possible
by mathematical syntheses in the first place. Cassirer distanced himself from
Helmholtz's naturalization of the foundations of mathematics, because a coherent
development of the argument requires that any ontological assumption be accounted
for in terms of its conceptual conditions. This is what distinguishes Kant's transcendentalism from a "copy" theory of knowledge, which is the view that knowledge
reproduces a mind-independent reality. This way to account for the correctness of
knowledge is doomed to failure, because, given that premise, it is impossible to find
reliable criteria to compare the copy with its allegedly transcendent model. Critical
idealists solved the problem by assuming that the criteria of knowledge are established in knowledge itself. The issue of the transcendental philosophy is to reconstruct the fundamental principles at work in the history of science, and it follows from
this formulation of the problem that the required principles are of a logical nature.

Thus, for Cassirer, it is not the case that the empirical origin of our conceptual apparatus warrants its applicability to the object of experience. On the contrary, mathematical concepts must be detached from any empirical representation. Otherwise, they could not play their foundational role in the formation of empirical concepts.

Is Helmholtz's epistemology committed to a copy theory of knowledge, as suggested by his designation of sensations as signs of external causes? How can Helmholtz overcome the previous objections against his conception of number? Cassirer's discussion of these questions suggests that a transcendental argument might provide us with the appropriate solution. Insofar as such a line of argument can be traced back to Helmholtz (as described in Section 4.2), I call Cassirer's objection to Helmholtz internal.

Cassirer's discussion of Helmholtz's and others' ordinal views of number is found in the second chapter of *Substance and Function* (1910). There, Cassirer maintained that recent developments in the theory of numbers, especially Dedekind's definition of 1888, provided us with a paradigmatic case of the general tendency Cassirer ascribed to the history of logic and of the exact sciences in the nineteenth century: scientific objects, once looked upon as bearers of relations, tended to be redefined in terms of complexes of relations. This was apparent in Dedekind's definition of number, because Dedekind himself emphasized that the nature of numbers consists of their relations to one another according to an order-setting function. Cassirer's point is that it is only the conceptual level reached by Dedekind's abstraction that warrants full precision in the distinction of any element of the series of natural numbers from the others. Cassirer's objection to Helmholtz is that numerical notations only work because of conceptual distinctions, which cannot be arbitrarily stipulated. Otherwise, it is hard to see how different signs could represent different things. A psychological viewpoint, according to which signs differ from each other because of their size and arrangement, would be out of place: it is not a question of the sensuous appearance of the signs, but of their meaning. Cassirer wrote:

> It is only the ambiguity in the concept of sign, only the circumstance that under it can be understood, now the bare existence of a sensuous content, and now the ideal object symbolized by the latter, which makes possible this reduction to the nominalistic schema. Leibniz, whose entire thought was concentrated upon the idea of a "universal characteristic," clearly pointed out in opposition to the formalistic theories of his time, the fact that is essential here. The "basis" of the truth lies, as he says, never in the signs but in the objective relations between ideas. If it were otherwise, we would have to distinguish as many forms of truth as there are ways of symbolizing. (Cassirer 1910, p.43)

The nominalistic schema is untenable, because a meaningful use of signs presupposes a logical basis or, as Leibniz put it, objective relations between ideas. In deeming such relations objective, Cassirer overcame Helmholtz's opposition between inner and outer experience.[29] Therefore, Cassirer reformulated Helmholtz's argument as follows:

---

[29] The passage above clearly foreshadows the concept of symbol that lies at the center of Cassirer's philosophy of symbolic forms. However, in 1910 and even in Cassirer (1929), Cassirer uses "sign" to refer to mathematical symbols in continuity with the mathematical tradition of his time and with Helmholtz's usage.

> The consideration of the "cardinal numbers" [...] occasions the discovery of no new prop-
> erty and no new relation, which could not have been previously deduced from the bare
> element of order. The only advantage is that the formulae developed by the ordinal theory
> gain a wider application, since they can henceforth be read in two different languages.
> (Cassirer 1910, p.42)

What is right in Helmholtz's argument is that the theory of cardinal numbers does not produce new objects, but realizes rather a new logical function: a finite sequence of unities earlier regarded as a series can now be regarded as a whole, namely, as a system, along with operations acting on it. Cassirer agreed with Helmholtz that, in this sense, the cardinal aspect presupposes the ordinal one. However, this consideration depends not so much on some psychological assumptions, as on the logical analysis of the system of natural numbers. Cassirer relied on Dedekind's definition of number, because it makes it clear that order is not intuitively given, but follows from the use of an ideal operation. This way of thinking enables us to study the structural properties of numbers independently of their allegedly intrinsic nature: the set of numbers is determined and coherently extended by the specification of the operations acting on it.

A similar consequence is characteristic of Helmholtz's approach. Helmholtz searched for a natural basis of arithmetic, because he believed that the extension of the laws of arithmetic to empirical manifolds cannot but proceed progressively. Cassirer's objection suggests the following consideration: one way to disentangle the ambiguity of Helmholtz's conception of numerical symbolism is to point out that in his search for the conditions of measurement, there is both a top-down direction from numerical structures to measurement situations and a bottom-up direction from empirical domains to mathematical laws.[30] It is because of the top-down reference of Helmholtz's inquiry that, notwithstanding his attempt to provide a physical interpretation, Helmholtz's conditions of measurement appear highly generalized and purely formal. At the same time, the bottom-up reference suggests that mathematical symbolism is open to further generalizations in view of possible empirical uses.[31] Critical idealism shows a similar tendency insofar as it makes the level of

---

[30] The dual character of Helmholtz's notions as both empirical and formal has been emphasized especially by DiSalle. However, I do not believe that Helmholtz's account of mathematical notions can be compared to Kant's notion of pure intuition in this respect (cf. DiSalle 2006). Not only did Helmholtz distance himself from Kant, but the main analogy from my point of view lies simply in the fact that the formal/empirical dichotomy was introduced only later. Such a dichotomy does not do justice to the dual direction of inquiry which is characteristic of Helmholtz's approach: it is because of this characteristic of his approach that some of his notions appear to us to be both formal and empirical, depending on the direction of inquiry. In fact, Helmholtz's definitions are neither formal nor empirical in the current understanding of these terms.

[31] For a clarification of the inductive aspect of Helmholtz's approach, see Schiemann (2009). Pulte (2006, p.199) refers to Schiemann's characterization of Helmholtz's approach as an inductive or bottom-up conceptualization to point out that, according to such an approach, there can be no sharp separation of intuition and conceptual knowledge. In the following chapters, I rely on Pulte also for the observation that Schlick tacitly presupposed such a separation in his interpretation of Helmholtz's epistemological writings on account of his own top-down approach to measurement. Cf. my former consideration about the formal/empirical dichotomy. For a thorough discussion of Schlick's reading of Helmholtz, see also Friedman (1997).

generalization dependent on scientific experience considered in its historical development. The range of hypotheses available at a given stage of knowledge cannot be fixedly delimited a priori. Nevertheless, mathematical thinking not only plays a foundational role in the context of specific theories, but also provides us with guiding principles for the explanation of new phenomena.

References

Biagioli, Francesca. 2013. Between Kantianism and empiricism: Otto Hölder's philosophy of geometry. *Philosophia Scientiae* 17.1: *La pensée épistémologique de Otto Hölder*, ed. Paola Cantù and Oliver Schlaudt, 71–92.

Biagioli, Francesca. 2014. What does it mean that "space can be transcendental without the axioms being so"? Helmholtz's claim in context. *Journal for General Philosophy of Science* 45: 1–21.

Campbell, Norman Robert. 1920. *Physics: The elements*. Cambridge: Cambridge University Press.

Cantù, Paola. 2008. Mathematik als Größenlehre. In *Wolffiana, 2: Christian Wolff und die europäische Aufklärung. Akten des 1. Internationalen Christian-Wolff-Kongresses, Halle (Saale), 4–8 April 2004*. Part 4. Section 8: *Mathematik und Naturwissenschaften*, ed. Jürgen Stolzenberg and Oliver-Pierre Rudolph, 13–24. Hildesheim: Olms.

Cassirer, Ernst. 1907. Kant und die moderne Mathematik. *Kant-Studien* 12: 1–49.

Cassirer, Ernst. 1910. *Substanzbegriff und Funktionsbegriff: Untersuchungen über die Grundfragen der Erkenntniskritik*. Berlin: B. Cassirer. English edition: Cassirer, Ernst. 1923. *Substance and Function and Einstein's Theory of Relativity* (trans: Swabey, Marie Collins and Swabey, William Curtis). Chicago: Open Court.

Cassirer. Ernst. 1929. *Philosophie der symbolischen Formen. Vol. 3: Phänomenologie der Erkenntnis*. Berlin: B. Cassirer.

Cassirer, Ernst. 1957. *Das Erkenntnisproblem in der Philosophie und Wissenschaft der neueren Zeit. Vol. 4: Von Hegels Tod bis zur Gegenwart: (1832–1932)*. Stuttgart: Kohlhammer.

Cohen, Hermann. 1888. Jubiläums-Betrachtungen. *Philosophische Monatshefte* 24: 257–291.

Darrigol, Olivier. 2003. Number and measure: Hermann von Helmholtz at the crossroads of mathematics, physics, and psychology. *Studies in History and Philosophy of Science* 34: 515–573.

Dedekind, Richard. 1872. *Stetigkeit und irrationale Zahlen*. Braunschweig: Vieweg. English edition in Dedekind (1901): 1–27.

Dedekind, Richard. 1888. *Was sind und was sollen die Zahlen?* Braunschweig: Vieweg. English edition in Dedekind (1901): 29–115.

Dedekind, Richard. 1901. *Essays on the theory of numbers*. Trans. Wooster Woodruff Beman. Chicago: Open Court.

Diez, José Antonio. 1997. Hundred years of numbers: An historical introduction to measurement theory 1887–1990. Part 1: The formation period. Two lines of research: Axiomatics and real morphisms, scales and invariance. *Studies in History and Philosophy of Science* 21: 167–181.

DiSalle, Robert. 1993. Helmholtz's empiricist philosophy of mathematics: Between laws of perception and laws of nature. In *Hermann von Helmholtz and the foundations of nineteenth-century science*, ed. David Cahan, 498–521. Berkeley: The University of California Press.

DiSalle, Robert. 2006. Kant, Helmholtz, and the meaning of empiricism. In *The Kantian legacy in nineteenth-century science*, ed. Michael Friedman and Alfred Nordmann, 123–139. Cambridge, MA: The MIT Press.

du Bois-Reymond, Paul. 1882. *Die allgemeine Functionentheorie. Vol. 1: Metaphysik und Theorie der mathematischen Grundbegriffe: Grösse, Grenze, Argument und Function*. Tübingen: Laupp.

Dummett, Michael. 1991. *Frege: Philosophy of mathematics*. Cambridge: Harvard University Press.

Elsas, Adolf. 1886. *Über die Psychophysik: Physikalische und erkenntnisstheoretische Betrachtungen*. Marburg: Elwert.

Erdmann, Benno. 1877. *Die Axiome der Geometrie: Eine philosophische Untersuchung der Riemann-Helmholtz'schen Raumtheorie*. Leipzig: Voss.

Ferrari, Massimo. 2009. Le forme della conoscenza scientifica: Cohen e Helmholtz. In *Unità della ragione e modi dell'esperienza: Hermann Cohen e il neokantismo*, ed. Gian Paolo Cammarota, 77–96. Soveria Mannelli: Rubettino.

Ferreirós, José. 1999. *Labyrinth of thought: A history of set theory and its role in modern mathematics*. Basel: Birkhäuser.

Frege, Gottlob. 1884. *Die Grundlagen der Arithmetik: Eine logisch mathematische Untersuchung über den Begriff der Zahl*. Breslau: Koebner.

Frege, Gottlob. 1893/1966. *Grundgesetze der Arithmetik, begriffsschriftlich abgeleitet*. Olms: Hildesheim.

Friedman, Michael. 1997. Helmholtz's *Zeichentheorie* and Schlick's *Allgemeine Erkenntnislehre*: Early logical empiricism and its nineteenth-century background. *Philosophical Topics* 25: 19–50.

Friedman, Michael. 1992. *Kant and the exact sciences*. Cambridge, MA: Harvard University Press.

Gray, Jeremy J. 2008. *Plato's ghost: The modernist transformation of mathematics*. Princeton: Princeton University Press.

Hamilton, William Rowan. 1835. *Theory of conjugate functions, or algebraic couples; with a preliminary and elementary essay on algebra as the science of pure time*. Dublin: Hardy.

Hatfield, Gary. 1990. *The natural and the normative: Theories of spatial perception from Kant to Helmholtz*. Cambridge, MA: The MIT Press.

Heidelberger, Michael. 1993. Fechner's impact for measurement theory. *Behavioral and Brain Sciences* 16: 146–148.

Heidelberger, Michael. 1994. Helmholtz' Erkenntnis- und Wissenschaftstheorie im Kontext der Philosophie und Naturwissenschaft des 19. Jahrhunderts. In *Universalgenie Helmholtz: Rückblick nach 100 Jahren*, ed. Lorenz Krüger, 168–185. Berlin: Akademie Verlag.

Heis, Jeremy. 2011. Ernst Cassirer's neo-Kantian philosophy of geometry. *British Journal for the History of Philosophy* 19: 759–794.

Helmholtz, Hermann von. 1855. Über das Sehen des Menschen. In *Vorträge und Reden*, vol. 1, 5th ed, 85–118. Braunschweig: Vieweg, 1903.

Helmholtz, Hermann von. 1862. Über das Verhältnis der Naturwissenschaften zur Gesammtheit der Wissenschaften. In *Vorträge und Reden*, vol. 1, 5th ed, 157–185. Braunschweig: Vieweg, 1903.

Helmholtz, Hermann von. 1876. The origin and meaning of geometrical sxioms. *Mind* 1: 301–321.

Helmholtz, Hermann von. 1878a. Die Tatsachen in der Wahrnehmung. In Helmholtz (1921): 109–152.

Helmholtz, Hermann von. 1878b. The origin and meaning of geometrical axioms. Part 2. *Mind* 3: 212–225.

Helmholtz, Hermann von. 1887. Zählen und Messen, erkenntnistheoretisch betrachtet. In *Philosophische Aufsätze: Eduard Zeller zu seinem fünfzigjährigen Doctor-Jubiläum gewidmet*, ed. Vischer, Friedrich Theodor, 70–97. Leipzig: Fues. Repr. in Helmholtz (1921): 70–97.

Helmholtz, Hermann von. 1903. *Vorlesungen über theoretische Physik*. Vol. 1.1: *Einleitung zu den Vorlesungen über theoretische Physik*, ed. Arthur König and Carl Runge. Leipzig: Barth.

Helmholtz, Hermann von. 1921. *Schriften zur Erkenntnistheorie*, ed. Paul Hertz and Moritz Schlick. Berlin: Springer. English edition: Helmholtz, Hermann von. 1977. *Epistemological writings* (trans: Lowe, Malcom F., ed. Robert S. Cohen and Yehuda Elkana). Dordrecht: Reidel.

Hölder, Otto. 1901. Die Axiome der Quantität und die Lehre vom Mass. *Berichten der mathematisch-physischen Classe der Königl. Sächs. Gesellschaft der Wissenschaften zu Leipzig* 53: 1–64.

Hölder, Otto. 1924. *Die mathematische Methode: Logisch erkenntnistheoretische Untersuchungen im Gebiete der Mathematik, Mechanik und Physik*. Berlin: Springer.

Husserl, Edmund. (1891/1970). *Husserliana*. Vol. 12: *Philosophie der Arithmetik: Psychologische und logische Untersuchungen, mit ergänzenden Texten (1890–1901)*, ed. Lothar Eley. Dordrecht: Springer. English edition: Husserl, Edmund. 2003. *Edmund Husserl: Collected Works*. Vol. 10: *Philosophy of arithmetic: Psychological and logical investigations with supplementary texts from 1887–1991* (trans: Willard, Dallas). Dordrecht: Springer.

Hyder, David. 2006. Kant, Helmholtz and the determinacy of physical theory. In *Interactions: Mathematics, physics and philosophy, 1860–1930*, ed. Vincent F. Hendricks, Klaus Frovin Jørgensen, Jesper Lützen, and Stig Andur Pedersen, 1–44. Dordrecht: Springer.

Hyder, David. 2009. *The determinate world: Kant and Helmholtz on the physical meaning of geometry*. Berlin: De Gruyter.

Kant, Immanuel. 1787. *Critik der reinen Vernunft*. 2nd ed. Riga: Hartknoch. Repr. in *Akademie-Ausgabe*. Berlin: Reimer, 3. English edition: Kant, Immanuel. 1998. *Critique of Pure Reason* (trans: Guyer, Paul and Wood, Allen W.). Cambridge: Cambridge University Press.

Köhnke, Klaus Christian. 1986. *Entstehung und Aufstieg des Neukantianismus: Die deutsche Universitätsphilosophie zwischen Idealismus und Positivismus*. Suhrkamp: Frankfurt am Main.

Krantz, David. H., R. Duncan Luce, Patrick Suppes, and Amos Tversky. 1971. *Foundations of measurement. Vol. 1: Additive and polynomial representations*. New York: Academic Press.

Krause, Albrecht. 1878. *Kant und Helmholtz über den Ursprung und die Bedeutung der Raumanschauung und der geometrischen Axiome*. Schauenburg: Lahr.

Kronecker, Leopold. 1887. Über den Zahlbegriff. In *Philosophische Aufsätze: Eduard Zeller zu seinem fünfzigjährigen Doctor-Jubiläum gewidmet*, ed. Vischer, Friedrich Theodor, 261–274. Leipzig: Fues.

Land, Jan Pieter Nicolaas. 1877. Kant's space and modern mathematics. *Mind* 2: 38–46.

Michell, Joel. 1993. The origins of the representational theory of measurement: Helmholtz, Hölder, and Russell. *Studies in History and Philosophy of Science* 24: 185–206.

Newton, Isaac. 1728. *Universal arithmetick: or, A treatise of arithmetical composition and resolution*, written in Latin by Sir Isaac Newton, translated by the late Mr. Ralphson, and revised and corrected by Mr. Cunn. London: Senex.

Patton, Lydia. 2009. Signs, toy models, and the a priori: From Helmholtz to Wittgenstein. *Studies in History and Philosophy of Science* 40: 281–289.

Poincaré, Henri. 1902. *La science et l'hypothèse*. Paris: Flammarion.

Potter, Michael. 2000. *Reason's nearest kin: Philosophies of arithmetic from Kant to Carnap*. Oxford: Oxford University Press.

Pulte, Helmut. 2006. The space between Helmholtz and Einstein: Moritz Schlick on spatial intuition and the foundations of geometry. In *Interactions: mathematics, physics and philosophy, 1860–1930*, ed. Vincent F. Hendricks, Klaus Frovin Jørgensen, Jesper Lützen, and Stig Andur Pedersen, 185–206. Dordrecht: Springer.

Reck, Erich. 2003. Dedekind's structuralism: An interpretation and partial defense. *Synthese* 137: 369–419.

Reck, Erich. 2013. Frege, Dedekind, and the origins of logicism. *History and Philosophy of Logic* 34: 242–265.

Riehl, Alois. 1925. *Der philosophische Kritizismus und seine Bedeutung für die positive Wissenschaft. Vol. 2: Die sinnlichen und logischen Grundlagen der Erkenntnis*, 2nd ed. Leipzig: Engelmann.

Russell, Bertrand. 1903. *The principles of mathematics*. Cambridge: Cambridge University Press.

Ryckman, Thomas A. 2005. *The reign of relativity: Philosophy in physics 1915–1925*. New York: Oxford University Press.

Schiemann, Gregor. 2009. *Hermann von Helmholtz's mechanism: The loss of certainty. A study on the transition from classical to modern philosophy of nature*. Trans. Cynthia Klohr. Dordrecht: Springer.

Tait, William W. 1996. Frege versus Cantor and Dedekind: On the concept of number. In *Frege: Importance and legacy*, ed. Matthias Schirn, 70–113. Berlin: De Gruyter.

Wolff, Christian. 1716. *Mathematisches Lexicon*. Leipzig: Gleditsch.

Chapter 5
Metrical Projective Geometry and the Concept of Space

5.1 Introduction

As noticed on several occasions in the previous chapters, Helmholtz's characterization of the relation between space, number, and geometry – especially in the writings that followed his dispute with Jan Pieter Nicolaas Land – can be interpreted as a generalization of the Kantian theory of space. Helmholtz's rule for coordinating geometry with empirical manifolds was the requirement that the points of a system in motion remain fixedly linked during displacement. He believed that it followed from the free mobility of rigid bodies that the form of the intuition of space in general coincides with that of a threefold extended manifold of constant curvature and includes both Euclidean and non-Euclidean geometries as special cases. After Sophus Lie's (1893, pp.437–471) critique of Helmholtz's inquiry into the foundations of geometry, such interpreters as Felix Klein, Henri Poincaré, and Ernst Cassirer, reformulated the argument by using Felix Klein's 1871 projective model of non-Euclidean geometry, which was Klein's first example of a group-theoretical classification of geometries. What are the reasons for ascribing to Helmholtz implicitly group-theoretical considerations? Are these considerations compatible with Helmholtz's claim about the empirical origin of geometrical axioms? Is it possible to provide a mathematical characterization of the form of intuition of space?

I discuss these questions in the next chapter. In order to provide an introduction to these problems, this chapter is devoted to Klein's work on projective geometry and its reception in the philosophical discussion about the Kantian theory of space. Such philosophers as Bertrand Russell argued that projective geometry is wholly a priori, "in so far as it deals only with the properties common to all spaces" (Russell 1897, pp.117–118). However, Russell did not attribute the same status to metrical properties or metrical projective geometry: the former depend on empirical factors; the latter rests upon a definition of distance that must be stipulated arbitrarily. According to Russell, it is in metrical properties alone that Euclidean and

© Springer International Publishing Switzerland 2016
F. Biagioli, *Space, Number, and Geometry from Helmholtz to Cassirer*,
Archimedes 46, DOI 10.1007/978-3-319-31779-3_5

non-Euclidean spaces differ. The axiom of distance is an exception, because – in Russell's formulation – this does not imply Euclid's proposition that two straight lines cannot enclose a space (i.e., cannot have more than one common point). Russell's axiom is that every point must have to every other point one, and only one, relation independent of the rest of space. This relation is the distance between the two points. A relation between two points can only be defined by a line joining them. Hence, a unique relation involves a unique line, that is, a line determined by any two of its points. Measurement is possible only in a space which admits such a line. Nevertheless, Russell's definition can be generalized to the case in which the straight line has a finite length, such as in the case of two antipodal points. His conclusion was that: "Distance and the straight line, as relations uniquely determined by two points, are thus a priori necessary to metrical Geometry. But further, they are properties which must belong to any form of externality" (p.173).

We turn back to Russell's definition of distance in Sect. 5.3.3. For now, it is worth noting that Russell contrasted his approach with Klein's classification of geometries, which he considered a merely technical result. By contrast, Ernst Cassirer attached great philosophical importance to this result for the clarification of the distinction between the general properties of space and the specific axiomatic structures. Following a line of argument that goes back to Helmholtz, Cassirer used Klein's classification to generalize the Kantian notion of space to a system of hypotheses, including both Euclidean and non-Euclidean geometries.

Reconsidering the philosophical reception of Klein will enable us to deepen the methodological aspect of the debate about Kant and non-Euclidean geometry. We mentioned already that one of the problems with the reconsideration of the Kantian theory of space after non-Euclidean geometry was that Kant apparently presupposed Euclid's synthetic method. Was Kant contradicted by the development of analytic methods in nineteenth-century geometry? First of all, it may be helpful to make a terminological remark. The term "analysis" (*Analyse*) indicated any kind of calculus and was introduced in opposition to the use of constructions in geometry. Analytic geometry differs from synthetic geometry because of its method of proof, which does not presuppose concrete constructions or even intuitions. Nevertheless, important mathematicians in the nineteenth century believed that analytical reasoning was compatible with the use of intuition as an overall view of geometric subject matters and, therefore, as a tool for exploring connections between different branches of mathematics. In this sense, Gauss (1880, p.365) maintained that Kant was not contradicted by later attempts to attain logical rigor in geometry. Riemann defended the role of intuition in a similar way in the quote from his Nachlass, which we mentioned in Sect. 3.2.2.

Klein's work on projective geometry offers an example of how analytic methods can be used in combination with geometrical constructions. The use of numerical representations and calculations, in his work, did not imply the view that geometrical concepts should be reduced to numbers. However, Klein attached much importance to numerical models in his search for a precise expression for the intuition that projective geometry includes Euclidean geometry as a special case. He used the algebraic theory of invariants to characterize and classify geometric properties as those relations that remain unchanged by ideal operations with a set of elements. The statement that the form of space can be specified in different ways resulted from the algebraic treatment. At the same time, Klein did not deny a role for intuition and geometrical constructions as psychological presuppositions for the development of mathematical theories.

Conversely, Kant's claim that mathematics is synthetic does not depend on a contraposition between analytic and synthetic methods in mathematics. He maintained that construction is required for the introduction of basic geometrical concepts (Kant 1787, p.203). However, for Kant, all mathematical judgments, including numerical formulas, are synthetic a priori (see Kant 1783, Sect. 2). Pure intuition plays the role of a mediating term between the concepts of the understanding and the empirical manifold. It is because of this role of intuition that Kant called mathematics synthetic. In 1910, Cassirer referred to the works of such mathematicians as Dedekind and Klein to reformulate Kant's claim by assuming that such mediation is carried out by the concept of function. Cassirer's approach enabled him to identify the form of space not so much as a specific axiomatic structure, but as Klein's classification. This example played an important role in Cassirer's interpretation of the notion of the a priori in hypothetical terms. He characterized the transcendental method using the analogy of Klein's definition of geometric properties as relative invariants of transformation groups and described his philosophical project as a universal invariant theory of experience.

5.2 Metrical Projective Geometry before Klein

Felix Klein was born in Düsseldorf in 1849. He studied mathematics and natural sciences at the University of Bonn in 1865. He became Julius Plücker's assistant at the age of 17, in 1866. At that time, Plücker was still teaching courses in experimental physics. Meanwhile, he had returned to his mathematical work in line geometry, the branch of projective geometry which goes back to Plücker's *System of the Geometry of Space in a New Analytical Treatment* (1846). When Plücker died in 1868, Alfred Clebsch invited Klein to Göttingen to complete and edit the second part of Plücker's *New Geometry of Space Founded on the Treatment of Line as Space Element* (1869). On that occasion, Klein became acquainted with the fundamental concepts of the theory of invariants formulated by the British mathematicians George Boole, Arthur Cayley, and James Joseph Sylvester, and introduced in Germany by Clebsch and by Paul Gordan. Klein developed the idea of a classification of geometries based on a projective metric after his stay in Berlin from 1869 to 1870. The idea reflects the various influences of his formation. In particular, Klein presented his projective models of non-Euclidean geometry as a synthesis between two different traditions. On the one hand, he was familiar with Arthur Cayley's 1859 analytic treatment of projective metric, on which Klein gave a presentation at Weierstrass's seminar in Berlin in 1870. On the other hand, he was introduced by his friend, Otto Stolz, the Austrian mathematician, to the autonomous, purely descriptive foundation of projective geometry by Christian von Staudt (1847,1856–1860). Klein (1921, pp.51–52) reported that in the summer of 1871, after his exchanges with Stolz, he arrived at the idea that there must be a connection between metrical projective geometry and non-Euclidean geometry.[1] Klein first mentioned the possibility of using a projective metric to classify geometrical hypotheses in his "Note on the Connection between Line Geometry and the Mechanics of Rigid Bodies" (1871a). Klein's detailed presentation of the projective models is found in a series of papers "On the So-Called Non-Euclidean Geometry," which appeared between 1871 and 1874.

We return to Klein's original commitment to the mechanics of rigid bodies and to the discussion about the geometry of physical space in the next section. This section provides an introduction to Klein's reception of the said traditions.

5.2.1   Christian von Staudt's Autonomous Foundation of Projective Geometry

Projective methods had been used in the seventeenth century by Girard Desargues and Blaise Pascal, but they tended to be abandoned after Descartes introduced the method of coordinates. Projective geometry flourished in the nineteenth century after the publication of Jean-Victor Poncelet's *Treatise on the Projective Properties of Figures* (1822). Poncelet's work was the first systematic study of the properties that are preserved by projections. His goal was to show that the descriptive treatment of these properties can attain the same generality and the same standards of rigor as Descartes's way of proceeding by formulating and solving equations between numbers representing points. On the one hand, Poncelet believed that geometry could be made independent of analysis. On the other hand, he pointed out that there are similarities between projective and algebraic methods. Firstly, both projective geometry and algebra look at quantities independently of the specific numerical values that can be assigned to them. Secondly, the projective treatment of figures can require the introduction of ideal elements for the interpretation of well-defined relations. In order to justify this way of proceeding, Poncelet extended the principle of continuity or principle of the permanence of formal laws from algebra to geometry (Poncelet 1822, pp.XIX–XXIV; 1864, pp.319, 338).

We return to the principle of continuity later on in this chapter. This section provides an introduction to the discussion about the place of the concept of magnitude in projective geometry. Poncelet's first remark about the general perspective on the projective treatment of figures refers to the fact that projections alter such magnitudes as distance and size. Therefore, specific numerical values associated with magnitudes in ordinary geometry do not usually express projective properties. The basic relation that remains unchanged by projections is not the distance between a

---

[1] For information about Klein's education and early geometrical works, see Rowe (1989) and Gray (2008, Ch.3). On the sources of Klein's classification of geometries, see also Wussing (1969, Ch.3); Birkhoff and Bennett (1988, pp.145–149) and Rowe (1992).

pair of points, but a proportion, which is called harmonic ratio, and whose definition depends on the circular order of the points on a projective line. The cross-ratio of four points, A, B, C, D, on a straight line, in this order, is given by:

$$\frac{CA}{CB} : \frac{DA}{DB}$$

Once given the coordinates, the harmonic ratio between four collinear points can be defined by using the arithmetical definition of the cross-ratio of four numerical values x_1, x_2, x_3, x_4:

$$\frac{x_3 - x_1}{x_3 - x_2} : \frac{x_4 - x_1}{x_4 - x_2}$$

The cross-ratio is harmonic if its value is -1.

The problem with this definition is that it presupposes both the use of numerical coordinates and the ordinary notion of segment, which is determined by two points A and B. The projective notion of segment requires at least three points because two points, A and B, divide the projective line through them into two segments. If we choose a third point, C, on the line, then a fourth point, D, lies in the same segment as C if A and B do not separate C and D, and in the other segment if A and B do separate C and D (see Enriques 1898, p.18). This example shows that projective notions may differ considerably from the corresponding notions in ordinary metrical geometry. Metrical notions only seemed to occur in the definition of harmonic ratio.

Christian von Staudt was the first to provide an autonomous foundation of projective geometry as a "science of position" rather than of quantity. In the *Geometry of Position* (1847), he introduced the concept of harmonic ratio as follows. Given three points, A, B and C, on a line u and a plane through u, he constructed a quadrangle, whose opposite sides intersect A and B, respectively, and one of whose diagonals intersects C. Then, the other diagonal intersects u at a point D, which is the same for every such quadrangle. D is called the harmonic conjugate of C relative to A and B or the fourth harmonic of the group $ABCD$.[2]

Forms of the same kind (e.g., projective lines and planes through a line) are called projective to each other if harmonic quadruples of elements of the first are in

---

[2] The construction described above goes back to Philippe de la Hire and is also found in Poncelet (1822, p.82). Von Staudt was the first to use it in the definition of the harmonic relation between two pairs of points on a projective line. In order to prove that there is one and only one point that is in such a relation to three given points, von Staudt repeated the construction of the quadrangle on another plane through the line. He proved that the second diagonals of the two quadrangles intersect with the line in the same point. The proof follows from a generalization of Desargues's theorem about perspective triangles to perspective quadrangles (i.e., distinct triangles or quadrangles, which are projections of the same figure). The theorem states that if the lines through the corresponding vertexes of the figures intersect with the same point, the points of intersection of the corresponding sides lie in the same line, and reciprocally (Staudt 1847, pp.40–43). For a modern presentation of the theorems of Desargues and of von Staudt, see Efimov (1970).

a one-to-one correspondence with those of the second. Von Staudt used the properties of harmonic groups to prove that, if two such forms have three elements in common, then they have all elements in common. Von Staudt's proof follows by reductio ad absurdum from the assumption that two projective lines coincide in more than two points, but not in all points. Suppose that the lines have two consecutive points A and B in common, and at least one point D subsequent to B. The said assumption leads to contradiction, because, in that case, there is a point C, which, along with D, separates AB harmonically (Staudt 1847, pp.49–51). This proposition is known as the fundamental theorem of projective geometry. It provides us with a definition of projectivity which can be extended to conics and to space (Staudt 1856–1860, p.5) and which is based on incidence relations, along with the assumption of continuity, regardless of relationships of congruence and of the ordinary notion of distance.[3]

The accomplishment of von Staudt's autonomous foundation of projective geometry is found in his *Contributions to the Geometry of Position*, which appeared in three volumes between 1857 and 1860. In the appendix to the second volume, von Staudt introduced the calculus of jets (*Würfe*), which offers one of the first examples of operations with entities different from both numbers and segments or angles. Von Staudt called a jet, and denoted $ABCD$, an ordered quadruple of elements A, B, C, D of a projective form. He called the jet proper if all its elements are distinct from each other; and called the jet improper if the distinct elements are three in number. There are three kinds of improper jets: (i) $ABCA$ or $BAAC$; (ii) $ABCB$ or $BABC$; and (iii) $ABCC$ or $CCAB$. Von Staudt indicated them with 0, 1 and ∞, respectively. Von Staudt (1856–1860, vol. 2, pp.166–182) used the properties of involutions to define equality, sum and the other operations with jets, and to prove that the sum of jets satisfies the same laws as the arithmetic sum.[4]

It is noteworthy that the symbols above do not indicate numerical values. Von Staudt introduced projective coordinates in order to determine the value of a jet as follows. Given three points of a projective line, C, A, A_1, he called the fourth har-

---

[3] This aspect of von Staudt's way of proceeding is apparent if one considers later axiomatizations of projective geometry. In this sense, Otto Hölder (1911, p.67, and note) maintained that in order to develop projective geometry in the manner of von Staudt (i.e., without metric foundations), one had to presuppose all axioms of plane linear geometry (connection, order and the axiom of parallel lines), except the axioms of congruence and the Archimedean axiom. Notice, however, that the first axiomatic treatment of (elementary) geometry goes back to Moritz Pasch (1882), and he did not provide an axiomatization in the modern sense. It was only Hilbert (1903) who formulated a set of axioms that are sufficient to characterize geometrical objects and relations up to isomorphism. Furthermore, von Staudt presupposed continuity as well. Later presentations of the proof of the fundamental theorem – beginning with Klein's (1874) – usually adopted an equivalent formulation of Dedekind's Archimedean continuity (see Darboux 1880; Pasch 1882, pp.125–127; Enriques 1898). Alternatively, Friedrich Schur (1881, p.253) used Thomae's (1873, p.11) definition of projectivity in terms of prospectivity to prove the fundamental theorem in a manner which is independent of the Archimedean axiom. On von Staudt's proof of the fundamental theorem of projective geometry and its development from 1847 to 1900, see Voelke (2008).

[4] On the use of involutions in von Staudt's calculus of jets, see Maracchia (1993) and Nabonnand (2008).

monic A_2. The indefinite repetition of the construction of the fourth harmonic generates a series C, A, A_1, A_2, A_3... such that CAA_1A_2 and $CA_1A_2A_3$, are harmonic jets and consecutive elements are not separated by any other elements in the series. The series thus generated corresponds to the series of the whole numbers, and it can be integrated so that the points of the line can be put into one-to-one correspondence with larger sets of numbers. The value of a proper jet $ABCD$ is negative if AC is separated by BD, enclosed between 0 and 1 if AB is separated by CD, and greater than 1 if AD is separated by BC. The value of an improper jet is 0, 1 or ∞, if the point D coincides with A, B or C, respectively. Given two points A and B, and set C at infinity, the jet $ABCD$ assumes every real value between $-\infty$ and $+\infty$ by varying the position of D in the sense ABC (Staudt 1856–1860, vol. 2, p.256).

Notice that the cross-ratio of four points corresponds to the value thus assigned to a jet. In fact there is no mention of the notion of cross-ratio in von Staudt's work. Not only did he avoid referring to the cross-ratio in the foundation of projective geometry, but the introduction of projective coordinates described above makes it superfluous to use this concept. Equivalent results can be obtained in a purely geometrical manner as consequences of the study of the projective properties of figures. The idea behind this approach is that the viewpoint of analytic geometry presupposes that of the geometry of position (see Nabonnand 2008, p.230). As we will see in the next section, projective coordinates can be introduced in a similar manner, even independently of the calculus of jets. However, in 1874, Klein showed that both von Staudt's proof of the fundamental theorem and his treatment of projective coordinates presupposed further assumptions about continuity, whose first analytic treatment goes back to Dedekind (1872). Since this fact seemed to call into question the feasibility of a purely synthetic foundation of geometry, Klein proposed a synthesis between analytic and synthetic methods.

5.2.2   Arthur Cayley's Sixth Memoir upon Quantics

Arthur Cayley's "Sixth Memoir upon Quantics" (1859) contains the first analytic treatment of a projective metric from the standpoint of the algebraic theory of invariants. Cayley's approach goes back to George Boole's "Researches on the Theory of Analytical Transformations" (1841) and "Exposition of a General Theory of Linear Transformations" (1843). Boole was the first to extend Joseph-Louis Lagrange's (1770–1771) study of unimodular transformations to the study of linear transformations in general. The British mathematician James Joseph Sylvester later introduced the term covariant to indicate a function of the coefficients and variables whose form does not change under such transformations. He called invariant a function only of the coefficients if it is endowed with the same property (Sylvester 1851).[5]

---

[5] For Cayley's sources, see Cayley (1889, pp.598–601). On the development of the algebraic theory of invariants, see Wussing (1969, pp.123–130).

From 1854 to 1859, Cayley dedicated a series of studies to algebraic forms, which he called quantics. In his sixth and last memoir on this subject, he used the theory of invariants to clarify the relationship between projective geometry and ordinary (Euclidean) metrical geometry. Cayley assumed a pair of imaginary points called the absolute. Any pair of real points can be taken with respect to the absolute. Cayley used this relation to define distance. He then showed that the formulas of metrical geometry can be derived from those of a projective metric based on his definition of distance. In the case of metrical geometry, the absolute degenerates into a pair of coincident points or a circle at infinity. Cayley concluded that projective geometry does not presuppose such metrical concepts as distance. On the contrary, the ordinary meaning of metrical notions can be derived from the more general viewpoint of projective geometry. He wrote:

> I remark in conclusion, that, *in my own point of view*, the more systematic course in the present introductory memoir on the geometrical part of the subject of quantics, would have been to ignore altogether the notions of distance and metrical geometry; for the theory in effect is, that the metrical properties of a figure are not the properties of the figure considered *per se* apart from everything else, but its properties when considered in connexion with another figure, viz. the conic termed the Absolute. The original figure might comprise a conic; for instance, we might consider the properties of the figure formed by two or more conics, and we are then in the region of pure descriptive geometry: we pass out of it into metrical geometry by fixing upon a conic of the figure as a standard of reference and calling it the Absolute. Metrical geometry is thus a part of descriptive geometry, and descriptive geometry is *all* geometry, and reciprocally; and if this be admitted, there is no ground for the consideration, in an introductory memoir, of the special subject of metrical geometry; but as the notions of distance and of metrical geometry could not, without explanation, be thus ignored, it was necessary to refer to them in order to show that they are thus included in descriptive geometry. (Cayley 1859, p.592)

Cayley's consideration suggests that his definition of distance was an accessory and could be ignored, once the formulas of metrical geometry were obtained from metrical projective geometry.

After Klein's classification of geometries of 1871, Cayley made it clear that he separated the idea of space sharply from the analytic theory of n-dimensional manifolds. He believed that Euclidean geometry captured the fundamental features of the idea of space lying at the foundation of external experience. Therefore, he ruled out a generalized notion of space, including non-Euclidean geometries as special cases (see Cayley's "Presidential Address to the British Association" from 1883 in Cayley 1896, pp.434–435). One of the goals of Klein's early writings on non-Euclidean geometry was to contradict such a view by attaching a spatial meaning to Cayley's notion of distance.

5.3 Felix Klein's Classification of Geometries

This section provides a brief account of Klein's first classification of geometries in 1871, with a special focus on the methodological issues at stake concerning the foundation of projective geometry. What can we learn from a projective coordinate system? Can metrical concepts be consistently avoided in the foundation of projective geometry? Is it possible to use a projective metric to obtain an overview of different hypotheses concerning the geometry of space? In order to establish the preconditions for the possibility of measurement, Russell (1897) generalized the Kantian notion of the form of externality to a structure endowed with homogeneity and allowing for the relativity of position. As already noted in the introduction, Russell identified the properties of such a structure as the projective properties that are common to Euclidean and non-Euclidean spaces. He considered projective coordinates and Cayley's definition of distance to be technical means to obtain mathematical results, which presuppose, but do not affect the idea of space. Such a view may well reflect Cayley's distinction between the idea of space and the analytic theory of n-dimensional manifolds. However, it is apparent from Klein's writings that he attached a spatial meaning to the notions under consideration. A Staudt-Klein coordinate system is something more than a mere labeling of points, because it entails a rule for constructing the points of a projective form. As a consequence of such a construction, the points of projective space can be ordered as numerical domains and proved to agree locally with the manifold of real numbers. In Klein's view, this way of proceeding was justified by the fact that when it comes to space, we may know very much less than we think we know. Therefore, any speculation about space presupposes a clarification of the conceptual postulates. One of the ideas behind Klein's classification of geometry was that the analytic treatment could provide us with a more general perspective on space than that of Euclidean geometry. His goal was to combine analytic methods with synthetic considerations in order to show that the general form of space can be specified in different ways.

5.3.1 A Gap in von Staudt's Considerations: The Continuity of Real Numbers

Klein was the first to point out that there was a gap in von Staudt's treatment of projective geometry. As we saw in the last section, the proof of the fundamental theorem followed by reductio ad absurdum from the assumption that the construction of a series of quadruples of harmonic points of a projective line can be interrupted at some point, which contradicts the assumption of the continuity of the line. However, von Saudt's assumptions do not suffice to obtain a contradiction: one must assume that a limit point of the sequence exists or that the continuity of the line corresponds to that of the real numbers.[6] Klein's remark is found in his second paper

---

[6] It can be hypothesized that Klein elaborated on a remark made by Weierstrass during Klein's stay in Berlin. Federigo Enriques (1907) reported that Weierstrass discussed the same subject in one of his lectures at the University of Berlin. Even though Enriques did not mention the date, Voelke (2008, p.288) supposes that the discussion might have taken place during the seminar attended by Klein in 1870. To support his conjecture, Voelke (2008, p.258) points out that it was Weierstrass who introduced the notion of a limit point in his proof that every bounded infinite set of real numbers have at least one limit point.

"On the So-Called Non-Euclidean Geometry" (1873, pp.139–140, 142, note). Soon after the publication of this paper, Jacob Lüroth and Hieronymus Georg Zeuthen, independently of each other, gave a rigorous proof of the fundamental theorem of projective geometry. Therefore, Klein reported the proof communicated to him by Zeuthen in a supplement to his paper one year later.[7]

We return to the assumption of continuity in the next section, after a brief presentation of Dedekind's (1872) analysis of the continuity of the real numbers. For now, it suffices to note that Klein's way to fill the gap in von Staudt's proof of the fundamental theorem entails the demand that quadruples of distinct points of a projective line be represented by real numbers. According to Klein, the same demand is implicit in the introduction of projective coordinates in the manner of von Staudt. Therefore, he made use of homogeneous coordinates, as first introduced by August Ferdinand Möbius, to include a point at infinity. Möbius (1827) identified these coordinates in his barycentric calculus as the center of gravity of a system of masses placed at the vertexes of an arbitrarily fixed triangle. In modern expositions, homogeneous coordinates are usually introduced as affine coordinates in three-dimensional space endowed with Cartesian coordinates in which sheaves of lines and planes (with center at the origin of the coordinate system) form a model of the projective plane (Yaglom 1988, p.40).

Klein acknowledged that the numerical representation can be varied arbitrarily – as long as each point of the projective plane (i.e., each of the lines of three-dimensional space passing through the origin) is described by the coordinates. Nevertheless, he maintained that von Staudt's construction of a numerical scale on a projective line enables the substitution of an arbitrary labeling of points with a specific rule. The harmonic sequence generated by three given points, one of which is set at infinity, by repetition of the construction of the fourth harmonic can be put into a one-to-one correspondence with the series of the whole numbers. The scale can be integrated with rational elements by specifying the further projections required for the construction of the middle points of any two given points. The construction of irrational elements follows from the demand of continuity. Klein sketched this construction in the second part of his paper from 1873, after discussing Plücker's extension of the notion of coordinate from points to lines in a three-dimensional space. A detailed description of the construction of the numerical scale on a projective line is found in Klein's lectures on "Non-Euclidean Geometry" (1893, pp.337–343; see also Hölder 1908; Efimov 1970, pp.257–269).

A simpler way to assign projective coordinates was introduced by Hilbert in the fifth chapter of "The Foundations of Geometry" (1899). Hilbert's coordinate system is based on his calculus of segments, which presupposes all of the axioms of linear and plane geometry (i.e., connection, order and the axiom of parallel lines), except the axioms of congruence and the Archimedean axiom (Hilbert 1899, p.55). Hilbert

---

[7] For a detailed presentation of this way of proving the fundamental theorem, see also Darboux (1880). This way of proceeding differs from Schur's (1881), because it presupposes Archimedean continuity: every point of the projective lines under consideration is thought of as a limit point of an infinite series of harmonic elements.

did not presuppose the fundamental theorem of projective geometry, because Desargues's theorem sufficed for the introduction of the calculus of segments.

Hilbert's way of introducing projective coordinates can be interpreted as a more coherent development of projective geometry in the manner of von Staudt – namely, without metric foundations – than Klein's (see Hölder 1911, p.67, and note). However, it remains true that a rigorous version of von Staudt's proof of the fundamental theorem of projective geometry presupposed the (Archimedean) continuity of the projective forms under consideration. Bearing in mind the fundamental role of the axiom of continuity both in the proof of the fundamental theorem and in the construction of a numerical scale on a projective line, Klein attached a lot of importance to the representation of space by the use of real numbers for the development of projective geometry. He reconsidered this aspect of von Staudt's work as follows:

> Analytic geometers did not pay much attention to von Staudt's researches. This may have been because of the widespread idea that the essential aspect of von Staudt's geometry lies not so much in the projective approach as in the synthetic form.
> Von Staudt's considerations have a gap, which can only be filled using an axiom, as described later on in the text. The same gap affects the extension of the method of von Staudt, as intended here. But our considerations concern not so much the extension as the original domain. The problem can be solved by specifying the analytical content of von Staudt's considerations, regardless of purely spatial ideas. Such content can be summarized in the demand *that projective space be represented by a numerical threefold extended manifold*. Besides, this is an assumption which lies at the foundation of any speculation about space. (Klein 1873, p.132, note)

The correlation between sets of projective points and subsets of real numbers sheds light on the method of projective geometry. The fact that projective geometry does not presuppose the metrical concepts of ordinary geometry does not mean that it should do without the numerical representation of space. Klein opposed the idea that the study of the projective properties of figures should be restricted to the synthetic form: owing to its generality, the projective approach entails analytic reasoning as well. This passage suggests that what Klein called the "analytical content" of von Staudt's considerations contributed to the idea of a classification of geometries from a general viewpoint. Not only is the demand of continuity implicit in von Staudt's considerations, but the possibility of representing space as a numerical manifold depends on the fact that projective space, as an abstract structure, is a more general concept than the three-dimensional Euclidean space of ordinary geometry. I believe that this is the reason why Klein maintained that any speculation about space presupposes the numerical representation.

In this connection, it is noteworthy that Klein (1873, p.114) agreed with Riemann that the theory of manifolds would prevent the empirical research from being hampered by views about the nature of space that were too narrow. It has been objected to that Klein's conception of manifold, being tied to the idea of a numerical manifold, is narrower than Riemann's (Norton 1999). The greater generality in Riemann's theory of manifolds comes from having no restriction to rigid bodies. A more detailed discussion of this point will require us to consider Klein's classification of geometries. For now, it is noteworthy that the significance of projective geometry for Klein depends not only on the fact that a projective space does not necessarily have a metric, but on the possibility of specifying metrical projective geometry in different ways. The introduction of projective coordinates in the manner of Staudt-Klein offers an important example of Klein's approach: the correlation between the points of a projective line and the real numbers is based on a projective rule of construction, which can be extended, in principle, to irrational elements. This way of proceeding suggests that arithmetic reasoning in geometry may prove useful in the quest for a general and rigorous study of the projective properties of figures, regardless of the metrical concepts of ordinary geometry. It is because of this level of generality that projective geometry provides us, at the same time, with the foundations of metrical geometry.

5.3.2   Klein's Interpretation of the Notion of Distance and the Classification of Geometries

Klein announced his classification of geometry based on the concept of a projective metric in his 1871a "Note on the Connection between Line Geometry and the Mechanics of Rigid Bodies." His starting point was Cayley's proof that metrical geometry can be obtained as a part of projective geometry. Klein was the first to explore the connection between Cayley's generalized metric and non-Euclidean geometry. He presented his classification of geometries in detail in another paper from 1871, "On the So-Called Non-Euclidean Geometry." He introduced the subject in this paper with the following remark about Cayley's work:

> Cayley's goal is to prove that the ordinary (Euclidean) metrical geometry can be conceived of as a particular part of projective geometry. Therefore, he articulates a general projective metric, and then he shows that its formulas can be used to obtain the formulas of ordinary metrical geometry, if the fundamental surface degenerates into a particular conic section, namely, the imaginary circle at infinity. The goal of the present paper instead is to clarify the *geometrical content* of Cayley's general metric, and to acknowledge that metrical projective geometry not only does include Euclidean metrical geometry as a special case, but fundamentally is in the same relation to the other metrical geometries that are derived from the theories of parallel lines. (Klein 1871b, p.574)

Klein's supposition that there is a geometrical content implicit in Cayley's metric clearly suggests a parallel with Klein's claim about von Staudt's synthetic approach: the axiom of continuity provides us with the analytical content of von Staudt's treatment of projective geometry. Klein's goal was to bridge the gap between these two traditions. In order to clarify the geometrical content of Cayley's metric, Klein defined distance as follows. He imagined the line intersecting two given points on a projective plane. The same line intersects the fundamental surface at two points. The distance between the former points is given by the logarithm of the cross-ratio they form with the points of the fundamental surface, multiplied by a constant. Klein's aim was to formulate a definition of distance that admitted a geometrical interpretation and that, at the same time, attained the generality of the definition of distance stipulated by Cayley.

Klein's interpreted the notion of distance by specifying the different types of fundamental surfaces that may occur in the construction described above. The fundamental surface can be: (i) an imaginary second-order surface; (ii) the inner points of a real, non-degenerate surface of second order; or (iii) the circle at infinity. According to the surface, Klein classified geometries into elliptic, hyperbolic and parabolic,[8] respectively. He defined spatial motions as the linear transformations that leave the fundamental surface unchanged. Metrical projective geometry offered the first example of a classification of geometric systems in group-theoretical terms. In 1872, Klein showed that geometric properties can be obtained and classified as relative invariants of groups of transformations, where the defining conditions for a set of transformations to form a group are, firstly, that the product of transformations of the group always gives a transformation of the group, and secondly, that for every transformation of the group, an inverse transformation exists in the group.

Klein used the theory of invariants to show that the said geometries are equivalent to the three classical cases of manifolds of constant curvature studied by Eugenio Beltrami in the "Fundamental Theory of Spaces of Constant Curvature" (1869). The equivalence of geometries follows from the identification of the corresponding groups of transformations according to the transfer principle that, given a manifold A and a group B on it, and $B \rightarrow B'$ when $A \rightarrow A'$, the B'-based treatment of A' can be derived from the B-based treatment of A (Klein 1893, p.72). The groups of transformation on a manifold of constant curvature differ from each other according to the measure of curvature, which can be positive, negative or equal to 0, and correspond to Klein's elliptic, hyperbolic and parabolic geometries, respectively. Beltrami proved that non-Euclidean geometries apply to manifolds of constant curvature, in the case that the curvature is positive or negative, whereas the measure of curvature in a Euclidean surface equals 0. Klein's classification showed that Euclidean geometry can be considered a limiting case of non-Euclidean geometry: the transformation group relative to a manifold of zero curvature corresponds to parabolic geometry, which is the degenerate case in which the points on the fundamental surface coincide.

Now we can see why the scope of Riemann's inquiry was broader than Klein's. Riemann's extension of Gauss's theory of surfaces to n-dimensional manifolds enabled him to consider the notion of manifold in the most general sense, as a set endowed with a topological structure. Therefore, his classification of the hypotheses which lie at the foundation of geometry also included manifolds of variable curvature. Torretti (1978, p.140) argued that Klein himself tended to presuppose the concept of a manifold in general in his way of proving the equivalence of geometries.

---

[8] Arguably, Klein's terminology relates to the fact that every linear transformation that maps a line onto itself can be associated with a characteristic quadratic equation. The transformation is elliptic, parabolic or hyperbolic, if the discriminant of this equation is less than, equal to or greater than 0 – namely, if the conic represented by this equation is an ellipse, a parabola or a hyperbola (see Torretti 1978, p.131).

The reliance of Klein's classification upon a projective metric and his use of the numerical representation of space, however, led him to restrict his attention to the more specific case of manifolds of constant curvature. For the same reason, Norton (1999, p.134) pointed out that Riemann's way of proceeding differs from Klein's because Klein's representation of space as a manifold of real numbers entails a specific notion of length. By contrast, Riemann's starting point was a continuous n-dimensional manifold free of metrical notions. He then added the notion of length on a localized basis.

Notwithstanding the difference between Riemann's approach and Klein's, it seems to me that Klein's considerations on the role of the manifold of real numbers in projective geometry need to be reconsidered in light of the fact that the analysis of the continuum of the real numbers, at the time, offered one of the clearest examples of how continuity can be defined rigorously, regardless of the notions of ordinary metrical geometry. This fact suggested that the representation of space as a numerical manifold could prove useful in the development of projective geometry, which was known to be a more general branch of geometry than metrical geometry. In fact imaginary elements play a fundamental role in Klein's interpretation of non-Euclidean geometry as well. Nevertheless, he emphasized the importance of the model provided by the real numbers, because it sufficed to show that the corresponding properties of geometrical figures admitted a variety of specifications, when considered from the more general viewpoint of projective geometry.

I believe that Klein's focus on manifolds of constant curvature was due to his commitment to classical mechanics. The possibility of representing spatial motions by using the theory of invariants provided a rigorous treatment of Helmholtz's (1870) thought experiment about the use of non-Euclidean geometry in measurement (see Klein 1898, p.588). This way of considering the geometry of physical space presupposed further developments in the group-theoretical treatment of geometry. Klein first presented the fundamental ideas of such an approach in his "Comparative Review of Recent Researches in Geometry." This paper circulated at the time Klein was appointed Professor at the University of Erlangen, in 1872, and became known as "Erlangen Program."[9] But it was only after the publication of Lie's work on continuous transformation groups, from 1888 to 1893, that Klein clarified the epistemological aspect of his earlier contributions to projective geometry. Most of his methodological and epistemological remarks are found in a series of lectures from the 1890s. In the same period, the "Erlangen Program" was translated into Italian (1890), French (1891) and English (1893), and in 1893 Klein published a revised version of it in the *Mathematische Annalen*.[10]

---

[9] The Erlangen Program is often mistaken for Klein's inaugural address as a newly appointed Professor at the University of Erlangen (see Rowe 1983). Klein's comparative review of the existing directions of geometrical research circulated as a pamphlet when he gave his inaugural address and became known as Erlangen Program, arguably because, after the second edition of 1893, Klein himself (e.g. in Klein 1921, pp.411–114) presented it as a retrospective guideline for his research (see also Gray 2008, pp.114–117).

[10] On the delayed reception of Klein's Erlangen Program, see Hawkins (1984, pp.451–463).

Before turning back to Klein's epistemological views, it may be helpful to take into account another objection to Klein's way of proceeding, namely, Bertrand Russell's discussion of the philosophical significance of metrical projective geometry and projective coordinates. The contrast between Russell's emphasis on the arbitrary character of the notions of a projective metric and Klein's considerations may shed some light on Klein's use of arithmetic reasoning in geometry.

5.3.3   A Critical Remark by Bertrand Russell

Russell's critical remark about Klein's work on projective metric is found in his early work, *An Essay on the Foundations of Geometry* (1897). Russell's objection concerned not so much Klein's mathematical results, as their significance to the debate about space and geometry. Russell divided the works on the foundations of geometry into three periods: the synthetic, the metrical and the projective. The first period was foreshadowed by Gauss's considerations concerning the consistency of non-Euclidean geometry and culminated in the development of non-Euclidean geometry by the denial of the parallel axiom in the works of János Bolyai and Nikolay Lobachevsky. Russell (1897, p.11) called this period synthetic, because he presumed that, once the parallel axiom is denied, the theorems of non-Euclidean geometry obtain by Euclid's synthetic method.[11] The second period includes Riemann's habilitation lecture of 1854 "On the Hypotheses Which Lie at the Foundation of Geometry" and Helmholtz's geometrical papers from 1868 and 1870. Russell called this period metrical because, despite the fact that Riemann's inquiry differed considerably from Helmholtz's, both Riemann and Helmholtz deemed space a kind of magnitude and based their inquiries on metrical notions, including the measure of curvature. According to Russell, this period was characterized by the significance of the inquiries into the foundations of geometry for the conception of space. These works showed that such properties as continuity and homogeneity do not suffice to attribute to space a single form that would imply the apriority of Euclidean geometry in Kant's sense. However, Russell distanced himself from Helmholtz's conclusion that all the properties of space are empirical. In order to defend a Kantian (broadly construed) theory of space, Russell maintained that at least some fundamental properties of space (i.e., continuity, homogeneity, having a finite number of dimensions) are common to both Euclidean and non-Euclidean spaces. These properties are a priori because they are necessarily presupposed in the perception of extended objects, and therefore, in measurement (Russell 1897, pp.60, 177).

Russell's conclusion is based not so much on metrical geometry, as on projective geometry. His argument is that:

---

[11] Russell seems to overlook the role of spherical trigonometry in the development of non-Euclidean geometry. On the importance of analytic methods and spherical trigonometry in the works of Bolyai and Lobachevsky, see Reichardt (1985); Rosenfeld (1988, Ch.6).

> [T]he distinction between Euclidean and non-Euclidean Geometries, so important in metri-
> cal investigations, disappears in projective Geometry proper. This suggests that projective
> Geometry, though originally invented as the science of Euclidean space, and subsequently
> of non-Euclidean spaces also, deals really with a wider conception, a conception which
> includes both, and neglects the attributes in which they differ. This conception I shall speak
> of as a form of externality. (Russell 1897, p.134)

Whereas metrical geometry for Russell contains empirical elements, projective
geometry is wholly a priori and provides us with the preconditions for the possibility of measurement, including the free mobility of rigid bodies (Russell 1897,
pp.118, 147).

Notwithstanding the importance of projective considerations in Russell's argument, he characterized the third, projective period in the inquiries into the foundations of geometry as a development of mathematical technicalities without
philosophical significance. This is because Russell deemed projective geometry a
priori only insofar as it is without a metric. Apparently, "projective geometry
proper," in Russell's sense, does not include metrical projective geometry, and this
affects the significance of Klein's classification of geometries. In this regard, Russell
wrote:

> Since these systems are all obtained from a Euclidean plane, by a mere alteration in the defi-
> nition of distance, Cayley and Klein tend to regard the whole question as one, not of the
> nature of space, but of the definition of distance. Since this definition, on their view, is
> perfectly arbitrary, the philosophical problem vanishes – Euclidean *space* is left in undis-
> puted possession, and the only problem remaining is one of convention and mathematical
> convenience. (Russell 1897, p.30)

Arguably, Russell bore in mind Cayley's separation between the (allegedly
Euclidean) idea of space and the analytic theory of manifolds. We already mentioned that Russell's form of externality includes the possibility of non-Euclidean
space, because he based the notion of distance on that of a straight line between two
points, where the properties of such a line depend on the points under consideration.
Euclid's proposition, for example, that two straight lines cannot enclose a space
(i.e., cannot have more than one common point) is not valid in the case of the lines
through antipodal points. Russell's approach is reminiscent of Cayley's, because
both of them distinguished a more fundamental idea of space from the analytic
theory of n-dimensional manifolds. Russell's goal was to show that the qualitative
notions of projective geometry (i.e., relations of order) provide us with necessary
presuppositions for the formation of such metrical notions as distance.

However, the claim that Klein's classification of geometry left Euclidean space
in undisputed possession was not Klein's view. Furthermore, this claim seems to
contradict Russell's own attempt to show that projective geometry provides us with
the properties of space that are common to both Euclidean and non-Euclidean systems. The disagreement with Klein is due to Russell's view about the use of arithmetic reasoning in geometry. We have already noticed that, for Klein, the
representation of a projective space as a manifold of real numbers lies at the foundation of any speculation about space in general. Russell's view of space differs from
Klein's, because Russell denied that there can be a proper numerical representation

of a continuous magnitude. He pointed out that the reduction of metrical notions (e.g., of distance) to projective forms depends on the assumption of imaginary elements. Therefore, he deemed such a reduction analytic or "purely symbolic" and, therefore, "philosophically irrelevant" (Russell 1897, p.28).

More specifically, Russell's disagreement with Klein concerned the meaning of projective coordinates. He wrote: "[Projective coordinates] are a set of numbers, arbitrarily but systematically assigned to different points, like the number of houses in a street, and serving only, from a philosophical standpoint, as convenient designations for points which the investigation wishes to distinguish" (Russell 1897, p.119). Since a Staudt-Klein construction generates an ordered series, the points of a projective line can obviously be put into a one-to-one correspondence with sets of numbers. The comparison with the number of the houses in a street emphasizes the fact that, notwithstanding the convenience of such a designation for the purpose of identifying distinct elements easily, their being distinct from each other is not a result of the designation, but its condition. Any coordinate assignment is arbitrary in that sense. The problem with projective coordinates is particularly that distance, as a function of projective coordinates, involves at least four elements. Russell's requirement that each element be distinguished from each other implies that any two elements should be in such a relation regardless of their relation to all others, as in the case of the ordinary notion of distance (p.35).

Russell's latter remark sheds light on his disagreement with Klein because the relevance of the numerical representation to the discussion about space in Klein's work lies precisely in the possibility of considering ordered sets of points as a system, without being committed to any specific assumption about the nature of the single elements. In current terms, a Staudt-Klein coordinate system is something more than a convenient way of designating points. In fact this coordinate system induces a topological structure in projective space which agrees locally with that of the set of real numbers (Torretti 1978, p.308).

In the interpretation proposed, this way of considering projective space and the classification of geometries also provided a classification of hypotheses concerning physical space. This way of proceeding, which Klein considered an example of arithmetization of mathematics, was related to his interests in physics. In Klein (1911), Klein used the concept of a projective metric to clarify the geometrical foundations of special relativity. In 1918, metrical projective geometry played an important role in Klein's intervention in the debate about the first cosmological models of general relativity.[12] Therefore, in a note to his *Collected Mathematical Papers*, Klein called his classification of geometries "the simplest way to clarify the

---

[12]The debate concerned the compatibility of the cosmological model developed by the Dutch astronomer Willem de Sitter with the principles of general relativity. Einstein argued for his own cosmological model by appealing to a principle borrowed from Mach. In a letter to de Sitter dated 24 March 1917, Einstein formulated the principle as follows: "In my opinion, it would be unsatisfactory, if a world without matter were possible. Rather, the $g_{\mu\nu}$-field should be determined by matter and not be able to exist without the latter." Klein and Hermann Weyl showed that De Sitter's model provided a counterexample to this principle. See the editorial note on "The Einstein-De Sitter-Weyl-Klein Debate," in Einstein (1998, pp.351–357).

newest physical (or even philosophical) ideas from a mathematical viewpoint" (Klein 1921, p.413).

In order to sketch the development of Klein's ideas, the next section gives a short account of his insights into the late nineteenth-century debate about the arithmetization of mathematics. To conclude, I contrast Russell's remarks on metrical projective geometry with the reception of Klein's work by Cassirer, who used Klein's classification of geometry to generalize the Kantian notion of the form of space.

5.4   The Arithmetization of Mathematics: Dedekind, Klein, and Cassirer

Klein's example of arithmetization of mathematics was Richard Dedekind's 1872 definition of irrational numbers. Whereas the continuity of the set of real numbers (i.e., the set formed by rational and irrational numbers) was usually inferred from that of spatial magnitudes, Dedekind's definition was based on the purely logical consideration of the mutual relations among the elements of a continuous set. We know from the previous chapter that Dedekind (1888) used the basic notions of set theory and the concept of function to define natural numbers. His earlier definition of irrational numbers was a crucial step in the development of his approach, because, in this case, he had to deal with mathematical objects (i.e., irrationals numbers) that do not admit a rigorous characterization other than in terms of relations between sets of rational numbers. A thorough reconstruction of the development of Dedekind's approach from 1872 to 1888 would require us to take into account the different versions of his theory of ideals.[13] Indeed, in his theory of algebraic integers of 1877, Dedekind himself compared his construction of the domain of ideal divisors of a ring of integers to his construction of the real numbers in 1872 (see Dedekind 1969, vol. 3, pp.268–269). For our present purpose, I restrict the consideration to numerical domains. In 1888, Dedekind maintained that even such familiar notions as that of natural number presuppose the more abstract concepts of set theory, including infinite sets, and proved that his axiomatization of the natural numbers is categorical. His way of proceeding suggested that the same standards of rigor could be attained in other branches of mathematics, beginning with projective geometry. Not only was an equivalent formulation of Dedekind's axiom of continuity required to give a rigorous proof of the fundamental theorem of projective geometry, but the study of the projective properties of figures presupposes the assumption of elements that – no less than irrational numbers – can be defined only as symbols for the relations among given elements. Consider, for example, the set of all points

---

[13] See Corry (1996); Avigad (2006). I refer to Avigad and Corry in particular, for a thorough account of how Dedekind's successive revisions of his theory of ideals shed light on his structuralist approach. According to Avigad (2006, p.168), the progression from Dedekind's first version of the theory of ideals to his last version represents a steady transition from Kummer's algorithmic style of reasoning to a style that is markedly more abstract and set-theoretic.

of a line u and an external point U. The lines of a bundle through U intersect u at all points, except for the line parallel to u. The points of intersection are thus divided into those that precede the turning point of the lines of the bundle and those that follow it. In order to fill the gap, one assumes in projective geometry that, even in that case, the lines intersect at infinity or a point designated by ∞.

Klein's considerations about Dedekind's axiom of continuity will introduce us to the foundational issues at stake with the arithmetization of mathematics. It is clear that the use of arithmetic reasoning in Klein's work did not imply a reduction of geometrical entities to arithmetical entities. "Arithmetic" here referred not so much to the theory of arithmetic, as to the method attributed to arithmetic by Dedekind. This method was supposed to be purely logical and independent of the intuitive content of mathematical statements. In fact, Klein did not deny the role of intuition in mathematics teaching and in mathematical practice. Nevertheless, he believed that foundational issues deserved a purely logical approach in order to account for all hypotheses which can occur in mathematical practice and in physics.

Insofar as Dedekind and Klein aimed at a purely logical or conceptual foundation of mathematics, their views can be considered a variant of logicism.[14] This was Cassirer's starting point for his understanding of the history of mathematics and of the exact sciences in the nineteenth century in terms of the tendency to replace concepts of substance with concepts of function. Owing to this tendency, Cassirer compared the mathematical method with the method of transcendental philosophy as to the generality of their perspectives on the objects of experience. At the same time, Cassirer's consideration implied a development of the transcendental method: the inclusion of hypotheses once deemed fundamental (i.e., the principles of Euclidean geometry) in more general systems of hypotheses suggests that the a priori in the sciences cannot be established once and for all and always relates to such systems. In conclusion, this section provides an interpretation of the logicist ideas, which Cassirer borrowed from Dedekind and Klein, as a development of the relativized conception of the a priori I drew back to Cohen.

---

[14] As referred to in Chap. 4, such scholars as Dummett (1991) and, more recently, Benis-Sinaceur (2015) draw logicism properly speaking back to Frege, and sharply distinguish the latter's logicism from Dedekind's view that "abstract objects are actually created by operations of our mind" (Dummett 1991, p.49). By contrast, Tait (1996) and Ferreirós (1999) reconsidered the logical aspect of Dedekind's abstraction from all the properties that have to do with a particular representation of mathematical domains (including spatial and temporal intuitions) in order to obtain a categorical characterization of mathematical structures. Cf. also Reck (2003) for a structuralist rather than psychological account of Dedekind's notions of abstraction and of creation. In this regard, Dedekind's view has been called logical structuralism. I especially rely on Ferreirós's broadening of logicism to include parallel versions of it, such as Dedekind's and Frege's.

5.4.1   Dedekind's Logicism in the Definition of Irrational Numbers

Dedekind's definition of irrational numbers is found in his 1872 essay on "Continuity and Irrational Numbers." This essay contains the first clarification of the distinction between continuity and such properties as infinite divisibility and density. In order to introduce these distinctions, Dedekind compared the set of rational numbers (i.e., numbers that can be expressed as the quotient of two integer numbers, with the denominator not equal to zero) with the set of points on a straight line. The problem is that there are infinite points on the line that do not correspond to rational numbers. These are the points that correspond to incommensurable lengths (e.g., the diagonal of a square whose side is the unit of length). The problem was known to ancient Greeks, and the usual solution was to introduce irrational numbers to fill the gaps. In the decimal system, these numbers are characterized by the fact that their decimal expansion continues without repeating. By contrast, the decimal expansion of a rational number either terminates after a finite number of digits or leads to a finite sequence of digits that repeats indefinitely. Dedekind's point is that the comparison with the points of a line does not provide us with an answer to the question: In what does continuity[15] consist? The answer to this question would provide what Dedekind called "a scientific basis" for the investigation of all continuous domains, including both magnitudes and continuous sets of numbers (Dedekind 1901, p.10).

Dedekind's emphasis lies in the fact that continuity cannot be grasped immediately. The continuity of a line, for example, depends on the following principle: If all points of the line fall into two classes such that every point of the first class lies to the left of every point of the second class, then one and only one point exists which produces this division of all points into two classes, this severing of the line into two portions. Dedekind wrote:

> The assumption of this property of the line is nothing else than an axiom by which we attribute to the line its continuity, by which we find continuity in the line. If space has at all a real existence it is not necessary for it to be continuous; many of its properties would remain the same even were it discontinuous. And if we knew for certain that space was discontinuous there would be nothing to prevent us, in case we so desired, from filling up its gaps, in thought, and thus making it continuous; this filling up would consist in a creation of new point-individuals and would have to be effected in accordance with the above principle. (Dedekind 1901, p.12)

---

[15] Dedekind's definition of irrational numbers can be considered a decisive step in the clarification of the mathematical notion of continuity. Notice, however, that especially in the introductory part of "Continuity and Irrational Numbers" he used "continuity" in a broader and more intuitive sense. The property of the line he was dealing with is not continuity but connectedness, which intuitively corresponds with the idea of having no breaks. A set is disconnected if it can be divided into two parts such that a point of one part is never a limit point of the other part; it is connected if it cannot be so divided. I am thankful to Jeremy Gray for pointing out to me that Dedekind wished to explain the connected character of the line.

Spatial intuition does not suffice to distinguish between continuous and discontinuous manifolds. The assumption that all the elements that produce a partition of the kind described above exist can only be postulated. Dedekind called this assumption an axiom because its correctness cannot be proved. Nevertheless, even under the hypothesis that space is discontinuous, the introduction of new point-individuals in order to fill the gaps follows logically from the said principle. The same holds true for the introduction of irrational numbers. Consider the set of rational numbers. Every rational number produces a division of all the others into two classes such that every number of the first class is less than every number of the second class. The number that produces the division is either the greatest number of the first class or the least number of the second class. This division shows the same characteristics of the division between the points of a line according to the premises of the principle of continuity. Dedekind called such a division a cut. He proved that every rational number corresponds to one and only one cut. However, there are infinitely many cuts not produced by any rational number. This is the characteristic that distinguishes a dense but discontinuous set (e.g., the set of rational numbers) from a continuous set. In order for continuity to be established, for every cut there must be one and only one number. In other words, irrational numbers must be introduced or "created" – as Dedekind put it – in correspondence with every cut that does not correspond to a rational number. As cuts are univocally determined, this way of proceeding provides us at the same time with an exact definition of irrational numbers (see Dedekind 1901, p.15).

As we saw in Sect. 4.3.2, in 1888, Dedekind adopted a similar way of proceeding in the definition of natural numbers. He started by abstracting from all the properties that do not have anything to do with numbers qua numbers, including the properties of spatiotemporal objects, and deemed numbers in general "free creations of the human mind." His emphasis in the use of this expression is not so much on the arbitrariness of some assumptions, as on the logical structure of the reasoning required for the definition of a numerical domain.[16] Irrational numbers offered an important example because, in that case, there are clearly no entities to which one may refer for the purpose of a univocal determination. Irrational numbers correspond not so much to particular entities, as to the particular divisions of ordered sets of rational numbers. Dedekind's view was that the existence of irrational numbers ought not to be inferred from that of other numbers or even non-numerical entities, because, in fact, every kind of number stands for a system of connections established by the mind.

Dedekind's definition of irrational numbers foreshadowed a view that became known as one of the first variants of logicism in the foundation of arithmetic. Summing up, Dedekind's view was characterized by the attempt to avoid all references to non-arithmetical elements and by the use of univocal correlation of the

---

[16] As already mentioned, Dedekind's use of "abstraction" and "creation" has sometimes been misunderstood as psychological (Cf. Dummett 1991, 49). In Chap. 4 and in the present chapter, I refer to more recent interpretations of the same operations as logical ones by Tait (1996), Ferreirós (1999), and Reck (2003).

elements of a system as the defining characteristic of a mathematical entity. Dedekind deemed his approach scientific because it promised generality and accuracy. This consideration introduces us to the next section about the arithmetization of geometry because it poses the problem whether using the same approach in other branches of mathematics would imply a reduction of mathematical entities to arithmetical ones. Insofar as Dedekind found an ontological foundation of numbers in the (naïve) notion of a set, he foreshadowed a set-theoretical kind of reductionism.[17] Nevertheless, I believe that Dedekind opened the door to a non-reductionist account of geometry. My emphasis is on the fact that the more general set-theoretic notions employed by Dedekind enabled him to consider both numerical and geometrical domains as different instantiations of the same mathematical structures. This aspect is evident in Klein's reception of Dedekind. In the following section, I refer to Klein's work to deal with the question as to whether, and to what extent, the same approach can be used in geometry.

5.4.2 Irrational Numbers, Axioms, and Intuition in Klein's Writings from the 1890s

Klein's most detailed discussion about arithmetization and the role of intuition in geometry is found in a series of writings and lectures from the 1890s. Klein seems to have developed his epistemological views in connection with his contributions on non-Euclidean geometry. His renewed interest in that subject in the 1890s was related to the appearance of Lie's work, *Theory of Transformation Groups*, which appeared in three volumes in 1888, 1890 and 1893. Lie's study of transformation groups provided essential requirements for the implementation of Klein's 1872 project of a general classification of geometries from the standpoint of group theory. Lie's work motivated Klein to develop and to promulgate his ideas (see Klein 1893, p.63, note; see also Hawkins 1984, pp.445–447). He revised his "Comparative Review of Recent Researches in Geometry" after the first Italian (1890) and French (1891) translations. The English translation appeared in 1893. Klein published a second version of his work in *Mathematische Annalen* in the same year. Klein's paper "On Non-Euclidean Geometry" appeared in the same journal in 1890. This paper includes the first detailed treatment of a class of surfaces in three-dimensional elliptic space that are locally isometric to the Euclidean plane. Klein heard about this class of surfaces in a lecture taught by William Kingdon Clifford in 1873. Klein's paper of 1890 offered a solution to the problem of determining the class of all surfaces in elliptic, hyperbolic, and parabolic space that are locally isometric to the Euclidean plane. The problem is known as the "Clifford-Klein problem" or "the problem of the form of space" (see Torretti 1978, p.151).

---

[17] Although Dedekind's conception appears to be naïve when compared to modern set theory, a set-theoretical approach is largely implicit in his logical foundation of arithmetic and became influential especially after Hilbert's reception of Dedekind (see Gray 2008, pp.148–151).

The existence of such a class of surfaces contradicted the view that the form of space, in a global sense, is endowed with a priori properties. But Klein denied a foundational role of spatial intuition for another reason: in Klein's psychological understanding of the notion of intuition, intuitive foundations were inexact and, therefore, compatible with different mathematical interpretations. For the same reason, Klein ruled out the definition of axioms as facts immediately known without need of proof. He characterized an axiom as "the postulate by which we read exact assertions into inexact intuition" (Klein 1890, p.572). He mentioned, for example, Dedekind's axiom of continuity. Klein wrote: "Since I do not attribute any precision to spatial intuition, I will not want the existence of irrationals to be derived from this intuition. I think that the theory of irrationals ought to be developed and delimited arithmetically, to be then transferred to geometry by means of axioms, and hereby enable the degree of precision that is required for the mathematical treatment" (p.572). Since spatial intuition does not enable us to distinguish between rational and irrational numbers, irrational numbers have to be introduced in the manner of Dedekind: one and only one number exists which produces a cut. At the same time, the quote above sheds some light on Klein's use of Dedekind's axiom of continuity. As discussed in Sect. 5.3.1, von Saudt's considerations presupposed the assumption of continuity. Klein pointed out that an equivalent formulation of Dedekind's axiom provided a rigorous way to prove the fundamental theorem of projective geometry and to introduce projective coordinates in the manner of Staudt-Klein.

It is clear from Klein's commitment to projective geometry that his interest lay not so much in a reduction of all of geometry to arithmetic or even to analytic geometry, as in a foundational issue to be solved in terms of Dedekind's logicism, namely, by abstracting from intuitively given contents and by adopting a rigorous method of proof. In this sense, the goal of arithmetization was to extend scientific criteria (i.e., generality and accuracy) from arithmetic to geometry. Arithmetization required the formulation of axioms or conceptual postulates.

The first axiomatic treatment of these subjects was due to Moritz Pasch. Pasch's *Lectures on Modern Geometry* (1882) include a proof of the fundamental theorem of projective geometry by means of the following formulation of the axiom of continuity: If the points $A_1 A_2 A_3 \ldots$ of a segment AB can be generated indefinitely and in such a way that A_1 lies between A and B, A_2 lies between A_1 and B, A_3 lies between A_2 and B, etc., then a point C (which can coincide with B) exists in the segment, such that, given a point D on AB between A and C, not all the points of the series $A_1 A_2 A_3 \ldots$ lie between A and D, and none of them lies between B and C. In order to introduce the notion of congruence, Pasch proved that this axiom, which is an equivalent formulation of Dedekind's continuity, entails that: Given the collinear points $AB_0 B_1 P$, if B_1 lies between A and P, and B_0 lies between A and B_1, then there is a positive integer λ, such that the $(\lambda+1)$th point of the series $AB_0 B_1$ follows every point of the segment AP. The point itself does not belong to the segment AP (Pasch 1882, p.125). In other words, Pasch showed that Dedekind's continuity entails the Archimedean property (see Hölder 1901).

Although Pasch's axiomatization was not complete, his work clearly showed the potential of a method of proof based on the explicit formulation of axioms. His deduction of the Archimedean property offered an example of the fact that propositions traditionally considered as basic notions (e.g., the so-called Archimedean axiom) can be obtained from more fundamental assumptions. Furthermore, Pasch clarified von Staudt's intuition that the notion of congruence in projective geometry depends on incidence relations or on what Pasch also called "graphic propositions" (Pasch 1882, p.118). His work also sheds light on the general notion of axiom. Whereas Pasch called all propositions that can be derived from other propositions theorems, he referred to unproven propositions as principles. Pasch maintained that the principles of geometry can only be learned by repeatedly occurring experiences and observations. However, the learning process remains unconscious, and it does not affect the mathematical development of geometrical systems. As Pasch put it: "The principles ought to include completely the empirical material to be elaborated in mathematics, so that, once the principles have been formulated, one needs not turn back to sense perception" (Pasch 1882, p.17).

Since Klein's considerations about arithmetization and the role of intuition in geometry were based mainly on examples drawn from projective geometry, he clearly profited from Pasch's work. At the same time, Klein distanced himself from Pasch's understanding of the notion of axioms for the following reason. While Klein believed that intuition and experience ought to be avoided in the foundation of mathematics, he attributed an important role in the evolution of mathematics to intuitive thinking. Intuition foreshadows the logical connections that have to be put in an exact formulation for the purposes of mathematical theories. Klein considered logical deduction necessary at that stage. However, he did not believe that logical deduction can ever include the empirical material provided by intuition completely. Therefore, intuition is necessary for discoveries both in pure and applied mathematics. Furthermore, Klein attached a lot of importance to intuitive thinking in mathematics teaching (see e.g., Klein 1894, pp.41–50; 1895).

Klein contrasted his definition of axiom with Pasch's in the concluding section of the paper "On Non-Euclidean Geometry" (Klein 1890, p.572, note).[18] A more detailed account of Klein's viewpoint is found in his lectures on non-Euclidean geometry from the academic year 1889–1890. Klein's lectures were elaborated by Friedrich Schilling and made available in printed version in 1893. A second edition by Walter Rosemann appeared posthumously in Klein (1928). Klein himself was

---

[18] Further developments in Klein's discussion with Pasch are found in their correspondence, which has been recently made available by Schlimm (2013). In particular, in a letter from October 19, 1891, Pasch fundamentally agreed with Klein's remarks about the notion of axiom in the concluding part of Klein (1890). However, Pasch distanced himself from Klein's defense of the role of intuition in mathematics. Although Pasch admitted that figures are commonly used in working on the axioms, he maintained that "the use of figures is merely a *facilitation* of the work; otherwise, the work would exceed our powers, or at least would progress much too slowly, or would not progress far enough. The consideration must be *possible* even without the figures, in other words: that which is derived from the figures must already be contained in the axioms, for otherwise the axioms are not complete" (Pasch in Schlimm 2013, p.193).

able to work on it with the collaboration of Rosemann until his death in 1925. The final version was completed by Rosemann in 1927.

The two editions of Klein's lectures differ considerably. Whereas in the first edition, Klein offers a historical reconstruction of knowledge about non-Euclidean geometry from the first projective perspectives on Euclidean metric to Lie's theory of transformation groups, the second edition provides a more comprehensive treatment of the same topics from a systematical viewpoint. Furthermore, several parts of the former edition were reworked completely in order to take into account later developments in the studies on non-Euclidean geometry and on related subjects. This is the case with the part of the first edition that concerns projective coordinates. Arguably, Klein recognized that Hilbert's calculus of segments enabled the introduction of projective coordinates in a simpler way than a Staudt-Klein coordinate system. More notably, Hilbert's calculus was a consistent development of the idea of doing geometry without metrical foundation, as it did not presuppose the fundamental theorem of projective geometry or the axiom of continuity, but only the theorem of Desargues. For my present purpose, I restrict my attention to the first edition of Klein's work, because his example of the different roles of axioms and intuitions in the 1890s was precisely the use of Dedekind's continuity in arithmetic and projective geometry. Klein related the discussion about the nature and the origin of geometrical axioms to that about the fundamental theorem of projective geometry. He distinguished between two different views about the axiom of continuity: the first, traditional view was that continuity is given by intuition; the second view goes back to Dedekind and looked at intuition as the trigger for the development of a completely different, logical characterization of irrational numbers. Correspondingly, there are two different ways to prove the fundamental theorem of projective geometry. The way of proof adopted by Klein presupposed Dedekind's continuity. The other way of proof goes back to Friedrich Schur (1881), who avoided Dedekind's continuity by obtaining the notion of projectivity from Johannes Thomae's (1873, p.11) definition of prospectivity (see Voelke 2008, pp.280–283).

Klein made it clear that both views about continuity lead to a rigorous proof of the fundamental theorem (Klein 1889–1890, p.311). His preference for the second view, and for the corresponding way of proving the theorem by using Dedekind's continuity, depended mainly on his general views about mathematical methodology and philosophy of science. Klein presented his view as a synthesis between empiricism and idealism.[19] On the one hand, he agreed with such empiricists as Pasch that

---

[19] This classification goes back to Paul du Bois-Reymond's *General Theory of Function* (1882). Regarding the fundamental notions of the calculus, du Bois-Reymond distinguished between idealism and empiricism as follows. Idealism is the view that limits exist as a logical presupposition of the calculus, although neither infinite nor infinitesimal quantities are imaginable in the sense of concrete intuition. By contrast, empiricism is the view that knowledge is grounded in immediate perception. Therefore, in the empiricist view, every representation in science must be referred to the objects of perception. In the case of such abstract concepts as the concept of limit, the representation can be obtained indirectly by the use of geometric constructions (see du Bois-Reymond 1882, pp.58–87). Du Bois-Reymond's approach differed from Klein's because the former did not propose a synthesis between two opposing views. The aforementioned sections of his work provided clarification on the assumptions of two equally possible views about the foundations of the calculus.

all concrete or even imaginary objects of intuition are distinguished from mathematical notions because of their vagueness. On the other hand, Klein required logical rigor in the foundation of mathematical theories, including geometry. Klein's stance led to a sharp distinction between mathematics and empirical science. At the same time, Klein believed that the purely logical development of mathematical theories was required to account for the connection between pure and applied mathematics. He wrote:

> If we ask ourselves what is it that makes some theories of pure mathematics applicable, we will have to make the following consideration: it is not so much a matter of inferring correct conclusions from correct premises as of obtaining the conclusions that follow with foreseeable correctness from approximately correct premises or of saying to what extent further inferences can be made. (Klein 1889–1890, p.314)

Logical rigor is required because the mathematical development of a theory consists of the clarification of the formal connections between the considered propositions. Since the correctness of assumptions concerning empirical contents cannot be ascertained beyond any doubt, this kind of reasoning is hypothetical. At the same time, owing to their deductive character, mathematical theories provide us with a precise formulation, classification and assessment of probability of the hypotheses occurring in other disciplines.

In this context, Klein did not discuss any particular example. Nevertheless, the quote above from his lectures on non-Euclidean geometry sheds some light on Klein's later considerations about the concept of a projective metric in his foundational inquiry into special relativity. Klein's classification of geometries shows both of the characteristics of purely mathematical theories that can be used in physics. The classification follows logically from such conceptual assumptions as Dedekind's axiom, and, at the same time, owing to its generality, it provides us with a clarification of the hypotheses involved in the interpretation of measurements. These hypotheses correspond only approximately to those of Euclidean geometry: they are equivalent to the assumption of a larger group of transformations, including both Euclidean and non-Euclidean groups as special cases.

5.4.3 Logicism and the A Priori in the Sciences: Cassirer's Project of a Universal Invariant Theory of Experience

In the third Chapter of *Substance and Function* (1910), "The Concept of Space and Geometry," Cassirer attributed a key role to projective geometry in the development of mathematical method. In particular, Cassirer regarded the generalization of the concept of form from concrete forms to a range of conceptual variations in the work of Poncelet as an important example of the general tendency to substitute concepts of substance with concepts of function in the history of mathematics. Cassirer wrote:

> Descartes charged ancient mathematics with not being able to sharpen the intellect without
> tiring the imagination by close dependence on the sensuous form, and Poncelet maintains
> this challenge throughout. The true synthetic method cannot revert to this procedure. It can
> only show itself equal in value to the analytic method, if it equals it in scope and universal-
> ity, but at the same time gains this universality of view from purely geometrical assump-
> tions. This double task is fulfilled as soon as we regard the particular form we are studying
> not as itself the concrete object of investigation but merely as a starting point, from which
> to deduce by a certain rule of variation a whole system of possible forms. The fundamental
> relations, which characterize this system, and which must be equally satisfied in each par-
> ticular form, constitute in their totality the true geometrical object. What the geometrician
> considers is not so much the properties of a given figure as the network of correlations in
> which it stands with other allied structures. (Cassirer 1910, p.80)

In Poncelet's study of the projective properties of figures, the dependence of laws on
the qualities of particular forms is reversed: the properties of particular forms
depend on the rule of variation of a whole system of possible forms. Therefore, from
the standpoint of projective geometry, figures that are distinct from each other in
ordinary geometry (e.g., circles, ellipses, hyperbolas, and parabolas) are classified
as the same kind of figures (i.e., the conics).

The generality of Poncelet's perspective depends on the principle of continuity
or principle of permanence of formal laws (Poncelet 1864, p.319). The principle
states that well-defined relations between the elements of a system subsist regardless of the existence of the related entities. It corresponded to the principle assumed
by Leibniz to introduce the so-called "infinitesimally small quantities" as limiting
points of convergent series in the calculus (see Leibniz 1859, p.106). The study of
the projective properties of figures requires the same principle, because projections
alter such absolute properties of figures as distance and the measurement of angles.
The properties that remain unchanged by projections do not always relate to elements that exist in the sense of ordinary geometry. Nevertheless, Poncelet believed
that the assumption of such properties was justified by the use of logic. He wrote:

> In some cases, the meaning of the properties [that follow from the principles of projective
> geometry] becomes purely ideal, since one or more of the elements involved have lost their
> absolute and geometrical existence. One may say that these properties are illusory, para-
> doxical in their object. Nevertheless, these properties are logical, and – if used correctly –
> they are appropriate to lead to incontestable and strict truths. (Poncelet 1865, p.68)

According to Cassirer, the permanence of formal laws justifies the creation of
new elements in correspondence with consistent variations of projective relations.
Cassirer paraphrased Poncelet's claim by saying that: "The new elements [...] are
paradoxical in their object but they are nevertheless thoroughly logical in their
structure, in so far as they lead to strict and incontestable truths" (Cassirer 1910,
p.85). The use of "creation" in this connection clearly suggests a parallel with
Dedekind's definition of irrational numbers. Indeed, Cassirer's remark was his starting point for his considerations about the use of Dedekind's continuity in the
introduction of projective coordinates. Not only did Dedekind's axiom of continuity
provide a necessary assumption for a Staudt-Klein coordinate system, but Dedekind's
way of introducing irrational numbers offered a model for a kind of reasoning that
is characteristic of the study of mathematical objects in terms of structures. In this

sense, Cassirer maintained that the introduction of projective coordinates in the manner of Staudt-Klein showed that the order of points in space can be conceived of in the same manner as that of numbers. In particular, Cassirer referred to Klein's interpretation of Cayley's definition of distance: the meaning of "distance" in projective metric depends not so much on the absolute properties of figures, as on the specific relation between four elements, two of which are ideal, since they belong to the fundamental surface. The fact that metric systems vary according to the kind of surface offered an argument for the plurality of geometric hypotheses concerning the form of space. Cassirer argued as follows:

> In this connection, projective geometry has with justice been said to be the universal "a priori" science of space, which is to be placed besides arithmetic in deductive rigor and purity. Space is here deduced merely in its most general form as the "possibility of coexistence" in general, while no decision is made concerning its special axiomatic structure, in particular concerning the validity of the axiom of parallels. Rather it can be shown that by the addition of special completing conditions, the general projective determination, that is here evolved, can be successively related to the different theories of parallels and thus carried into the special "parabolic," "elliptic" or "hyperbolic" determinations. (Cassirer 1910, p.88)

Cassirer borrowed from Russell's work of 1897 the idea that projective geometry could be used to reformulate the Kantian conception of geometry as a priori science of space. At the same time, Cassirer's approach enabled him to appreciate the aspect of Klein's work overlooked by Russell: the correlation between the points of space and subsets of real numbers depends on the properties of projective space. Spatial and numerical domains remain distinct. However, the correlation established by Klein enabled him to abstract from the intuitive character attributed to space in the Euclidean tradition and to redefine spatial notions in terms of a system of hypotheses. In opposition to Russell, Cassirer maintained that the development of a projective metric , besides mathematical convenience, has great philosophical importance. In Cassirer's view, metrical projective geometry shows that the a priori role of geometry is strictly related to the hypothetical character of geometrical assumptions: projective geometry can be called a "universal a priori science of space" insofar as this includes the specification of the conditions for a comparison of geometric systems. Cassirer related this result to his interpretation of the Kantian theory of space in continuity with Leibniz's relational notion of space as order of coexistence in the second volume of *The Problem of Knowledge in Modern Philosophy and Science* (1907).

Another aspect of Klein's work that emerges from Cassirer's reconsideration is Klein's conviction that a clarification of the foundations of mathematical theories was necessary to account for the connection between pure and applied mathematics. In particular, Cassirer (1910, pp.108–111) referred to Klein's idea of a comparative review of geometrical researches from the standpoint of group theory. According to Cassirer, Klein's characterization of geometric properties as relative invariants of groups of transformations offered a rational ground for a choice among hypotheses in physics. Cassirer's argument is discussed the next chapter. For now, it is noteworthy that, in this sense, Cassirer compared Klein's way of proceeding with that of transcendental philosophy. In the fifth chapter of *Substance and Function*, which is devoted to the problem of induction, Cassirer wrote:

> Since we never compare the system of hypotheses in itself with the naked facts in themselves, but always can only oppose one hypothetical system of principles to another more inclusive, more radical system, we need for this progressive comparison an ultimate constant standard of measurement of supreme principles of experience in general. Thought demands the identity of this logical standard of measurement amid all the change of what is measured. In this sense, the critical theory of experience would constitute the universal invariant theory of experience, and thus fulfill a requirement clearly urged by inductive procedure itself. The procedure of the "transcendental philosophy" can be directly compared at this point with that of geometry. Just as the geometrician selects for investigation those relations of a definite figure, which remain unchanged by certain transformations, so here the attempt is made to discover those universal elements of form, that persist through all change in the particular material content of experience. (Cassirer 1910, pp.268–269)[20]

Referring to the same quote, Friedman (2001) denies that Cassirer's conception of the a priori is a relativized one, because, as argued by him before (Friedman 2000, Ch.7), Cassirer (and the Marburg School more generally) defended a purely regulative conception of the a priori, according to which what is absolutely a priori are simply those principles that remain throughout the ideal limiting process. Therefore, Friedman finds that Cassirer has not much to say when it comes to the question: "How is it possible to venture a transformation of our present constitutive principles resulting in a genuine conceptual change or shift of paradigm"? According to Friedman, Cassirer suggests in the above quote that:

> [N]ot only we are never in a position to take our present constitutive principles as ultimate, we are also never in a position to know how the future evolution of constitutive principles will actually unfold. The best we can do, at any given time, is make an educated guess, as it were, as to what these ultimate, maximally general and adequate constitutive principles might be. (Friedman 2001, pp.65–66)

I discuss this problem and Cassirer's solution in Chap. 7. Regarding Cassirer's conception of the a priori, it is worth noting that a few lines before, on p.168, he introduced the reader to the inductive aspect of his philosophical project by saying that even the principles of Newtonian mechanics need not be taken as absolutely unchanging dogmas; he rather regarded these principles as the "temporarily simplest intellectual hypotheses, by which we establish the unity of experience." This claim shows that an invariant theory of experience in Cassirer's sense implied a rela-

---

[20] Ihmig (1997, pp.306–326) refers to Cassirer's comparison between the transcendental and the mathematical method to indicate a series of analogies between critical idealism and Klein's Erlangen Program. Notwithstanding the significance of this comparison for reconsidering Cassirer's relationship to Klein, it seems to me to be reductive to restrict the consideration to Klein's general idea of a group-theoretical treatment. My suggestion is to reconsidered the importance of Klein's projective model throughout his writings on non-Euclidean geometry. Not only did metrical projective geometry offer the first example of a classification of geometries by the use of the theory of invariants, but Klein used this example to support his epistemological views about the relationship between pure and applied mathematics. In this regard, I believe that there are more substantial points of agreement between Cassirer and Klein than the analogies between the Erlangen Program and critical idealism.

tivized and historicized conception of the a priori. As we saw in Chap. 2, the inductive aspect of Cassirer's approach goes back to Cohen. Similar to Cohen's interpretation of the Kantian theory of experience, the results of a universal theory of experience in Cassirer's sense depend on the history of science. This aspect of Cassirer's philosophical project clearly distinguishes it from Kant's. At the same time, Cassirer's account of mathematical method reflects an important characteristic of the method of transcendental philosophy: ontological assumptions are disregarded, so that hypotheses regarding the objects of experience can be classified from the most general viewpoint. Notwithstanding Cassirer's commitment to logicism, this consideration depends not so much on a formal-logical approach toward mathematical theories, as on Cassirer's insights into the potential of Dedekind's way of proceeding in foundational inquiries. As we saw in the previous section, even Klein believed that logicism in this sense was compatible with the demands of an empiricist theory of knowledge. Cassirer's goal was to use a Kantian architectonic of knowledge to answer the question: What is it that makes some theories of pure mathematics applicable? Cassirer's view was that the mathematical concept of function, which for Cassirer instantiated Kant's notion of the synthetic a priori, offered a model for the formation of concepts in natural science.

References

Avigad, Jeremy. 2006. Methodology and metaphysics in the development of Dedekind's theory of ideals. In *The architecture of modern mathematics*, ed. José Ferreirós and Jeremy Gray, 159–186. Oxford: Oxford University Press.

Beltrami, Eugenio 1869. Teoria fondamentale degli spazi a curvatura costante. *Opere Matematiche* 1: 406–429. Milano: Hoepli, 1902.

Benis-Sinaceur, Houria. 2015. Is Dedekind a logicist? Why does such a question arise? In *Functions and generality of logic: Reflections on Dedekind's and Frege's logicisms*, ed. Houria Benis-Sinaceur, Marco Panza and Gabriel Sandu, 1–57. Heidelberg: Springer.

Birkhoff, Garrett, and Mary Katherine Bennett. 1988. Felix Klein and his *Erlanger Programm*. In *History and philosophy of modern mathematics*, ed. William Aspray and Philip Kitcher, 144–176. Minneapolis: University of Minnesota Press.

Boole, George. 1841. Researches on the theory of analytical transformations with a special application to the reduction of the general equation of the second order. *Mathematical Journal* 2: 64–73.

Boole, George. 1843. Exposition of a general theory of linear transformations. *Mathematical Journal* 3(1–20): 106–119.

Cassirer, Ernst. 1907. *Das Erkenntnisproblem in der Philosophie und Wissenschaft der neueren Zeit*, vol. 2. Berlin: B. Cassirer.

Cassirer, Ernst. 1910. *Substanzbegriff und Funktionsbegriff: Untersuchungen über die Grundfragen der Erkenntniskritik*. Berlin: B. Cassirer. English edition: Cassirer, Ernst. 1923. *Substance and Function and Einstein's Theory of Relativity* (trans: Swabey, Marie Collins and Swabey, William Curtis). Chicago: Open Court.

Cayley, Arthur. 1859. A sixth memoir upon quantics. In Cayley (1889): 661–592.

Cayley, Arthur. 1889. *Collected mathematical papers*, vol. 2. Cambridge: University Press.

Cayley, Arthur. 1896. *Collected mathematical papers*, vol. 11. Cambridge: University Press.

Corry, Leo. 1996. *Modern algebra and the rise of mathematical structures*. Boston: Birkhäuser.

Darboux, Jean Gaston. 1880. Sur le théorème fondamental de la géométrie projective. *Mathematische Annalen* 17: 55–61.

Dedekind, Richard. 1872. *Stetigkeit und irrationale Zahlen*. Braunschweig: Vieweg. English edition in Dedekind (1901): 1–27.

Dedekind, Richard. 1888. *Was sind und was sollen die Zahlen?* Braunschweig: Vieweg. English edition in Dedekind (1901): 29–115.

Dedekind, Richard. 1901. *Essays on the theory of numbers*. Trans. Wooster Woodruff Beman. Chicago: Open Court.

Dedekind, Richard. 1969. *Gesammelte mathematische Werke*, ed. Robert Fricke, Emmy Noether, and Öystein Ore. Bronx: Chelsea.

du Bois-Reymond, Paul. 1882. *Die allgemeine Functionstheorie. Vol. 1: Metaphysik und Theorie der mathematischen Grundbegriffe: Grösse, Grenze, Argument und Function*. Tübingen: Laupp.

Dummett, Michael. 1991. *Frege: Philosophy of mathematics*. Cambridge: Harvard University Press.

Efimov, Nikolaj V. 1970. *Höhere Geometrie. Vol. 2: Grundzüge der projektiven Geometrie*. Braunschweig: Vieweg.

Einstein, Albert. 1998. *Collected papers. Vol. 8: The Berlin years: 1914–1918*, ed. Robert Schulmann, A.J. Kox, Michel Janssen, and József Illy. Princeton: Princeton University Press.

Enriques, Federigo. 1898. *Lezioni di geometria proiettiva*. Bologna: Zanichelli.

Enriques, Federigo. 1907. Prinzipien der Geometrie. *Enzyklopädie der Mathematischen Wissenschaften*, 3a.1b: 1–129.

Ferreirós, José. 1999. *Labyrinth of thought: A history of set theory and its role in modern mathematics*. Basel: Birkhäuser.

Friedman, Michael. 2000. *A parting of the ways: Carnap, Cassirer, and Heidegger*. Chicago: Open Court.

Friedman, Michael. 2001. *Dynamics of reason: The 1999 Kant lectures at Stanford University*. Stanford: CSLI Publications.

Gauss, Carl Friedrich. 1880. *Werke*. Ed. Königlichen Gesellschaft der Wissenschaften di Göttingen, 4.

Gray, Jeremy J. 2008. *Plato's ghost: The modernist transformation of mathematics*. Princeton: Princeton University Press.

Hawkins, Thomas. 1984. The *Erlanger Program* of Felix Klein: Reflections on its place in the history of mathematics. *Historia Mathematica* 11: 442–470.

Helmholtz, Hermann von. 1870. Über den Ursprung und die Bedeutung der geometrischen Axiome. In Helmholtz (1921): 1–24.

Helmholtz, Hermann von. 1921. *Schriften zur Erkenntnistheorie*, ed. Paul Hertz and Moritz Schlick. Berlin: Springer.

Hilbert, David. 1899. Grundlagen der Geometrie. In *Festschrift zur Feier der Enthüllung des Gauss-Weber-Denkmals in Göttingen*, 1–92. Leipzig: Teubner.

Hilbert, David. 1903. *Grundlagen der Geometrie*. 2nd revised ed. Leipzig: Teubner.

Hölder, Otto. 1901. Die Axiome der Quantität und die Lehre vom Mass. *Berichten der mathematisch-physischen Classe der Königl. Sächs. Gesellschaft der Wissenschaften zu Leipzig* 53: 1–63.

Hölder, Otto. 1908. Die Zahlenskala auf der projektiven Geraden und die independente Geometrie dieser Geraden. *Mathematische Annalen* 65: 161–260.

Hölder, Otto. 1911. Streckenrechnung und projektive Geometrie. *Berichte über die Verhandlungen der Königl. Sächs. Gesellschaft der Wissenschaften zu Leipzig* 63: 65–183.

Ihmig, Karl Norbert. 1997. *Cassirers Invariantentheorie der Erfahrung und seine Rezeption des "Erlanger Programms"*. Hamburg: Meiner.

Kant, Immanuel. 1783. *Prolegomena zu einer jeden künftigen Metaphysik die als Wissenschaft wird auftreten können*. Riga: Hartknoch. Repr. In *Akademie-Ausgabe*, 4: 253–384.

Kant, Immanuel. 1787. *Critik der reinen Vernunft*, 2nd ed. Riga: Hartknoch. Repr. in *Akademie-Ausgabe*, 3.

Klein, Felix. 1871a. Notiz, betreffend den Zusammenhang der Liniengeometrie mit der Mechanik starrer Körper. *Mathematische Annalen* 4: 403–415.

Klein, Felix. 1871b. Über die sogenannte Nicht-Euklidische Geometrie. *Mathematische Annalen* 4: 573–625.

Klein, Felix. 1872. *Vergleichende Betrachtungen über neuere geometrische Forschungen*. Erlangen: Deichert.

Klein, Felix. 1873. Über die sogenannte Nicht-Euklidische Geometrie, 2 Teil. *Mathematische Annalen* 6: 112–145.

Klein, Felix. 1874. Nachtrag zu dem zweiten Aufsatz über Nicht-Euklidische Geometrie. *Mathematische Annalen* 7: 531–537.

Klein, Felix. 1889–1890. *Vorlesungen über Nicht-Euklidische Geometrie*, elaborated by Fr. Schilling. Göttingen 1893.

Klein, Felix. 1890. Zur Nicht-Euklidischen Geometrie. *Mathematische Annalen* 37: 544–572.

Klein, Felix. 1893. Vergleichende Betrachtungen über neuere geometrische Forschungen. *Mathematische Annalen* 43: 63–100.

Klein, Felix. 1894. *Lectures on mathematics at Northwestern University*. Evanston, reported by A. Ziwet. New York: Macmillian.

Klein, Felix. 1895. Die Arithmetisierung der Mathematik. *Nachrichten von der Königl. Gesellschaft der Wissenschaften zu Göttingen. Geschäftliche Mitteilungen*: 82–91.

Klein, Felix. 1898. Gutachten, betreffend den dritten Band der Theorie der Transformationsgruppen von S. Lie anlässlich der ersten Vertheilung des Lobatschewsky-Preises. *Mathematische Annalen* 50: 583–600.

Klein, Felix. 1911. Über die geometrischen Grundlagen der Lorentzgruppe. *Physikalische Zeitschrift* 12: 17–27.

Klein, Felix. 1921. *Gesammelte mathematische Abhandlungen*, vol. 1, ed. Robert Fricke and Alexander Ostrowski. Berlin: Springer.

Klein, Felix. 1928. *Vorlesungen über nicht-euklidische Geometrie*. 2nd ed, ed. Walter Rosemann. Berlin: Springer.

Lagrange, Joseph-Louis. 1770–1771. Réflexions sur la résolution algébrique des équations. *Nouveaux mémoires de l'Académie royale des sciences et belles-lettres de Berlin, années 1770 et 1771*. Repr. in *Oeuvres*, 3: 203–421. Paris: Gauthier-Villars, 1869.

Leibniz, Gottfried Wilhelm. 1859. *Leibnizens mathematische Schriften. Vol. 4: Briefwechsel zwischen Leibniz, Wallis, Varignon, Guido Grandi, Zendrini, Hermann und Freiherrn von Tschirnhaus*, ed. Carl Immanuel Gerhardt. Halle: Schmidt.

Lie, Sophus. 1893. *Theorie der Transformationsgruppen*. Vol. 3. Leipzig: Teubner.

Maracchia, Silvio. 1993. *Dalla geometria euclidea alla geometria iperbolica: Il modello di Klein*. Napoli: Liguori.

Möbius, August Ferdinand. 1827. *Der barycentrische Calcul: Ein neues Hülfsmittel zur analytischen Behandlung der Geometrie*. Leipzig: Barth.

Nabonnand, Philippe. 2008. La théorie du *Würfe* de von Staudt – Une irruption de l'algèbre dans la géométrie pure. *Archive for History of Exact Sciences* 62: 201–242.

Norton, John. 1999. Geometries in collision: Einstein, Klein, and Riemann. In *The symbolic universe: Geometry and physics 1890–1930*, ed. Jeremy Gray, 128–144. Oxford: Oxford University Press.

Pasch, Moritz. 1882. *Vorlesungen über neuere Geometrie*. Leipzig: Teubner.

Plücker, Julius. 1869. *Neue Geometrie des Raumes, gegründet auf die Betrachtung der geraden Linie als Raumelement*. Leipzig: Teubner.

Poncelet, Jean-Victor. 1822. *Traité des propriétés projectives des figures: Ouvrage utile à ceux qui s'occupent des applications de la géométrie descriptive et d'opérations géométriques sur le terrain*. Paris: Bachelier.

Poncelet, Jean-Victor. 1864. *Applications d'analyse et de géométrie qui ont servi de principal fondement au Traité des propriétés projectives des figures*. Paris: Gauthier-Villars.

Poncelet, Jean-Victor. 1865. *Traité des propriétés projectives des figures*, vol. 1, 2nd ed. Paris: Gauthier-Villars.

Reck, Erich. 2003. Dedekind's structuralism: An interpretation and partial defense. *Synthese* 137: 369–419.

Reichardt, Hans. 1985. *Gauß und die Anfänge der nicht-euklidischen Geometrie*, 2nd ed. Leipzig: Teubner.

Rosenfeld, Boris A. 1988. *A history of non-Euclidean geometry: Evolution of the concept of a geometric space*. Trans. Abe Shenitzer. New York: Springer.

Rowe, David E. 1983. A forgotten chapter in the history of Felix Klein's "Erlanger Programm". *Historia Mathematica* 10: 448–454.

Rowe, David E. 1989. The early geometrical works of Sophus Lie and Felix Klein. In *The history of modern mathematics. Vol. 1: Ideas and their reception*, ed. David E. Rowe and John McCleary, 209–273. Boston: Academic.

Rowe, David E. 1992. Klein, Lie, and the *Erlanger Program*. In *1830–1930: A century of geometry, epistemology, history and mathematics*, ed. Luciano Boi, Dominique Flament, and Jean-Michel Salanskis, 45–54. Berlin: Springer.

Russell, Bertrand. 1897. *An essay on the foundations of geometry*. Cambridge: University Press.

Schlimm, Dirk. 2013. The correspondence between Moritz Pasch and Felix Klein. *Historia Mathematica* 40: 183–202.

Schur, Friedrich. 1881. Über den Fundamentalsatz der projectivischen Geometrie. *Mathematische Annalen* 18: 252–254.

von Staudt, Karl Georg Christian. 1847. *Geometrie der Lage*. Nürnberg: Korn.

von Staudt, Karl Georg Christian. 1856–1860. *Beiträge zur Geometrie der Lage*. 3 Vols. Nürnberg: Korn.

Sylvester, James Joseph. 1851. On the general theory of associated algebraical forms. *Mathematic Journal* 6: 289–293.

Tait, William W. 1996. Frege versus Cantor and Dedekind: On the concept of number. In *Frege: Importance and legacy*, ed. Matthias Schirn, 70–113. Berlin: De Gruyter.

Thomae, Johannes. 1873. *Ebene geometrische Gebilde erster und zweiter Ordnung vom Standpunkte der Geometrie der Lage betrachtet*. Halle: Nebert.

Torretti, Roberto. 1978. *Philosophy of geometry from Riemann to Poincaré*. Dordrecht: Reidel.

Voelke, Jean Daniel. 2008. Le théorème fondamental de la géométrie projective: évolution de sa preuve entre 1847 et 1900. *Archive for History and Exact Sciences* 62: 243–296.

Wussing, Hans. 1969. *Die Genesis des abstrakten Gruppenbegriffes: Ein Beitrag zur Entstehungsgeschichte der abstrakten Gruppentheorie*. Berlin: VEB Deutscher Verlag der Wissenschaften.

Yaglom, Isaak M. 1988. *Felix Klein and Sophus Lie: Evolution of the idea of symmetry in the nineteenth century*. Trans. Sergei Sossinsky. Boston: Birkhäuser.

Chapter 6
Euclidean and Non-Euclidean Geometries in the Interpretation of Physical Measurements

6.1 Introduction

Klein's classification of geometries by the use of group theory inaugurated a new phase in the debate on the geometry of space. On the one hand, the conclusion of Riemann's and Helmholtz's inquiries into the foundations of geometry appeared to be confirmed: Euclidean geometry does not provide us with the necessary presuppositions for empirical measurement, because both Euclidean and non-Euclidean assumptions can be obtained as special cases of a more general system of hypotheses. On the other hand, Helmholtz had believed that he had shown that the free mobility of rigid bodies implied and was implied by a metric of constant curvature, which includes spherical and elliptic geometries. Sophus Lie criticized Helmholtz and addressed the problem of characterizing the form of space in a completely different way. Even in 1854, Riemann pointed out that metrical relations in observable phenomena do not imply that the free mobility of rigid bodies holds true at the infinitesimal level (Riemann 1996, p.661). Lie was the first to formulate a set of necessary and sufficient conditions for a Riemannian metric of constant curvature, including free mobility at the infinitesimal level (see Lie 1893, pp.437–471). As Lie's proof did not refer to physical bodies, his demand did not provide a condition of measurement in Helmholtz's sense. Nevertheless, Lie's correction of Helmholtz's inquiry into the foundations of geometry of 1868 called into question Helmholtz's claim that geometrical axioms depend on the existence of rigid bodies.

The most challenging argument against Helmholtz's empiricism, however, was formulated by Henri Poincaré: observation and experiment cannot contradict geometrical assumptions because the application of geometrical concepts to empirical objects, including the characterization of solid bodies as "rigid," already presupposes these kinds of assumptions. On the one hand, Poincaré ruled out the view of geometrical axioms as empirical propositions. On the other hand, he maintained that geometrical axioms, unlike synthetic a priori judgments, are not necessary and

© Springer International Publishing Switzerland 2016
F. Biagioli, *Space, Number, and Geometry from Helmholtz to Cassirer*,
Archimedes 46, DOI 10.1007/978-3-319-31779-3_6

have to be stipulated: the choice among hypotheses, in the case of equivalent geometries, can only be guided by considerations of simplicity and conformity to experience.

The present chapter is devoted to the reception of Poincaré's argument in neo-Kantianism. Poincaré's objections to Helmholtz enabled neo-Kantians such as Bruno Bauch and Richard Hönigswald to reconsider Alois Riehl's argument that, in the case when observation contradicted our expectations regarding the motion of solid bodies, one should postulate a physical cause, instead of revising the Euclidean hypotheses. By contrast, owing to the relativized conception of the notion of a priori which goes back to Hermann Cohen's interpretation of Kant, Ernst Cassirer acknowledged the possibility of revising geometrical assumptions. However, Cassirer maintained that conventional criteria do not provide us with sufficient conditions for measurement: the interpretation of measurements depends on conceptual rules and, ultimately, on rational criteria. Cassirer relied on the group-theoretical analysis of space to infer such criteria from the relations of geometrical systems to one another.

6.2   Geometry and Group Theory

The debate about the relationship between geometry and experience after Poincaré's criticism of Helmholtz presupposed a group-theoretical approach to geometry. The first part of this section provides basic information on the development of the idea of using group theory to classify geometries in the works of Klein and Poincaré. The second part is devoted to the discussion about the possibility of using group theory to deal with the problems posed by Helmholtz. On the one hand, Klein (1898), Poincaré (1898b) and Cassirer (1950) maintained that the group-theoretical approach was implicit in Helmholtz's 1870 thought experiments about non-Euclidean metrical relations. On the other hand, Moritz Schlick (in Helmholtz 1921) called into question such an interpretation and interpreted Helmholtz's general properties of space as psychological qualities. Schlick's critical remarks lead us to the question whether it is appropriate to use mathematics to express the properties of space. Poincaré's and Cassirer's answers to this question shed some light on their disagreement about the conventional nature of geometrical assumptions. Whereas Poincaré's reasons for the use of the concept of group in this connection depend on his account of spatial perception, Cassirer's argument presupposes his reformulation of the Kantian claim that mathematics is synthetic in terms of the logic of the concept of function.

6.2.1   Klein and Poincaré

Klein first presented the project of a unified treatment of geometry by using group theory in Klein (1872), when he became Christian von Staudt's successor as Professor of Geometry at the University of Erlangen. During his inaugural address on that occasion, which was devoted to the relationship between pure and applied mathematics in mathematics teaching, another written text was distributed and appeared in the proceedings of the University of Erlangen under the title "A Comparative Review of Recent Researches in Geometry." This work was republished in a revised version in 1893 in *Mathematische Annalen* and became known as the "Erlangen Program."[1] Klein's goal was to compare the various branches of nineteenth-century geometry from a general viewpoint. In the introductory remarks of his paper, Klein (1893, pp.63–65) reported that he arrived at a solution after his classification of geometries into elliptic, hyperbolic, and parabolic in Klein (1871). Klein's classification presupposed the application of the algebraic theory of invariants to the study of the projective properties of figures. Thereby, projective properties are characterized as those properties that remain invariant under projective transformations. In 1872, Klein reformulated his argument by using the notions introduced by Evariste Galois and Camille Jordan: operations form a group if the product of any two operations of the group also belongs to the group and if, for every operation of the group, there exists in the group an inverse operation.[2]

---

[1] The Erlangen Program is often mistaken for Klein's inaugural address (see Rowe 1983). It is only after the second edition of Klein's "Comparative Review of Recent Researches in Geometry" that this work, also called the Erlangen Program, became known as a retrospective guideline for Klein's research (see Gray 2008, p.117).

[2] Klein specified the second condition in the 1893 version of the paper. In the first version of 1872, he adopted Jordan's (1870, p.22) definition, which referred to finite groups of permutations. In that case, the closure of a set of elements relative to a fundamental operation (i.e., the first of the said conditions) is a sufficient condition for the set to form a group. Afterwards, Lie drew Klein's attention to the fact that the existence of an inverse operation is required in the case of infinite groups (see Wussing 1969, p.139). Furthermore, it is noteworthy that both Galois and Jordan dealt with groups of permutations. A set G of permutations forms a group if: (i) G contains the (unique) identity permutation $\varepsilon = \begin{pmatrix} 1 & 2 \ldots n \\ 1 & 2 \ldots n \end{pmatrix}$; (ii) together with every two permutations $\sigma = \begin{pmatrix} 1 & 2\ldots & n \\ a_1 & a_2 \ldots & a_n \end{pmatrix}$ and $\tau = \begin{pmatrix} 1 & 2\ldots & n \\ b_1 & b_2 \ldots & b_n \end{pmatrix}$ (where the equality $\tau = \sigma$ is not excluded), G contains the product $\tau\sigma = \begin{pmatrix} 1 & 2\ldots & n \\ b_1 & b_2 \ldots & b_n \end{pmatrix} \cdot \begin{pmatrix} 1 & 2\ldots & n \\ a_1 & a_2 \ldots & a_n \end{pmatrix} = \begin{pmatrix} 1 & 2\ldots & n \\ b_{a_1} & b_{a_2} \ldots & b_{a_n} \end{pmatrix}$; (iii) together with every permutation σ, the set G also contains the inverse permutation $\sigma^{-1} = \begin{pmatrix} a_1 & a_2 \ldots & a_n \\ 1 & 2\ldots & n \end{pmatrix} = \begin{pmatrix} 1 & 2\ldots & n \\ \alpha_1 & \alpha_2 \ldots & \alpha_n \end{pmatrix}$, where $\alpha_{a_i} = i$, $i = 1, 2, \ldots, n$ (see Yaglom 1988, pp.12–13). The group-theoretical treatment of geometry showed that the same conditions generally apply to groups of operations. In this sense, Klein's Erlangen Program can be considered a fundamental step in the development of the abstract concept of group (Wussing 1969, pp.132–143).

Klein defined motion as a transformation performed on the whole of space. He observed that the totality of motions forms a group. Klein called the group of motions the "principal group" of space transformations and established that geometric properties are characterized by their remaining invariant under the transformations of the principal group. At the same time, Klein emphasized that geometry is not restricted to the theory of invariants relative to the principal group. Space in mathematics is subsumed to the more general concept of n-dimensional manifold and can be defined as a three-dimensional manifold. Klein pointed out that the transformations of space, in this more general sense, also form groups. Unlike the case of space transformations in the former sense, however, there is not one group distinguished above the rest by its significance; each group is equally admissible. Therefore, Klein generalized the problem of characterizing geometric properties as follows: "Given a manifold and a group of transformations of the same; to investigate the configurations belonging to the manifold with regard to such properties as are not altered by the transformations of the group" (Klein 1893, p.67). Or, to put it in terms of the theory of invariants: "Given a manifold and a group of transformations of the same; to develop the theory of invariants relating to that group" (p.67).

In order to clarify the relations of one group to another, Klein adopted the following principle of transfer. Suppose that a manifold A has been investigated with reference to a group B. If, by any transformation, A is converted into a second manifold A', the group B of transformations, which transformed A into itself, will become a group B', whose transformations are performed upon A'. The principle states that every property of a configuration contained in A obtained by means of the group B corresponds to a property of a configuration in A' to be obtained by the group B'. According to Klein, this principle sheds light on the arbitrariness of the choice of a particular space element (e.g., the point or the line). Klein wrote:

> As there is nothing at all determined at the outset about the number of arbitrary parameters upon which these configurations shall depend, the number of dimensions of our line, plane, space, etc., may be anything we like, according to our choice of the element. *But as long as we base our geometrical investigation upon the same group of transformations, the substance of the geometry remains unchanged.* (Klein 1893, p.73)

The choice of a particular group is arbitrary in the same sense. Klein, for example, mentioned the fact that the specifications of a projective metric according to his 1871 classification correspond to the three classical cases of manifolds of constant curvature in Beltrami's theory of 1869. Klein used the transfer principle above to prove that elliptic, hyperbolic and parabolic geometries are equivalent to the geometry of manifolds of constant positive, negative and zero curvature, respectively. Klein's classification showed that Euclidean geometry – which corresponds to a parabolic metric – can be obtained as a limiting case of a more general system of hypotheses. From the standpoint of group theory, the same conclusion follows from the fact that the principal group is included in the group of projective transformations. Relative invariants of the principal group (e.g., absolute distance and parallelism) are not invariant relative to the larger group. The group-theoretical view of geometry shows that the projective group is included itself in the continuous transformation group. Klein identified the study of the relative invariants of this group as the domain of the science foreshadowed by Leibniz and called by him analysis situs.[3]

As we mentioned in the previous chapter, Klein's ideas did not receive much attention before the first translations of "A Comparative Review of Recent Researches in Geometry" into Italian (1890), French (1891), and English (1892–1893) and Klein's revised version of 1893.[4] Klein (1893, p.63, note) reported that he was motivated to promulgate his earlier project by Lie's *Theory of Transformation Groups*, which appeared in three volumes in 1888, 1890 and 1893. Lie's study of transformation groups provided essential requirements for the implementation of the project of a unitary treatment of geometry from the standpoint of group theory.[5] Meanwhile, related ideas had been developing in Paris. In 1880, Poincaré defined geometry as "the study of the group of operations formed by the displacements to which one can subject a body without deforming it" (Poincaré 1997, p.11). Over the course of the following years, Poincaré and other leading figures of the Paris school of mathematics had fruitful exchanges with Lie during his research into the theory

---

[3] Leibniz's goal was to develop a general science of situational relations in order to represent non-spatial relations (i.e., relations between monads). The reception of Leibniz's ideas in the nineteenth century differed considerably from his original project and took place in very different contexts. Hermann Grassmann (1847) compared the analysis situs to his *Theory of Extension* (Grassmann 1844). Giuseppe Peano and his school contributed both to the rediscovery of Grassmann's work in the second half of the nineteenth century and to the connection between Leibniz's analysis situs and the vector calculus. On the other hand, "geometry of position" was used to designate projective geometry as well. It was in that context that Johann Benedikt Listing, in a letter to his old school teacher dated 1836, introduced the term "topology," which was substituted for "analysis situs" in the twentieth century. Listing introduced a new term because the phrase "geometry of position" had been used by Lazare Carnot (1803). The fundamental ideas of algebraic topology go back to a fragment from Riemann's discussions with Enrico Betti. The fragment appeared in Riemann (1876) with the title "Fragment Belonging to Analysis Situs." The development of these ideas is due mainly to Poincaré and lies at the origin of the discipline now known as topology. Klein was arguably acquainted with Riemann's fragment. However, in the Erlangen Program, Klein specifically referred to the continuous transformation group. On Leibniz's project and its reception in the nineteenth century, see De Risi (2007, pp.XII, 111–114). On Poincaré's work on topology, see Gray 2013, Ch.8.

[4] The question of the influence of Klein's Erlangen Program is controversial. I rely on Hawkins (1984) for a historically well-documented reconstruction of the delayed reception of the Erlangen Program. Furthermore, Hawkins gives evidence of the role of other mathematicians, including Sophus Lie, Henri Poincaré and Gino Fano, in the implementation of a research program in line with Klein's view. Cf. Birkhoff and Bennett (1988) for the view that Klein had a major influence on later mathematical researches, including his own. More recently, Gray points out the different backgrounds of these readings: whereas mathematicians, such as Birkhoff, often consider Klein's Erlangen Program very influential, a number of historians, including Hawkins and Erhard Scholz, showed that the solid work establishing group theory between 1870 and 1890 was done by Camille Jordan, Sophus Lie and Henri Poincaré, among others. Klein himself did not implement or advance the view of the program (Gray 2008, p.117).

[5] On Klein's relationship to Lie, see Hawkins (1984) and Rowe (1989; 1992).

of continuous functions. The idea shared by these mathematicians was that the concept of group provides us with the highest principle of mathematics as a whole.[6]

In line with the remarks above, Poincaré (1882) developed a model of non-Euclidean geometry. His considerations about the foundations of geometry are found in a series of writings from the 1890s. In 1902, Poincaré collected his earlier writings on that subject in the first part of *Science and Hypothesis*. His goal was to use the group-theoretical approach to clarify the relationship between geometry and experience. Poincaré's argument for the conventionality of geometry in that connection is discussed in Sect. 6.3. For now, it is noteworthy that both Poincaré's and Klein's classifications of geometry culminated in the analysis situs. However, whereas Klein referred to the group of continuous transformations, Poincaré's analysis situs lies at the foundation of modern topology. This fact led to an important difference in their approaches to measurement. For Klein, the group-theoretical analysis shows that spatial notions can be interpreted in various ways according to the transformation group under consideration. In order to justify this way of proceeding, Klein observed that there are no constraints imposed by sense perception upon the conception of space. Since we know nothing about nonmeasurable quantities, nothing precludes us from introducing numerical representations of space and purely conceptual postulates, such as Dedekind's continuity (Klein 1898, pp.593–594). Klein identified the foundations of metrical geometry as the concept of a projective metric, because projective geometry depends on such postulates, rather than on the consideration of specific quantities, and because his model of non-Euclidean geometry showed that metrical geometries can be derived from a projective metric as special cases.

Poincaré discussed the status of projective geometry in his 1899 review of Bertrand Russell's *Essay on the Foundations of Geometry* (1897). We know from the previous chapter that Russell's work contains a defense of a Kantian theory of space, according to which only the axioms that are common to both Euclidean and non-Euclidean spaces are a priori; the axioms that distinguish Euclidean space from non-Euclidean spaces, on the other hand, are empirical and depend on the metrical structure of actual space. Therefore, Louis Couturat called Russell's view "a revised and integrated version of Kant's Transcendental Aesthetic, reconsidered in light of metageometry" in a review from 1898 (Couturat 1898, p.355). Couturat's discussion with Russell was followed by Poincaré's review, which contains several critiques of both of Russell's claims. From 1899 to 1900, Couturat promoted the discussion between Poincaré and Russell as a member of the editorial board of the *Revue de métaphysique et the morale*.[7]

---

[6] After a journey to Paris in 1882, Lie reported to Klein on his meeting with Poincaré and on the latter's view of mathematics as a "tale about groups" (*Gruppengeschichte*). On that occasion, Lie informed Poincaré about Klein's Erlangen Program. After that, Poincaré and Darboux remained in touch with Lie – who was based at the University of Leipzig – and promoted studies in the theory of continuous functions at the École normale supérieure in Paris (see Hawkins 1984, p.448).

[7] For a thorough reconstruction of the debate between Russell and Poincaré over the status of geometrical axioms, see Griffin (1991, pp.171–181); Nabonnand (2000).

Although Russell reconsidered his axiomatization after Poincaré's objections, there remained a fundamental disagreement about the definition of distance. We have already mentioned that Russell took distance as a fundamental relation between two points. By contrast, Poincaré urged clarification of the foundations of measurement from the general viewpoint of group theory. It followed that distance can be defined as an invariant relation between pairs of points relative to a transformation group. Equality and diversity between the elements can likewise vary according to the transformation group under consideration. Projective geometry, for example, does not imply an essential distinction between ideal and ordinary points. Poincaré agreed with Russell that projective metrical geometry was based on a stipulation about the definition of distance. However, he emphasized the conventional aspect of any way to define distance. As pointed out by Griffin (1991), Nabonnand (2000) and Gray (2013), the disagreement with Russell concerns definitions in mathematics, and it is mainly a philosophical one.[8] Whereas for Russell, our intuitions regarding the fundamental concepts provide mathematical definitions with their true meaning and distinguish them from mere conventions, Poincaré's point is that there is no other way of defining terms in mathematics except in relation to other terms. Some stipulation is required, because existence in mathematics does not depend on external reference, but on the consistency of the propositions admitted.

The conventional aspect of Poincaré's notions posed the problem of clarifying the physical meaning of geometry. In fact, some of Poincaré's objections against Russell specifically regard the impossibility of carrying out an experimentum crucis to make a choice between Euclidean and non-Euclidean geometry. However, as Griffin pointed out, Russell was not committed to the view that the truth of Euclidean geometry can be put to the test. He believed that experiment might establish that non-Euclidean geometries are false, since the value of curvature might deviate from 0 sufficiently to be detected. Insofar as Poincaré relies on the underdetermination of theory by evidence, his argument about the conventionality of geometry is not conclusive. It might be argued that, although Euclidean geometry was true or false, it was impossible to determine it empirically. However, it does not follow that the truth or falsity of Euclidean geometry could not be determined empirically because Euclidean geometry was neither true nor false (see Griffin 1991, pp.180–181).

We turn back to Poincaré's argument for the conventionality of geometry in the next section, as this will require us to consider the whole argument as presented by Poincaré in his 1898b essay on the foundations of geometry and subsequently incor-

---

[8] For a reconstruction of how Russell's and Poincaré's different mathematical conceptions influenced their disagreement on philosophical matters, see Nabonnand (2000, p.259): "In presupposing the models of Euclidean and non-Euclidean geometries, Russell is led to restrict his consideration to the metrical concepts. Poincaré, who endorses the viewpoint of the transformation groups – which in the eyes of Russell is nothing more than a change in formulation without philosophical importance –, considers distance to be essentially an invariant of a group, whose content is equivalent to that of this group. The choice of the distance depends, therefore, on a convention no less than the choice of the group."

porated in *Science and Hypothesis*. Therefore, I consider Poincaré's understanding of mathematical definitions an important premise of his argument: the argument being that conventions are required for the formation of spatial concepts and for the use of geometry in physics. In this regard, Poincaré adopted a view of mathematical objects that can be traced back to Riemann and Dedekind, namely, the view that mathematical objects are defined by their relations to one another independently of our prior intuitions. I have already mentioned that similar views were shared by Helmholtz and Klein. Notwithstanding the abstract character of mathematical definitions, Poincaré believed that conventionalism provided him with a theoretical framework for rethinking the problem of the applicability of geometry.[9]

Regarding the dispute with Russell, my focus is on the a priori side of the matter. As mentioned in Chap. 5, Russell identified the a priori properties of space as projective properties. However, he denied that metrical projective geometry can provide the foundations of metrical geometry. In other words, Russell granted projective geometry the status of an a priori science of space only insofar as it dealt with qualities, because any quantitative relation presupposes some qualitative identity of the objects to be measured. Given the essential difference between qualities and quantities, metrical projective geometry, according to Russell, is a purely technical development, and a reduction of metrical geometry to projective geometry remains impossible. Regarding the foundation of metrical geometry, Poincaré observed that the projective transformation group contains the Euclidean group as a subgroup. This fact enabled him to account for the properties that Russell considered common to all spaces by identifying the invariants of the larger group. Not only does the group-theoretical approach clarify the relations between projective geometry and the systems of metrical geometries from a mathematical viewpoint, but it affects the philosophical implications of Russell's argument: the size of the group does not determine a distinction between qualities and quantities. Nevertheless, it is noteworthy that Poincaré considers this a fundamental distinction as well. The only "qualitative" geometry in Poincaré's sense is analysis situs, because it enables us to consider space independently of its form. In order to characterize the amorphous space of analysis situs, Poincaré introduced such properties as continuity and the number of dimensions as topological properties (see Poincaré 1899, pp.276–277).

To sum up, Poincaré's considerations about the status of projective geometry show that the properties of space according to Poincaré are fundamentally different from the corresponding properties in Klein's works. My suggestion is that the difference with Klein does not depend solely on Poincaré's wider conception of analysis situs, which Klein could not have considered at the time he wrote his "Comparative Review;" Poincaré's distinction between topological and metrical properties determines a different approach to the problem of measurement. Correspondingly, he distinguished between different levels in the formation of spatial concepts. Whereas

---

[9] On the explanatory power of Poincaré's talk of conventions when it comes to measurement, see Gray (2013). The contrast with Russell is illuminating: "Russell sought to define distance, or perhaps to elucidate our familiar concept of distance by saying what it *is*, while Poincaré could only do so by saying how the concept if distance is *used*, that is, measured" (Gray 2013, p.81).

Poincaré's account of space in topological terms presupposed spatial intuition, his view about the principles of measurement was that the specification of metrical properties, beginning with the introduction of Dedekind's continuity, is a matter of convention. By contrast, Klein started from the fact that our intuitions about space, in contrast with geometrical knowledge, are inherently imprecise. Therefore, he emphasized the need of conceptual postulates, such as Dedekind's continuity, both in foundational inquiries into geometry and in empirical research.

Bearing in mind the difference between Klein's and Poincaré's approaches, we now deal with the use of group theory in later interpretations of Helmholtz's thought experiments of 1870 by Klein, Schlick and Cassirer.

6.2.2 Group Theory in the Reception of Helmholtz's Work on the Foundations of Geometry: Klein, Schlick, and Cassirer

The third volume of Sophus Lie's *Theory of Transformation Groups* (1893, pp.471–523) contains a solution to the so-called "Riemann-Helmholtz problem" of establishing a set of necessary and sufficient conditions for obtaining a Riemannian metric of constant curvature, namely, the form that characterizes both Euclidean geometry and the two cases of non-Euclidean geometry that became known as the geometries of Bolyai-Lobachevsky and of Riemann. Lie presented his treatment as a correction of Helmholtz's inquiry into the foundations of geometry of 1868, because he showed with several examples that a group may act freely at the finite level and not act freely at the infinitesimal level. These examples made it clear that the free mobility of rigid bodies did not suffice to obtain a metric of constant curvature. What was required was an equivalent formulation of free mobility at the infinitesimal level. Lie's formulation showed that another of Helmholtz's conditions (i.e., the monodromy of space) was superfluous. Lie provided the first mathematically sound solution to the Riemann-Helmholtz's problem of space. However, his inquiry differed significantly from Helmholtz's because Lie's conditions, unlike the free mobility of rigid bodies, did not apply to empirical measurement straightforwardly.[10]

In 1898, owing to his work on the theory of continuous transformation groups, Lie was awarded the first Lobachevsky prize. In a review of the third volume of Lie's work delivered by Klein on that occasion, Klein addressed the problem of establishing the conditions of measurement from the standpoint of projective geometry and of group theory. Notwithstanding the difference between metrical and projective geometry, the connection lies in the fact that projective geometry enables us to make metrical considerations dependent on projective relations. Klein referred to his proof that projective geometry is independent of the theory of parallel lines and

---

[10]For a thorough comparison between Helmholtz and Lie, see Torretti (1987, pp.158–171).

to the related classification of geometries into elliptic, hyperbolic and parabolic. In 1898, Klein maintained that metrical projective geometry and group theory provided a suitable mathematical account of Helmholtz's views about the foundations of geometry, if one considered not so much Helmholtz's (1868) treatment of the problem of space – for which Klein relied on Lie – as Helmholtz's (1870) thought experiments about measurement in a convex mirror. According to Klein, Helmholtz foreshadowed the possibility of a projective treatment of non-Euclidean geometry, insofar as he used Beltrami's model to establish the correspondence between operations with rigid bodies in our world and the operations carried out by a hypothetical inhabitant of the world in the mirror. Klein identified these operations as collineations to point out that the result of Helmholtz's inquiry into the foundations of metrical geometry coincided with that of Klein's projective metric: the Euclidean transformation group is a subgroup of the projective group. This fact affects empirical measurement, because it shows that non-Euclidean geometries had to be considered approximate special cases that can occur in the interpretation of measurements in limited regions of space (Klein 1898, pp.598–600). Regarding the form of space as a whole, Klein (1898, p.597) referred to his essay "On Non-Euclidean Geometry" (1890) to point out that there is an even larger variety of possible hypotheses. This remark made it clear that the significance of Helmholtz's inquiry regarding the problems concerning measurement did not depend on a global representation of space, but on the possibility of representing spatial motions by means of transformations and transformation groups (cf. Coffa 1991, pp.49–51).[11]

A similar conclusion is found in Schlick's comments to the centenary edition of Helmholtz's *Epistemological Writings* (1921) by Schlick and Paul Hertz. Meanwhile, Hilbert had proved that there is no twice-differentiable isometric embedding of all of non-Euclidean space in Euclidean three-dimensional space (Hilbert 1903, pp.162–172). This result made it clear that Helmholtz's thought experiments cannot be considered models of non-Euclidean geometry. Schlick (in Helmholtz 1921, p.29, note 18) observed that the lack of a mathematical model did not affect Helmholtz's epistemological argument: non-Euclidean hypotheses cannot be ruled out as approximations of the relations of measure found in physical space. Schlick's remark differs from Klein's, because it depends not so much on the use of group theory in the interpretation of Helmholtz's thought experiments, as on Schlick's distinction between intuitive and physical space. According to Schlick, this distinction was implicit in Helmholtz's (1878a, p.121) distinction between the general form of space and its narrower specifications, namely the axioms of geometry. Owing to this distinction, Helmholtz maintained that space can be considered a form of intuition in Kant's sense (i.e., the order of appearance as opposed to the matter), and yet not necessarily imply the axioms, in the following sense:

---

[11] Klein's account differs from Coffa's, because Klein does not presuppose an extension of Beltrami's two-dimensional model to the three-dimensional case. On the problems of interpreting Helmholtz's thoughts experiments as a three-dimensional model of non-Euclidean geometry, see Sect. 3.2.3, and note 8.

> To cite a parallel instance, it undoubtedly lies in the organisation of our optical apparatus that everything we see can be seen only as a spatial distribution of colours. This is the innate form of our visual perceptions. But it is not in the least thereby predetermined how the colours we see shall co-exist in space and follow each other in time. (Helmholtz 1878b, p.213)

The parallel with color mixtures suggests that Helmholtz's distinction was motivated, firstly, by his knowledge of the theory of manifolds[12] and, secondly, by his psychological interpretation of Kant's theory of pure sensibility.

Schick (in Helmholtz 1921, p.128, note 33) addressed the question whether the general characteristics of space in Helmholtz's sense admitted a mathematical expression by using axioms other than the axioms of congruence. Arguably bearing in mind Lie's and others' axiomatizations, Schick observed that modern mathematicians seem to be inclined to answer this question affirmatively. However, Schick argued against the use of projective geometry or group theory for the purpose of reformulating Helmholtz's distinction, because such an interpretation seemed to call into question the intuitive character of the form of space. In order to uphold Helmholtz's distinction, Schick characterized the general properties of space as indescribable, psychological factors in spatial perception. He referred not so much to Klein as to Poincaré's interpretation of the qualitative aspects of space in topological terms. However, Schick distanced himself from Helmholtz's psychological conception of Kant's forms of intuition. Given the empirical origin of the representation of space, Schick ruled out the idea that intuitive space provides us with a form of intuition in Kant's sense: whereas space and time are forms according to Kant, the qualities of sensation are contents given in intuition. Schick goes on to say that sensations "have a wholly different significance for cognition from what the forms have (only the latter are namely for [Kant] sources of synthetic judgements a priori." Schick concludes that Helmholtz's use of the notion of form as referred to the qualities of sensation is not Kantian; Helmholtz's remarks only indicates that the "qualities of sensation are purely subjective" (Schlick in Helmholtz 1921, p.166, note 16). Regarding Helmholtz's claim that "space can be transcendental without the axioms being so" (p.149), Schlick wrote:

> If "transcendental" – in accordance with Helmholtz's linguistic usage – means the same as a priori, and if the latter is understood in Kant's meaning, then the a priori nature of the axioms is no different from the a priori nature of space. In Kant the assertion of the a priori nature of the latter has indeed only the purpose of explaining the apodictic validity of the axioms. But Helmholtz's a priori is precisely not Kant's, and instead has only the meaning that the spatiality of perception is something purely subjective in the same sense as are the qualities of sensations. Under this presupposition, of course, the "transcendental" nature of spatial intuition need yield no ground for the validity of any synthetic a priori axioms about space. (Schlick in Helmholtz 1921, p.182, note 65)

---

[12] Color was one of Riemann's examples of continuous manifolds in Section I.1. of "On the Hypotheses Which Lie at the Foundation of Geometry." Another important source for Helmholtz's considerations was Hermann Grassmann's work on manifolds and its application to color mixtures (see Hatfield 1990, p.218, note 106; Hyder 2009, Ch.4).

Schlick's distinction between sense qualities and geometrical concepts poses the problem of correlating the purely subjective space of intuition with the physical-geometrical meaning of space. This is the goal of Schlick's method of coincidences, which he presented as foreshadowed in Helmholtz's conception of congruence. We discuss Schlick's method in the next chapter. For now, it is worth noting that Schlick's interpretation of Helmholtz is problematic, for two reasons.

Firstly, Schlick's reference to Poincaré suggests that Helmholtz's distinction between the general properties of space and the axioms of geometry can be reformulated as a distinction between topological and metrical properties. The problem with such a formulation is that the general characteristics of space would admit an axiomatic expression as well.[13] A related problem of Schlick's reliance on the topological/metrical distinction is that it does not seem to take into account the metrical aspect of Helmholtz's conception of the general form of intuition. Helmholtz's general characteristics of space include constant curvature, which, for Helmholtz, follows from the requirement of the free mobility of rigid bodies.[14]

Secondly, as pointed out by Friedman (1997), there is no place for a distinction between acquaintance with or purely subjective intuition of space and discursive knowledge in Helmholtz's theory of local signs. The parallel with Kant lies precisely in Helmholtz's use of conceptual rules for providing local signs with their meaning. In particular, Friedman draws attention to the following passage from Helmholtz's *Physiological Optics* (1867): "The fundamental principle of the empiricist view is [that] sensations are signs for our consciousness, where learning to understand their meaning is left to our understanding" (Engl. trans. from Friedman 1997, p.31). Obviously, Helmholtz's empiricist explanation of the "transcendental" nature of space differs from Kant's aprioristic one, because Helmholtz's free mobility of rigid bodies and, therefore, constant curvature is supposed to be inferred from observation of the behavior of solid bodies in space. Insofar as the law thus inferred provides us with a univocal interpretation of all series of sensations that may occur in experience, intuitive space retains, nonetheless, the status of a form of intuition in Kant's sense. According to Friedman, even in the *Physiological Optics*, "Helmholtz comes close to the view that lawlike relations among our sensations – arrived at by inductive inferences in accordance with the principle of causality or the lawlikeness

---

[13] Torretti (1978, pp.166–167) points out that the general properties of space may not have counted as axioms in the Euclidean tradition. This does not mean that they cannot be axiomatized at all. Following Helmholtz's analogy with color mixtures, Torretti suggests that the characteristics required for the interpretation of the general properties of space may be specified in terms of of the theory of manifolds. He describes Helmholtz's space as a differentiable, three-dimensional manifold. The axioms of quantity, on the other hand, are not determined by the form of space because their formulation presupposes the existence of the solid bodies we experience. Following a similar line of reasoning, Lenoir (2006, pp.201–202) describes Helmholtz's form of spatiality as a differentiable, n-fold extended manifold.

[14] This problem affects even later interpretations of Helmholtz's notion of space in topological terms (see note 13). On the metrical aspect of Helmholtz's concept of space, cf. Friedman (1997), Ryckman (2005) and Pulte (2006).

of nature – are *constitutive* of their relationship to an external world" (Friedman 1997, p.33).

In Friedman's reading, the passages from "The Facts in Perception" commented on by Schlick lend plausibility to Helmholtz's naturalization of the forms of intuition. The localization of the objects in space according to the free mobility of rigid bodies also provides us with a construction of the concept of space as a three-dimensional manifold of constant curvature. Helmholtz's claim that space can be transcendental without the axioms (of specifically Euclidean) geometry being so indicates that the specific value of the measure of curvature is not determined a priori and depends on empirical considerations. Referring to Friedman (1997), Ryckman (2005) contrasts "Schlick's Helmholtz" with Helmholtz's original account of the form of spatial intuition. Ryckman summarizes his account by saying that "Helmholtz argued against the Kantian philosophy of geometry while retaining an inherently Kantian theory of space" (Ryckman 2005, pp.73–74). Helmholtz's theory can be considered inherently Kantian, insofar as his notion of rigid body receives a consistent interpretation as constitutive of the concept of congruence on which geometrical measurement rests. In distancing himself from the Kantian philosophy of geometry, Helmholtz is not blurring the distinction between form and contents; rather, the borderline is set somewhere else for the form of intuition to include all possible displacements in physical space. Related to the problem of space, this again means that whereas the notion of a three-dimensional space of constant curvature can be inferred from free mobility, the choice between the specific cases of such a manifold depends on the structure of actual space.

As we saw in the previous chapter, Cassirer's goal in 1910 was to show that such a way to reformulate Kant's form/content distinction was compatible with a Kantian architectonic of knowledge. Therefore, he reinterpreted Kant's thesis about the synthetic character of mathematics in terms of the use of the mathematical concept of function in mathematics and physics. The appreciation of Klein's approach to geometry was Cassirer's starting point for his account of the relationship between space and geometry. Group theory makes the characterization of spatial notions dependent on the formal analysis of ideal operations and of their relations to one another. At the same time, the group-theoretical analysis of space provides us with a possible interpretation of the Kantian notion of the a priori form of space and with a clarification of the relationship between this form and a variety of specifications.

Cassirer clarified his motivations for generalizing Kant's form of spatial intuition by using group theory in the 1940s. In the fourth volume of *The Problem of Knowledge in Modern Philosophy and Science*, which was written in 1940 and appeared posthumously in 1950 in English translation and 1957 in the German original, Cassirer characterized Klein's approached by saying that Klein clearly distinguished between the problems regarding mathematical structures and the existence of mathematical objects. Klein's inquiries into the foundations of geometry suggest that immanent developments in mathematics depend not so much on the solution of ontological problems, as on the study of mathematical structures. In this sense, Cassirer maintained that the success of the group-theoretical approach was due to the idea of obtaining propositions about specific domains – including both numbers and quantities – from the analysis of a system of operations (Cassirer 1950, pp.37–38). This way of proceeding shows that there can be levels of increasing generality in the characterization of mathematical structures. Cassirer mentioned, for example, the fact that the affine group is broader than the Euclidean one, and does not make a difference between a circle and an ellipse, but only between finite and infinite conic sections. From the point of view of the projective group, all conic sections are classified as the same figure. The group of continuous transformations makes no difference between a cone-shaped figure and a cube (p.41).

Regarding the origin of these ideas, Cassirer (1950, pp.49–50) prized Helmholtz for having put the Kantian view that the concept of space as a form of intuition cannot be referred to as an absolute object in connection with more recent insights into the formation of mathematical concepts. Notwithstanding the problems of Helmholtz's claim that spatial notions, including the free mobility of rigid bodies, have an empirical origin, Cassirer referred to the fact that Helmholtz, instead of presupposing some specific magnitudes, explored the possibility of reconceptualizing spatial notions in terms of displacement and coordinating rules for their empirical use. Following a similar line of argument, Klein and Poincaré explained distance by describing what measurement is. Therefore, Cassirer maintained that the group-theoretical approach was implicit in Helmholtz's inquiries into the foundation of geometry: Helmholtz foreshadowed the view that the concept required for a unified account of various geometrical hypotheses was not so much the concept of space, as that of group (Cassirer 1857, pp.49–50). According to Cassirer, this view followed from Helmholtz's distinction between general properties and narrower specifications of space. Helmholtz's claim that space can be "transcendental" with regard to its general properties does not imply that the specific axiomatic structure of space can be determined a priori. Therefore, Helmholtz distanced himself from Kant and maintained that the choice among geometrical hypotheses is a matter for empirical science. Nevertheless, Helmholtz's use of the notion of transcendental suggests that such coordinating principles as the free mobility of rigid bodies play the role of conditions of experience in Kant's sense, insofar as they are necessary presuppositions for the possibility of measurement.

Summing up, Cassirer arrived at the same conclusion as the more recent literature about Helmholtz's epistemological writings and their significance for a renewal of the Kantian theory of space. At the same time, we know from the previous chapters that Cassirer, in line with the Marburg reading of Kant, distanced himself from Helmholtz's naturalized version of the theory of pure intuitions. Cassirer believed that Helmholtz's considerations deserved a more precise formulation in light of more recent developments in the logical foundation of mathematics. The group-theoretical analysis of space confirmed that this could be done independently of Helmholtz's psychologism in the philosophy of mathematics. It might be objected to that Cassirer's perspective on the problem of space is broad enough to synthetize views that might be incompatible with each other, such as Kantianism, empiricism and conventionalism.

In order to characterize Cassirer's view better, the next section offers a comparison between Cassirer and Poincaré. To conclude this section, I think that Cassirer arguably bore in mind Schlick's objections against a mathematical characterization of intuitive space and was very clear about the suitability of group theory for reformulating Helmholtz's argument in mathematical terms. Cassirer based his reading of Helmholtz in an article of 1944 on "The Concept of Group and the Theory of Perception," on the following consideration:

> Modern geometry endeavors to attain progressively to more and more fundamental strata of spatial determination. The depth of these strata depends upon the comprehensiveness of the concept of group; it is proportional to the strictness of the conditions that must be satisfied by the invariance that is a universal postulate with respect to geometrical entities. Thus the objective truth and structure of space cannot be apprehended at a single glance, but have to be *progressively* discovered and established. If geometrical thought is to achieve this discovery, the conceptual means that it employs must become more and more universal. (Cassirer 1944, p.30)

Despite the fact that Cassirer considered Helmholtz's attempt to find a common measure between mathematical thought and perception unsuccessful, Cassirer believed that the concept of group offered a possible solution to the problem of mediating between different levels of knowledge, insofar as it led to progressive generalizations regarding the structure of space. In this sense, Cassirer captured an important aspect of Helmholtz's approach. As mentioned in Chap. 4, Cassirer agreed with Helmholtz that there is both a bottom-up direction – from experience to mathematical concepts – and a top-down direction – from mathematical structures to physical interpretations – in the theory of measurement. Cassirer's remark about Helmholtz did not receive much attention in the literature. However, it offered a plausible response to Schlick's objection by clearly distinguishing the group-theoretical analysis of the problem of space from a purely formal account of spatial notions.

Furthermore, Cassirer and Helmholtz agreed in the conclusion that, owing to a synthesis of both directions in the inquiry into the principles of knowledge, the possibility of further generalizations must be left open and scientific truths cannot be fixed once and for all.

In *Substance and Function* (1910), Cassirer drew such a conclusion from his comparison between Klein's geometry and the method of transcendental philosophy. Cassirer called his philosophical project a universal invariant theory of experience to express the fact that there is also an inductive aspect in the search for the conditions of knowledge. He concluded:

> The goal of critical analysis would be reached, if we succeeded in isolating in this way the ultimate common element of all possible forms of scientific experience; i.e., if we succeeded in conceptually defining those moments, which persist in the advance from theory to theory because they are the conditions of any theory. At no given stage of knowledge can this goal be perfectly achieved; nevertheless it remains as a demand, and prescribes a fixed direction to the continuous unfolding and evolution of the systems of experience. (Cassirer 1910, p.269)

By 1944, it became clear to Cassirer that his stance enabled him to integrate a Kantian perspective on knowledge with the demands of an empiricist theory of knowledge such as Helmholtz's.

6.3   The Relationship between Geometry and Experience: Poincaré and the Neo-Kantians

The leading idea of Poincaré's inquiries into the foundations of geometry was that the concept of group provided the highest principle of mathematics. In a series of writings from the 1890s, Poincaré used the concept of group to clarify the relationship between geometry and experience. On the one hand, Poincaré rejected Helmholtz's view that geometrical axioms have empirical origins and pointed out the role of stipulations in the formation of spatial notions. On the other hand, Poincaré excluded the case that geometry provides us with conditions of experience in Kant's sense because of the possibility of revising stipulations about the form of space under empirical circumstances. For Poincaré, only the "latent idea of a certain number of groups" (i.e., of Lie groups) "pre-exists in our mind." Which of the subgroups should we take to characterize a point in space? Poincaré's answer is that experiment "guides" us "by showing us what choice adapts itself best to the properties of our body" (Poincaré 1902, p.88).

There have been different interpretations of Poincaré's philosophy of mathematics both at that time and in the literature. Folina (1992) and Crocco (2004) paid special attention to the Kantian aspects of Poincaré's considerations. They showed in different ways that for Poincaré, as well as for Kant, infinite domains can only be constructed according to a priori laws.[15] By contrast, Gray (2013) argued that Poincaré's way of thinking more clearly emerges from the interactions between his philosophical views and his contributions to different scientific disciplines, including the theory of functions, topology and mathematical physics. Gray agrees with other authors that Poincaré was guided in his mathematical and scientific work by his philosophical views, as is often the case in times when the nature of mathematics changes profoundly. However, Gray emphasizes that what holds Poincaré's life work together to a remarkable degree is "the tight hold his epistemology had on his ideas of ontology, on what constitutes an answer to a mathematical or physical problem (and what does not), and on what the practice of mathematics and physics consists of" (Gray 2013, p.7). Gray showed that Poincaré's critical attitude towards the set-theoretical foundations of mathematics that were coming out of Germany depended largely on these motivations, rather than on Poincaré's reliance on a full-blown Kantian philosophy of mathematics.

It may be helpful to notice that Gray's account differs from Folina's in particular. According to Crocco (2004), the parallel with Kant depends on the constitutive role of mathematics in the theory of knowledge and might shed light on the relation between intuition, construction and convention in Poincaré's epistemology. On the

---

[15] However, there are important differences between Folina's (1992) and Crocco's (2004) readings. Whereas Folina emphasizes the parallel with Kant regarding for the idea that arithmetic and geometry have different kinds of synthetic judgments a priori (i.e., formulas and axioms, respectively), Crocco points out the deeper connection between space and time in Poincaré's conception of synthesis a priori. The latter account has the advantage of offering a unitary perspective on Poincaré's epistemology.

other hand, Poincaré clearly distanced himself from Kant regarding the philosophy of geometry and in his commitment to the problem of measurement. This led Poincaré to conceive of space as more closely related to time than is the case in Kant's *Critique of Pure Reason*. Furthermore, it is noteworthy that Poincaré's relationship to Kant was mediated by French neo-Kantianism. The interrelations with German neo-Kantianism remain largely unexplored and very much worth being taken into account, as the idea of a renewal of the Kantian philosophy of space arguably depended on similar motivations. It goes beyond the scope of the present study to undertake such a comparison. The following section offers a reconstruction of Poincaré's argument for the conventionality of geometry, with a special focus on his remarks on Kant. This will introduce us to the reception of Poincaré's argument by German neo-Kantians. Poincaré's objection to Helmholtz lent plausibility to some of the earlier arguments against geometrical empiricism by Cohen and Riehl. At the same time, Poincaré's argument for the conventionality of geometry motivated neo-Kantians, such as Bruno Bauch, Richard Hönigswald and Cassirer, to reconsider Kant's view that mathematics is synthetic as an alternative to both empiricism and conventionalism.

6.3.1 The Law of Homogeneity and the Creation of the Mathematical Continuum

The most detailed presentation of Poincaré's argument for the conventional character of geometry is found in his article of 1898b "On the Foundations of Geometry," which was incorporated into the first part of his 1902 collection of his earlier wrings on that subject under the title *Science and Hypothesis*.[16] Poincaré's argument entails compelling objections against Helmholtz. Nevertheless, it is noteworthy that Poincaré's reception of Helmholtz played an important role in the development of his view. Poincaré agreed with Helmholtz that sensations per se do not possess a spatial character. The representation of space depends not so much on single sensations, as on movement and association of ideas. Therefore, Poincaré distanced himself from Kant's definition of space as the form of outer intuition a priori. According to Poincaré, Kant's definition caused confusion between sensible and geometric space. In order to distinguish clearly between the meanings of the concept of space, Poincaré called sensible space representative and posed the problem of clarifying the conditions for the formation of geometrical concepts. He identified geometric

---

[16] Despite the fact there were developments in Poincaré's thought throughout the 1890s, his choice to collect his writings on the foundations of mathematics in a comprehensive exposition suggests that his argument for the conventionality of geometry should be considered as a whole (see Ben-Menahem 2006, p.40). The proposed reconstruction of the argument is mainly based on Poincaré (1902), as this appears to have been most popular in Germany and was translated into German by Ferdinand Lindemann in 1904. References to earlier writings are given in those cases in which there were significant changes.

space as a form of the understanding, because its definition requires an active intervention of the mind and the creation of the mathematical continuum.

Poincaré's goal was to provide a natural basis for the introduction of the concept of group. According to Poincaré, the formation of geometrical concepts is occasioned by the observation of external changes. These are distinguished from internal changes, because they are involuntary and they do not presuppose muscular sensations. External changes are divided into changes of state and changes of position: only after the latter kind of changes, can the initial situation be reproduced by internal changes. In such a case, Poincaré said that external changes are subject to compensation, which is a fundamental condition for the development of geometry. The said circumstances lead us to assume that both the parts of the objects that undergo external changes and our sense organs can be brought back to the initial disposition. Poincaré recognized that solid bodies are indispensable for the formation of geometrical concepts, because they are those that better satisfy this condition (Poincaré 1902, p.61).

Helmholtz's influence is apparent in Poincaré's emphasis on the role of experiences with solid bodies in the formation of geometrical concepts. At the same time, Poincaré made it clear that solid bodies cannot be identified as rigid bodies in Helmholtz's sense. Poincaré pointed out that geometry is not grounded so much in a collection of facts, as in some rules for the interpretation of sense perception. More precisely, the starting point for the formation of geometrical concepts is the exact formulation of geometric laws. Helmholtz's free mobility of rigid bodies corresponds to the following formulation of the law of homogeneity:

> Suppose that by an external change we pass from the aggregate of impressions A to the aggregate B, and that then this change α is corrected by a correlative voluntary movement β, so that we are brought back to the aggregate A. Suppose now that another external change α' brings us again from the aggregate A to the aggregate B. Experiment then shows us that this change α', like the change α, may be corrected by a voluntary correlative movement β', and that this movement β' corresponds to the same muscular sensations as the movement β which corrected α. (Poincaré 1902, pp.63–64)

As we saw in Sect. 4.2.3, Helmholtz's explanation for the repeatability of measurements presupposed the existence of rigid bodies. However, his theory of measurement seemed to face a circularity problem, as Helmholtz maintained that the objective meaning of quantitative notions depends on the repeatability of measurement operations. In order to avoid circularity, Poincaré acknowledged the ideal character of the law: Poincaré's compensation – as an operation of the mind – corresponds only approximately to actual experiences with solid bodies. In the case that experiment should not confirm our expectations regarding the possibility of compensation, it would follow that the phenomena under consideration are changes of state rather than changes of position. In the case of an approximate confirmation, it would be possible to hypothesize that some change of state occurred during displacement. In no case does experiment compel us to revise the law.

To sum up, representative space differs from geometric space because of its heterogeneity. Poincaré observed that visual space is not homogeneous, because not all the points on the retina play the same role in the formation of images. Since tactile space is more complicated than visual space, it differs even more widely from geometric space (Poincaré 1902, pp.52–55). The turning point is provided by the law of homogeneity and the remaining, defining conditions of the transformation group of external changes, including the demand that for every operation of the group, there is an inverse operation in the group. Poincaré's considerations about the formation of geometrical concepts enabled him to reformulate the problem of space in group-theoretical terms.

Before considering Poincaré's argument for the conventionality of geometry, it may be helpful to add some details about the related problem of formulating general conditions of measurement. Poincaré addressed the problem of defining suitable conditions for assigning measure to any magnitude in one of the papers incorporated in *Science and Hypothesis*, "The Mathematical Continuum" (1893). In order to adopt a general viewpoint, Poincaré assumed that the magnitudes under consideration must be incommensurable and considered the conditions for introducing the concept of measure. In this sense, he referred to Paul du Bois-Reymond's definition of mathematical magnitudes in *The General Theory of Function* (1882) and to Helmholtz's "Counting and Measuring from an Epistemological Viewpoint" (1887). Poincaré (1893, p.33) considered Helmholtz's essay an excellent treatment of that subject. At the same time, Poincaré proposed a new explanation for the assumption of continuity. For Poincaré, the assumption was motivated by the need to eliminate a contradiction of sense perception. Common sense suggests that a magnitude is continuous if sensations caused by it are such that it is possible to distinguish two sensations from each other, but not from a third sensation. Poincaré expressed this fact as follows:

$$A = B, B = C, A < C;$$ where the capital letters indicate sensations.

This formula was Poincaré's expression for the physical continuum. In order to eliminate the contradiction, one assumes Euclid's first axiom that two magnitudes that are distinct from a third magnitude are distinct from each other. The axiom enables us to say that, in the case of a physical continuum, A is slightly different from B, as B from C, even though the difference is not perceptible to our senses. Therefore, Poincaré maintained that the contradiction of the physical continuum compels us to "invent" the mathematical continuum (Poincaré 1902, p.22).

Poincaré distinguished two stages in the creation of the mathematical continuum. In the first stage, one has to intercalate new terms between the terms distinguished already. The condition for this operation to generate a continuum of the first order is that this operation can be pursued indefinitely. In order to justify this requirement, Poincaré made the following comparison between the generation of a continuum of the first order and the generation of the series of integer numbers:

> Here everything takes place just as in the series of the integers. We have the faculty of conceiving that a unit may be added to a collection of units. Thanks to experiment, we have had the opportunity of exercising this faculty and are conscious of it; but from this fact we feel that our power is unlimited, and that we can count indefinitely, although we have never had to count more than a finite number of objects. In the same way, as soon as we have

intercalated terms between two consecutive terms of a series, we feel that this operation may be continued without limit, and that, so to speak, there is no intrinsic reason for stopping. (Poincaré 1902, pp.24–25)

On the one hand, recurrent reasoning differs from analytic inferences, because the operation required entails an infinite number of inferences. On the other hand, experiment does not suffice either, because of the finite number of the terms presently under consideration. Poincaré attributed the intermediate status of a priori synthetic intuition to recurrent reasoning or proof by recurrence and defined it as "the power of the mind which knows it can conceive of the indefinite repetition of the same act when the act is once possible. The mind has a direct intuition of this power. Experiment can only be for it an opportunity for using it, and thereby of becoming conscious of it" (Poincaré 1902, p.13). Poincaré seems to adopt a Kantian view of mathematics in opposition to logicism. Notice, however, that Poincaré's notion of intuition differs considerably from Kant's. Goldfarb (1988, p.63) contrasted these notions as follows: whereas for Kant sensibility provides us with the necessary conditions for the constitution of empirical objects, Poincaré seems to refer to the fact that some mathematical truths are intuitive and do not require further justification. Despite Poincaré's use of Kant's terminology, his psychological understanding of the notion of intuition would prevent him from reconsidering the structure of Kant's argument for the applicability of mathematics.[17] Folina, among others, pointed out that Poincaré's notion of intuition cannot be reduced to a psychological concept, insofar as he identified the foundations of mathematics as laws instead of facts.[18] In this sense, intuition in Poincaré's philosophy of mathematics plays some normative role and it has a constitutive function in the construction of infinite domains. This interpretation offers a plausible account of Poincaré's intentions in his use of Kant's notions. However, there is another aspect of his notion of intuition that can hardly be traced back to Kant: not only does experience provide the opportunity of exercising the faculty of intuition for Poincaré, but it enables us to becoming conscious of it by using it.[19]

Furthermore, Poincaré's construction in intuition requires conventions for its completion. The series generated as described above corresponds to the set of rational numbers and is characterized by density. Given the difference between density and continuity, the creation of a continuum of the second order requires us to intercalate new terms according to the laws of irrational numbers. In order to give an intuitive representation of this stage in the creation of the mathematical continuum, Poincaré introduced the concepts of a line without breadth and of a point without size. The line can be imagined as the limit towards which a band tends that is getting thinner and thinner, and the point as the limit towards which an area tends that is getting smaller and smaller. Now suppose that a line is divided into two half lines. Each of these half lines will appear to us to be a band of a certain breath. These

---

[17] On the constitutive role of mathematics in Poincaré's epistemology, cf. Crocco (2004).

[18] See especially Folina (2006, p.288).

[19] For a discussion of this aspect of Poincaré's notion in contrast to Kant, see MacDougall (2010, p.140).

bands will fit closely together, because there must not be any interval between them. The common part will appear to us to be a point which will still remain if we imagine that the bands become thinner and thinner. The assumption of continuity corresponds to the intuitive truth that if a line is divided into two half lines, the common limit of these half lines is a point.

Poincaré contrasted this description of the creation of the mathematical continuum with Dedekind's definition of irrational numbers. Poincaré characterized the mathematical perspective on the continuum as follows: "Mathematicians do not study objects, but the relations between objects; to them it is a matter of indifference if these objects are replaced by others, provided that the relations do not change. Matter does not engage their attention, they are interested by form alone" (Poincaré 1902, p.20). Owing to such an approach, Dedekind introduced irrational numbers as symbols that stand not so much for incommensurable quantities, as for well-defined partitions of rational numbers. In other words, Dedekind characterized the mathematical continuum as a system of symbols. Poincaré's objection was that Dedekind did not account for the origin of these symbols. Therefore, Poincaré distinguished between the mathematical and the physical continuum. According to him, the formation of the concept of number presupposes the infinite divisibility of some physical continuum. He agreed with Dedekind that the notion of continuity is a creation of the mind. At the same time, Poincaré emphasized that it is experiment that provides the opportunity.

A related problem regards Poincaré's worries about the set-theoretic reconstruction of the continuum. These are explicit in Poincaré's later papers on the relationship between mathematics and logic.[20] Since Dedekind's definition of irrational numbers provides us with an example of non-predicative definition,[21] it may be supposed that geometrical intuition, for Poincaré, was the only mean to fill the gaps between classes of rational elements (see Folina 1992, pp.122–127). This solution is not necessarily Kantian if one considers Cassirer's view that the transcendental inquiry does not impose any constraints upon the set-theoretic reconstruction of the continuum.[22] Cassirer explored more the possibility of reformulating Kant's

---

[20] Poincaré (1906, p.307) referred to Russell's objection that the definition of irrational numbers as upper limits of sets of real numbers is non-predicative. Intuitively, the definition of two objects A and A' is non-predicative, if A occurs in the definition of A', and vice versa. In formal terms, this kind of definition can be derived from the schema of the axioms of comprehension: $\exists Y \forall X (X \in Y \leftrightarrow \varphi (X))$. Russell and Poincaré believed that the use of non-predicative definitions caused the antinomies of set theory. The problem lies in the fact that these definitions seem to contradict some conditions for quantification. Poincaré indicated two possible ways to avoid antinomies. The domain under consideration can be restricted to those elements that could be specified independently of the quantification. Alternatively, in the case of indefinite domains, one can require that the classification of proper subsets of the set remain unvaried under quantification (see Heinzmann 1985, pp.72–73).

[21] For an illustration of this example, see Heinzmann 1985, p.42.

[22] Regarding Poincaré's objection, it must be noticed that the later axiomatization of set theory offered a possible solution to the problems concerning non-predicative definitions. In order to avoid antinomies, it suffices to integrate the schema of the axioms of comprehension with the schema of the axioms of separation. Given a set A, the elements that satisfy the property φ are separated and reunited in Y, which is a subset of A, according to the following schema: $\exists Y \forall X (X \in Y \leftrightarrow X \in A \land \varphi (X))$. See Heinzmann 1985, pp.10–11.

argument for the synthetic character of mathematics in terms of Dedekind's and Klein's logicism.

We have already noticed that Poincaré introduced the problem of the origin of symbols by referring to Helmholtz's "Counting and Measuring." Notice, however, that Poincaré's solution differed considerably from Helmholtz's.[23] As we saw in Chap. 4, Helmholtz showed that the laws of addition hold true for ordinal numbers by definition and proved that the same laws can be extended to cardinal numbers. His theorem provided the basis for Helmholtz's formulation of the conditions for the application of the same laws to physical magnitudes. By contrast, Poincaré introduced the mathematical continuum in ordinal terms. The definition of arithmetical operations requires further, conventional assumptions, because the consideration of equidistant terms or divisibility into equal parts presupposes the choice of some unit of measure. This way of proceeding corresponds to the fact that Poincaré characterized space via the concept of group, independently of the reference to numerical models. The concept of group provided a geometrical formulation of the law of homogeneity and of Euclid's first axiom, which Helmholtz considered to be the first law of addition or the formal definition of equality. According to Poincaré, Euclid's first axiom provided a definition of equality, because of the possibility of transforming one geometrical figure into another. Given a transformation α of a figure A into a figure B and another transformation of the same kind, β, of B into C, the axiom states that A equals C if the product $\alpha\beta$ – which is a transformation of A into C – belongs to the same kind as α and β when considered separately (Poincaré 1899, p.257). This formulation of the axiom corresponds to another of the defining conditions for the transformations under consideration to form a group, namely, the closure of a set of elements under a fundamental operation.

To borrow a phrase from Heinzmann (1985), Poincaré's goal was to reconcile analysis with intuition. He combined internal consistency – as a formal criterion of definitions – with consistency relative to intuitive constraints on experience. Given the empirical origin of intuitive truths, Poincaré avoided Kant's notion of pure intuition. Nevertheless, his account of the formation of geometrical concepts retains the structure of Kant's metaphysical exposition of the concept of space, insofar as Poincaré's geometric space presupposes the more fundamental concept of a three-dimensional continuum. Poincaré defined it as "a sort of non-measurable magnitude analogous to magnitudes concerning which we may say that they have grown larger or smaller, but not that they have become twice or three times as large" (Poincaré 1898b, 40). Therefore, Poincaré criticized Helmholtz for not having being able to account for the origin of spatial notions: Helmholtz's way of proceeding clearly presupposes his knowledge of analytic geometry (p.40). These considerations were Poincaré's starting point for the development of geometric conventionalism. His argument for the conventionality of geometry presupposed the view that conventional assumptions are required in order for the concept of measure to apply to the more general concept of a three-dimensional continuum.

---

[23] It is revealing that Poincaré omitted his reference to Helmholtz in the revised version of the paper on the mathematical continuum as found in *Science and Hypothesis*.

6.3.2   Poincaré's Argument for the Conventionality of Geometry

Poincaré presented geometrical conventionalism as a plausible alternative to both Kantianism and empiricism about the origin and meaning of geometrical axioms. In order to understand the notion of axioms better, Poincaré started by clarifying the relationship between axioms and definitions. He noticed that mathematical objects differ from empirical objects because existence in mathematics only depends on the consistency of the definition of the supposed entity both in itself and with the propositions previously admitted. Therefore, the meaning of mathematical definitions cannot depend on the existence of empirical objects. Take, for example, the following definition of equality:

> Two figures are equal when they can be superposed. To superpose them, one of them must be displaced until it coincides with the other. But how must it be displaced? If we asked that question, no doubt we should be told that it ought to be done without deforming it, and as an invariable solid is displaced. The vicious circle would then be evident. (Poincaré 1902, pp.44–45)

Poincaré bore in mind Helmholtz's definition of spatial notions by the use of the free mobility of rigid bodies. Because of circularity, this definition has no proper meaning and it appears clear to us only because we are accustomed to the properties of natural solids, which do not differ much from those of ideal solids.[24] However, Poincaré pointed out, for example, that the same definition would appear to be meaningless to beings living in a world in which there were only fluids. Therefore, Poincaré maintained that the meaning of such definitions as that of the equality of two figures depends on the explicit formulation of the mathematical properties of the operations involved in the comparison. The example shows that mathematical definitions differ from empirical definitions because they imply some axiom, namely, some assumption that does not depend on experience.

As in the case of the mathematical continuum, postulates can be occasioned by the need to solve a contradiction of sense perception. In this sense, Poincaré acknowledged the importance of experiences with solid bodies for the formation of geometrical concepts. But it is only the formulation of the axioms that provide definitions with their meaning. The other way around, axioms are only "definitions in

---

[24] The reference to Helmholtz remains implicit, but it emerges clearly from Poincaré's example. We turn back to Poincaré's remark in the next chapter, as it appears to have influenced Schlick's objection against Helmholtz's definition of congruence: "This definition reduces congruence (the equality of two extents) to the coincidence of point pairs in rigid bodies 'with the same point pair fixed in space', and thus presupposes that 'points in space' can be distinguished and held fixed. This presupposition was explicitly made by Helmholtz [...], but for this he had to presuppose in turn the existence of 'certain spatial structures which are regarded as unchangeable and fixed'. Unchangeability and fixity (the term 'rigidity' is more usual nowadays) cannot for its own part again be specified with the help of that definition of congruence, for one would otherwise clearly go round in a circle. For this reason the definition seems not to be logically satisfactory" (Schlick in Helmholtz 1921, pp.192, 31, note 31). Schlick goes on to argue for Poincaré's geometrical conventionalism as the only way to escape the circle.

disguise" (Poincaré 1902, p.50), because their formulation is free and it is only limited by the necessity of avoiding contradiction. On the one hand, postulates remain true even if the experimental laws which have determined their adoption are only approximate. On the other hand, Poincaré rejected the view that geometrical axioms are synthetic a priori judgments, because of the possibility of choosing between equivalent hypotheses.

In order to highlight this point, Poincaré (1902, p.41) enunciated Lie's theorem as follows. Suppose that: (1) Space has n dimensions, (2) the movement of an invariable figure is possible and (3) p conditions are necessary to determine the position of this figure in space. The theorem states that the number of geometries compatible with these assumptions is limited. Furthermore, if n is given, an upper limit can be assigned to p: the choice is limited to the three classical cases of manifolds of constant curvature. As the essay from 1898 makes clear, Poincaré believed that the group-theoretical approach provides us with a criterion for making a choice as well. An invariant subgroup of which all the displacements are interchangeable, namely, the translation group, exists only in Euclidean geometry. "It is this that determines our choice in favor of the geometry of Euclid, because the group that corresponds to the geometry of Lobatchévski does not contain such an invariant sub-group" (Poincaré1898b, p.21). Therefore, he argued that the Euclidean group is the simplest. It is preferable because, "all other things being equal, the simplest is the most convenient" (p.42).[25]

In 1902, Poincaré summarized his argument by saying that the choice of Euclidean geometry depends on two reasons. The first is that the formulas of Euclidean geometry are simpler than those of non-Euclidean geometry, just as a polynomial of the first degree is simpler than a polynomial of the second degree. The second is that Euclidean geometry agrees sufficiently with the properties of natural solids. The first criterion may seem to be merely formal. Notice, however, that Poincaré had already limited the choice to those manifolds whose geometry was supposed to provide conditions of measurement. However, Poincaré did not take into consideration the hypothesis of a manifold of variable curvature considered by Riemann. For Poincaré, Riemannian geometries, insofar as they are incompatible with the movement of a variable figure, can never be but purely analytical (Poincaré 1902, pp.47–48). At that time, Poincaré could not know about Einstein's use of Riemann's geometry in general relativity. We return to this point in the next chapter. For now, it is noteworthy that Poincaré's argument referred not so much to

---

[25] As pointed out by Gray (2013, pp.48–57), the earlier formulation of the argument in Poncaré (1898b) sheds light on the priority of the concept of group over that of space in Poincaré's philosophy of geometry. The same idea lies behind Poincaré's definition of distance in terms of measurement. I argue in what follows that it was this aspect of Poincaré's approach that especially influenced Cassirer in his reconstruction of the problem of space from Helmholtz to Poincaré. The leading idea of Cassirer's account is that the construction of the concept of space required the mathematicians to take a step back and look at the more general concept of group.

logically equivalent geometries, as to geometries that could be considered equivalent representations of physical reality.[26]

Poincaré's commitment to the problems concerning measurement is apparent in the second of the said criteria. The claim that Euclidean geometry agrees sufficiently with empirical laws about solid bodies followed from Poincaré's description of a supposedly non-Euclidean world along the lines of his 1882 model of non-Euclidean geometry. Imagine that both measured objects and our standards of measurement undergo some dilatation or contraction in proportion to their distance from the center of a world enclosed in a sphere. Poincaré supposed that these changes depend on a non-uniform distribution of temperature: this is greatest at the center, and gradually decreases as one moves towards the circumference of the sphere, where it is absolute zero. The inhabitant of the non-Euclidean world may adopt Bolyai-Lobachevsky geometry. The thought experiment shows that the choice among geometrical hypotheses is not determined a priori. Nevertheless, Poincaré emphasized that the choice is not determined by experiment either. Assuming that inhabitants of our world would find themselves in the non-Euclidean world, Poincaré believed that they would base their measurements on Euclidean geometry, rather than change their habits. Poincaré's conclusion is that geometrical axioms are neither synthetic a priori judgments nor empirical judgments. They are conventions, because one geometry cannot be "more true" than another; it can only be more convenient according to the said criteria (Poincaré 1902, pp.64–58).

Poincaré's explanation of the formation of spatial notions via the concept of group played a fundamental role in his argument in favor of geometric conventionalism, as it enabled him to present the opposition between Kantianism and empiricism as an impasse, which could be overcome from a more comprehensive viewpoint. To conclude his considerations about the non-Euclidean world, Poincaré wrote:

> The object of geometry is the study of a particular "group"; but the general concept of group pre-exists in our minds, at least potentially. It is imposed on us not as a form of our sensitiveness, but as a form of our understanding; only, from among all possible groups, we must choose one that will be the standard, so to speak, to which we shall refer natural phenomena.

---

[26]This again emerges most clearly in Poincaré's 1898 formulation of the argument: "Have we the right to say that the choice between geometries is imposed by reason, and, for example, that the Euclidean geometry is alone true because the principle of the relativity of magnitudes is inevitably imposed upon our mind? [...] Unquestionably reason has its preferences, but these preferences have not this imperative character. It has its preferences for the simplest because, all other things being equal, the simplest is the most convenient. Thus our experiences would be equally compatible with the geometry of Euclid and with a geometry of Lobachévski which supposed the curvature of space to be very small. We choose the geometry of Euclid because it is the simplest. If our experiences should be considerably different, the geometry of Euclid would no longer suffice to represent them conveniently, and we should choose a different geometry" (Poincaré 1898b, p.42). On Poincaré's commitment to the problems concerning measurement, see also Hölder (1924, p.400). On the physical equivalence of the geometries considered by Poincaré, see also Torretti (1978, pp.327, 336) and Ben-Menahem (2006, p.58).

Experiment guides us in this choice, which it does not impose on us. It tells us not what is the truest, but what is the most convenient geometry. (Poincaré 1902, pp.70–71)

The characterization of the general concept of group as a form of the understanding enabled Poincaré to account for the possibility of revising the principles of measurement according to the group-theoretical analysis of space: geometrical hypotheses can be compared in terms of transformation groups` and of their relations to one another. At the same time, conventionalism provided a more plausible account than empiricism, because the introduction of the concept of measure in the continuum and the definition of metrical concept always requires conventions.

6.3.3   The Reception of Poincaré's Argument in Neo-Kantianism: Bruno Bauch and Ernst Cassirer

Poincaré's argument for the conventionality of geometry inaugurated a new phase in the debate about the foundations of geometry and it was influential in the later debate about the philosophical consequences of general relativity. However, it is worth noting that Poincaré's argument served different, even opposed purposes when it was received. On the one hand, Schlick used Poincaré's distinction between representative and geometric space to argue against the Kantian theory of space. On the other hand, the reception of Poincaré in neo-Kantianism was mainly due to his objections to Helmholtz's geometrical empiricism. Neo-Kantians such as Cassirer, Natorp, Bauch, and Hönigswald read Poincaré's philosophy of geometry in connection with the earlier discussions of geometrical empiricism by Cohen and Riehl. We deal with Cassirer's discussion with Schlick on the relationship between geometry and physics after general relativity in the next chapter. The present section offers a brief account of the reception of Poincaré's argument by Bauch (1907) and Cassirer (1910). The comparison between Bauch and Cassirer shows that, even in neo-Kantianism, the interpretation of Poincaré's philosophy of geometry was controversial and led to different consequences regarding the legacy of the Kantian theory of space.

Bruno Bauch dealt with the status of geometrical axioms in a paper entitled "Experience and Geometry Considered in Their Epistemological Relation" and published in 1907 in *Kant-Studien*. Bauch used Poincaré's interpretation of non-Euclidean geometry to argue for the aprioricity of Euclidean geometry: notwithstanding the fact that the "translation," as Poincaré (1902, p.43) put it, corresponds to the theorems of non-Euclidean geometry, Euclidean geometry for Bauch provides us with the "fundamental language" (Bauch 1907, p.227). As we saw in the third chapter, Riehl drew a similar conclusion from the fact that the consistency of non-Euclidean systems can be proved only relative to Euclidean models. Bauch seems to refer to the same fact. Furthermore, he relied on the distinction between formal and transcendental logic. Whereas from the standpoint of formal logic, all non-self-contradictory systems of geometry are equivalent, Bauch pointed out that they differ regarding their relation to experience. Poincaré's argument shows that all geometries are independent of experience. Nevertheless, Bauch maintained that from the standpoint of transcendental logic, there can be only one geometry that lies at the foundation of experience, because knowledge is possible only with regard to human experience. Bauch contrasted the viewpoint of mathematics with that of the theory of knowledge as follows:

> No philosopher who followed geometrical speculations with some interest in and understanding of logic will deny the value of free geometry, as well as that of free mathematics in general. But, as a theoretician of knowledge, he is mainly interested in that branch of geometry, whose extent at least does not cross the limits of a modest inhabitant of the earth. So we assist to the curious fact that the theory of knowledge, despite its bad reputation of being speculative, cannot compete in speculative height with the highest geometrical speculations. Compared with geometry, the theory of knowledge keeps its feet on the ground. (Bauch 1907, p.234)

Bauch's "modesty" enabled him to maintain that the foundational role of Euclidean geometry for experience is compatible with the freedom of mathematics in the speculation about non-Euclidean spaces. Since his argument rules out other geometrical hypotheses in the representation of actual space, however, Bauch is committed to the less modest conclusion that the theory of knowledge provides us with immutable principles of empirical knowledge.

Richard Hönigswald agreed with Bauch and presented a similar argument at the third International Congress of Philosophy, which took place in Heidelberg in 1908. In a paper entitled "On the Difference and the Relationship between the Logical and the Knowledge-Theoretical Aspects of the Critical Problem of Geometry," Hönigswald summarized the argument by saying that "geometries differ from each other only in their epistemological relation to experience; for this is positive only in the case of Euclidean geometry" (Hönigswald 1909, p.891). According to Hönigswald, Euclidean geometry is singled out by its validity for experience, despite the fact that this geometry – similar to any other geometry – is independent of experience. This formulation of the argument is clearly reminiscent of Riehl's characterization of the transcendental inquiry as the proof that a priori concepts, despite their being subjective, determine objective features of the things we experience (Riehl 1904, p.267).[27]

---

[27] In 1904, Hönigswald wrote his Dissertation *On Hume's Doctrine of the Reality of the External World* at the University of Halle under the supervision of Riehl. Riehl's influence is also apparent in Hönigswald's Habilitation Thesis, *Contributions to the Theory of Knowledge and Methodology* (1906). In my opinion, both Hönigswald's paper on geometry of 1908 and his essay *On the Discussion about the Foundations of Mathematics* (1912) show the importance of Riehl's teaching for the development of Hönigswald's philosophy of mathematics and of science, even though, after 1906, Hönigswald distanced himself from Riehl's views about the possibility of knowledge of things in themselves. Arguably, Riehl influenced Bauch as well, when they met in Halle between 1903 and 1904. On that occasion, Bauch collaborated with both Riehl and Hönigswald. Bauch's friendship with Hönigswald lasted until Bauch adhered to Nazism. In 1933, Hönigswald lost his professorship in Munich according to racist legislation. In 1939, he was deported to the Dachau concentration camp for 5 weeks and was only freed following international protests. He then emigrated to the United States (see Ollig 1979, pp.73–81; 88–93).

The problematic aspect of this use of Poincaré's argument is that Bauch overlooked an important premise: a choice between hypotheses, in Poincaré's sense, did not occur among all of the possibilities that can be considered from a formal-logical viewpoint. Poincaré restricted the range of the hypotheses considered to those geometries that can provide equivalent interpretations of empirical measurements. Cassirer's study of Klein's Erlangen Program enabled him to appreciate the group-theoretical approach to the problem of physical space better. The possibility of considering a variety of hypotheses depends on the definition of geometrical properties as relative invariants of transformation groups. Invariants can vary according to the group under consideration. At the same time, the group-theoretical treatment presupposes the ideal and, therefore, exact nature of geometrical knowledge: the meaning of geometrical concepts relative to a specific transformation group is univocal (Cassirer 1910, p.91).

Cassirer reformulated Poincaré's argument as follows. Owing to the relativity of the theory of invariants, Euclidean geometry is not a necessary presupposition of measurement. Nevertheless, the use of the concept of group in the classification of spatial movements provides us with a plausible account of the conditions for making a choice among hypotheses. According to Lie's theorem, both classical cases of non-Euclidean geometries are compatible with the assumption of Euclidean metric at the infinitesimal level. Poincaré made it clear that the criteria required for making a choice in such a case cannot be purely empirical. Even though the range of possibilities is restricted to the three cases of manifolds of constant curvature, geometrical hypotheses cannot be put to the test: given the approximate character of empirical measurements, the results of measurements are equally compatible with the geometric systems considered. Therefore, Cassirer adopted Poincaré's criteria of simplicity and convenience in the formulation of empirical predictions. However, Cassirer argued for the rationality of such criteria. Even Poincaré described simplicity not so much in psychological terms, as in terms of an objective feature of mathematical structures when compared to one another. Regarding the second criterion, Cassirer did not refer to Poincaré's considerations about the formation of the representation of space. According to Cassirer, it sufficed to point out the priority of homogeneity over heterogeneity in the formation of geometrical concepts: the systems of non-Euclidean geometry – insofar as they are in local agreement with Euclidean metric – are more complex than Euclidean geometry and require additional assumptions (Cassirer 1910, p.109).

To sum up, Cassirer did not exclude the possibility of giving an empirical interpretation of the more complex systems of non-Euclidean geometry (Cassirer 1910, p.111). Nevertheless, in order to account for the assumption of Euclidean geometry in the physical theory of his time, he argued for a combination between rational criteria and empirical evidence in the formulation of hypotheses. Therefore, Cassirer contrasted rational criteria with arbitrary factors and distanced himself from Poincaré's characterization of geometrical hypotheses as conventions. Since this consideration regards the principles of measurement, Cassirer's stance on Poincaré's terminology in *Substance and Function* is found not in the chapter about space and geometry, but in the subsequent chapter about the concepts of natural science. Cassirer wrote:

> The distinction between an "absolute" truth of being and a "relative" truth of scientific knowledge, the separation between what is necessary from the standpoint of our concepts and what is necessary in itself from the nature of the facts, signifies a metaphysical assumption, that must be tested in its right and validity before it can be used as a standard. The characterization of the ideal conceptual creations as "conventions" has thus at first only one intelligible meaning; it involves the recognition that thought does not proceed merely receptively and imitatively in them, but develops a characteristic and original spontaneity (*Selbsttätigkeit*). Yet this spontaneity is not unlimited and unrestrained; it is connected, although not with the individual perception, with the system of perceptions in their order and connection. It is true this order is never to be established in a single system of concepts, which excludes any choice, but it always leaves room for different possibilities of exposition; in so far as our intellectual construction is extended and takes up new elements into itself, it appears that it does not proceed according to caprice, but follows a certain law of progress. This law is the ultimate criterion of "objectivity;" for it shows us that the world-system of physics more and more excludes all the accidents of judgment, such as seem unavoidable from the standpoint of the individual observer, and discovers in their place that necessity that is universally the kernel of the concept of the object. (Cassirer 1910, p.187)[28]

Cassirer referred to Poincaré to support the view that a description of scientific laws as empirical generalizations from the comparison of particular phenomena would imply circularity: measurements confirm scientific laws insofar as laws have been presupposed in the formulation of hypothetical predictions. For the same reason, one can consider different hypotheses. In order to avoid circularity, Cassirer emphasized the constructive character of theoretical thinking. On the one hand, he avoided reference to absolute reality in the search for criteria of physical objectivity. On the other hand, he avoided arbitrariness in theory change, because the relative truth of scientific theories can be accounted for in terms of increasingly general validity.

Cassirer captured an important aspect of Poincaré's conception of scientific laws. In order to distance himself from the extreme version of conventionalism advocated by Édouard Le Roy, in the introduction to *The Value of Science* (1905), Poincaré wrote:

> Does the harmony the human intelligence thinks it discovers in nature exist outside of this intelligence? No, beyond doubt a reality completely independent of the mind which conceives it, sees or feels it, is an impossibility. A world as exterior as that, even if it existed, would for us be forever inaccessible. But what we call objective reality is, in the last analysis, what is common to many thinking beings, and could be common to all; this common part [...] can only be the harmony expressed by mathematical laws. (Poincaré 1913, p.209)

---

[28] Cassirer (1910, p.186) referred in particular to Poincaré's considerations regarding the measurement of time. As in the case of geometry, Poincaré maintained that one way to measure time cannot be "more true" than another; it can only be more convenient. He referred to the fact that the definition of time presupposes the laws of mechanics. Poincaré's requirement was that time be defined in such a way that the equations of mechanics result as simply as possible (Poincaré 1898a, p.6). The difference with geometry lies in the fact that Poincaré did not consider geometry a part of mechanics. However, this difference does not affect Cassirer's argument insofar as this regards the conditions of measurement in general.

Both Cassirer and Poincaré identified the objectivity of scientific theories with the regularity of nature expressed by mathematical laws. Such a view of objectivity does not presuppose the existence of a mind-independent structure of the world, because the fact that this structure would be inaccessible to us would make it impossible to claim for the truth of science. At the same time, Cassirer argued for the rationality of Poincaré's "conventions," because Cassirer's emphasis lay on the role of mathematics in the definition of objective reality across theory change. Objectivity from Cassirer's dynamical perspective of knowledge depends on the exclusion of subjective factors: insofar as the individual viewpoint does not affect the general validity of mathematical laws, the objectivity of physical theories can be conceptualized in terms of a univocal correlation between mathematical structures and the phenomena. In *Einstein's General Relativity* (1921), Cassirer described the correlation between the physical continuum and the mathematical as follows: "In so far as physics is an objectifying science working with the conceptual instruments of mathematics, the physical continuum is conceived by it as related to and exactly correlated with the mathematical continuum of pure numbers" (Cassirer 1921, p.453).

It is clear from this consideration that Cassirer, unlike Poincaré, did not attach much importance to the distinction between the topological continuum of intuition and the introduction of metrical concepts. Cassirer attributed a mediating role to mathematics in the relationship between different levels of knowledge, firstly, because metrical concepts are no less constructive than topological concepts. Secondly, the possibility of considering a variety of geometrical hypotheses does not rule out the truth of geometry in Cassirer's relativized sense, as exact determinateness of geometrical concepts relative to transformation groups. In Cassirer's view, "modern group-theory is far from denying the truth of any geometrical system; but it declares that no single system has a claim of definitiveness. Only the *totality* of possible geometrical systems is really definitive" (Cassirer 1945, p.194). Cassirer's remark suggests that the disagreement with Poincaré's geometrical conventionalism depends not so much on its consequences for the relationship between geometry and space, but on the conception of truth. Conventions for Poincaré are not necessarily arbitrary and do not contradict the truth of scientific theories.[29] However, he distinguishes sharply conventions from supposedly true propositions in order to make it clear that the choice between different conventions is free. Both Poincaré and Cassirer – insofar as their philosophy of geometry led to the same consequences regarding the possibility of revising the principles of measurement – belong to the tradition Coffa called semantic.[30] Cassirer's view differs from Poincaré's because, for Cassirer, freedom is a defining characteristic of reason.

---

[29] On the distinction between conventionality and arbitrariness, see Ben-Menahem (2006, p.54). Ben-Menahem (2006, Preface) interprets Poincaré's conventionalism as an attempt to disentangle the concepts of convention and truth. The goal of such an approach is to avoid the tendency to consider some theoretical assumptions (e.g., the principles of geometry) necessary when in fact they are free.

[30] Consider the following description of the idea of the semantic tradition: "Much of the most interesting philosophy of science developed in the past few decades has been inspired by the [...] idea: Many fundamental scientific principles are not necessarily thought – indeed, it takes great effort to

In order to highlight these points, the next section provides a brief account of Cassirer's view in 1910. I suggest that Cassirer's view goes back to his clarification of the concept of the symbol in the context of the debate about the origin of numerals.

6.4   Cassirer's View in 1910

As we saw in the previous chapters, Cohen was one of the first philosophers to call into question Helmholtz's psychological explanations concerning the origin of mathematical notions. It is worth noting, however, that Cohen, contrary to Poincaré, appreciated Helmholtz's approach precisely because of Helmholtz's use of analytic methods. Not only did the development of analytic geometry offer an argument for Cohen's revision of the Kantian theory of space by identifying space with a form of pure thought, but it also played an important role in Cohen's clarification of the relationship between the concept of number and that of magnitude. Cohen used Helmholtz's theory of measurement to support the view that numbers, owing to the principles of addition, have a constitutive function in the definition of physical magnitudes. Furthermore, Cohen agreed with Helmholtz that analytic geometry provides us with necessary presuppositions for the scientific treatment of spatial magnitudes. Cohen believed that Kant's arguments for the homogeneity of space as an a priori property of the same was confirmed by Helmholtz's idea of a physical geometry, insofar as aprioricity for Cohen depends not on independence of experience in a psychological sense, as on the role of the mathematical method in the interpretation of physical phenomena.

Cassirer borrowed from Cohen the conviction that the development of the deductive method in mathematics sheds light on the connection between mathematics and physics. On the one hand, mathematics offers the clearest example of the fact that the meaning of symbols depends on the conceptual connections established in the definition of mathematical structures. On the other hand, owing to the unifying power of mathematical reasoning, symbols are suitable for a variety of possible interpretations independently of the reference to specific intuitive contents. Therefore, the mathematical method plays a fundamental role in the abstraction from subjective factors which is required for physical objectivity. Cassirer's first example of this way of proceeding is the subordination of the concept of magnitude to that of order in the theory of numbers. He referred not so much to the idea of nonmeasurable magnitudes, as to Dedekind's definition of numbers as free creations of the human mind. In the case of natural numbers, this corresponds to the fact that, given an initial element, the series generated by indefinitely mapping a number into its successor coincides with the series of ordinal numbers. The consideration of the

---

develop the systems of knowledge that embody them; but their denial also seems oddly impossible – they need not be thought, but if they are thought at all, they must be thought as necessary" (Coffa 1991, p.55).

cardinal aspect depends on Dedekind's proof that the operations with the elements thus generated apply to numbers in general (see Dedekind 1901, pp.67–70). In the case of the definition of irrational numbers, the generative power of the concept of function emerges from the possibility of establishing a one-to-one correspondence between divisions of rational numbers and the set of real numbers. The extension of the original set does not presuppose the reference to such entities as intensive quantities, because it finds its justification in the idea of a univocal correlation[31] between series.

As pointed out by Ryckman (1991), Cassirer's reception of Helmholtz played a decisive role in the extension of the concept of series from pure mathematics to theoretical physics. In particular, Ryckman draws attention to Helmholtz's distinction between signs and images in the theory of local signs, which he sketches as follows: "While our sensations are not purely 'arbitrary' designations of the objects of the outer world in the sense that, e.g., *chien* and *Hund* are purely conventional designations of 'dog', they are 'mere designations' nonetheless" (Ryckman 1991, p.68). This distinction posed the problem of accounting for the meaning of sensations. According to Ryckman, Helmholtz introduced the idea of coordination as the mode of relation between sign and designatum in the following quote from Helmholtz's "Facts in Perception" (1878a, p.122): "The relation between the two is restricted to the fact that the same object, exerting an influence under the same circumstances, calls forth the same sign, whereas unlike signs always correspond to unlike influences" (Engl. Trans. in Ryckman 1991, p.68). The epistemological significance of this idea lies in the fact that it enables a characterization of the lawful character of natural processes. Helmholtz postulated on the presupposition of the regularity of nature that an unambiguous correspondence exists between series of sensations and causal relations in nature.

I believe that the previous chapters provided new evidence of Helmholtz's influence on Cassirer. One of Cassirer's motivations is found in Helmholtz's (1887) problem concerning the possibility of representing magnitudes by using numbers. The objective meaning of numerical representations for Helmholtz and Cassirer depends on the possibility of unambiguously interpreting the results of measurement, and, ultimately, on the postulate of the comprehensibility of nature.[32] Helmholtz's ordinal conception of number was especially relevant to Cassirer's epistemological

---

[31] The German term for "correlation" is *Zuordnung*, and, in that context, indicates a mathematical function. "Univocal correlation" or "coordination" in Cassirer's sense refers both to the one-to-one correspondence between series that occurs, for example, in Dedekind's definition of irrational numbers and to the general idea of functional dependence between different elements of a theory. On mathematical and epistemological uses of "Zuordnung" between the nineteenth century and the first decades of the twentieth century, see Ryckman (1991, pp.58–60). As pointed out by Ryckman, the idea of a univocal correlation occupies a central place in such different epistemologies as Cassirer's and Schlick's.

[32] As pointed out in Chap. 4, the comprehensibility of nature is presupposed in Helmholtz's physical interpretation of the homogeneity of the sum and the summands. The equivalent proposition for the composition of magnitudes to be considered in terms of arithmetical sum depends on the repeatability of measurements.

use of the notion of series. At the same time, Cassirer believed that the notion of symbol deserved a clarification according to more recent developments in mathematics. Bearing in mind Dedekind's characterization of the mathematical continuum independently of the example of the continuity of a line, Cassirer maintained that the guiding idea for the development of geometry from infinitesimal geometry to the group-theoretical classification of geometries was that of a transposition of spatial concepts into serial concepts (Cassirer 1910, p.73). In particular, Cassirer emphasized the importance of the concept of a projective metric as follows:

> The evolution of modern mathematics has approached the ideal, which Leibniz established for it, with growing consciousness and success. Within pure geometry, this is shown most clearly in the development of the general concept of space. The reduction of metrical relations to projective realizes the thought of Leibniz that, before space is defined as a *quantum*, it must be grasped in its original qualitative peculiarity as an "order of coexistence" (*ordre des coexistences possibles*). The chain of harmonic constructions, by which the points of projective space are generated, provides the structure of this order, which owes its value and intelligibility to the fact that it is not sensuously presented but is constructed by thought through a succession of relational structures. (Cassirer 1910, p.91)

Similar to Cohen (1896/1984, p.75) before him, Cassirer referred to Leibniz for the idea that spatial order presupposes an intellectual construction. In this sense, the reduction of metrical relations to projective relations corresponds to the exclusion of sense data and immediate intuition from the consideration of geometrical figures. It does not follow that metrical considerations are ruled out. In this regard, Cassirer recalled Leibniz's dispute with Clark, who advocated Newton's theory of absolute space and absolute time. One of Clark's objections against Leibniz's definitions of space as an order of coexistence and of time as an order of succession was that space and time are first of all quantities, which position and order are not. Leibniz replied that determinations of magnitude are also possible within pure determinations of order, insofar as the members in the series are distinguished and the distance between them can be conceptually defined. Therefore, Leibniz claimed: "Relative things have their magnitudes just as well as absolute things; thus, e.g., in mathematics, relations or proportions have magnitudes, which are measured by their logarithms; nevertheless they are and remain relations" (Leibniz 1904, pp.189–190). According to Cassirer, a similar remark applies to the projective definition of distance as the logarithm of a certain cross-ratio (Cassirer 1910, p.91, note).

Cassirer maintained that the transposition of geometrical concepts into serial concepts culminated in the idea of a general classification of geometries from the standpoint of group theory, because it becomes clear from that standpoint that spatial notions can be determined and varied only relative to groups of ideal operations. For Cassirer, this way of proceeding is justified by the fact that the most diverse forms of calculus belong to one logical type, and also agree in their "fruitfulness" for the problems of mathematical natural science (Cassirer 1910, p.99). Cassirer's idea is that the issue at stake in mathematics, as well as in physics, is not to point out the ultimate substantial constitution of magnitudes, but to find a logical viewpoint for their determination (p.100). This corresponds to the fact that such concepts as matter, atom and energy do not refer immediately to sense data. The detachment from sense perception is particularly evident in the case of the concept of energy: "What is given us are qualitative differences of sensation: of warm and cold, light and dark, sweet and bitter, but not numerical differences of quantities of work" (Cassirer 1910, p.189). Therefore, Cassirer maintained that the reference of sensations to such magnitudes and their mutual relations or the measurement of perception depends on a set of theoretical assumptions that might be subject to revision. He believed, for example, that the advance of energetics over classical mechanics lay in a clarification of the relation between number and magnitude. The same constitutive function which Cohen attributed to the concept of number in the definition of physical magnitudes emerges from the following remark:

> [E]nergism contains a motive from the beginning, which protects it more than any other physical view from the danger of an immediate hypostatization of abstract principles. Its fundamental thought, from an epistemological point of view, does not go back primarily to the concept of space, but to that of number. It is to numerical values and relations that the theoretical and experimental inquiry are alike directed, and in them consists the real kernel of the fundamental law. Number, however, can not be misunderstood as substance, unless we are to revert to the mysticism of Pythagoreanism, but it signifies merely a general point of view, by which we make the sensuous manifold unitary and uniform in conception. (Cassirer 1910, p.189)[33]

The first step in the objectification of the phenomena consists of the arrangement of the members of individual series according to a numerical scale. The fundamental law is the principle of the conservation of energy, and it presupposes a second step, because it provides us with a coordinating principle for the series thus obtained. Cassirer mentions, for example, the fact that the equivalence of motion and heat first presented itself as an empirical statement. The law of energy enables one to extend the connection to the totality of possible physical manifolds, as it directs us to correlate every member of a manifold with one and only one member of any other manifold (p.193).

To sum up, Cassirer agreed with Poincaré that the concept of group provides us both with a general principle of mathematics and a possible solution to the problem of mediating between geometry and experience. However, owing to his reliance on the logist tradition of Dedekind and Klein, on the one hand, and to Cohen's teaching, on the other, Cassirer advocated the aprioricity of the concept of group for a different reason. Whereas Poincaré's reconstruction of the origin of mathematical symbols led to a sharp distinction between the topological intuition of spatial continuum and the concept of measure, Cassirer looked for a general viewpoint for the classification of geometrical hypotheses. The concept of group presented itself as the most suitable candidate, not because of the assumption of nonmeasurable magnitudes as a starting point for the definition of space, but because it offered a clarification of the foundations of metrical geometry as suggested by Klein: metri-

---

[33] The reference to Cohen is implicit, but it emerges clearly from Cassirer's contrast between the epistemological approach to the problem of measurement and the mystical view of numbers as substances (see, e.g., Cohen 1888, p.272, already quoted in Chap. 4)

cal notions can be characterized and compared as relative invariants of transformation groups. In this sense, the group-theoretical analysis of space confirmed the general tendency to substitute concepts of substance with concepts of function, which for Cassirer, goes back to Dedekind's definition of numbers as free creations of the human mind. The connection with theoretical physics depends on the idea of a univocal correlation between mathematical structures and sense data as a condition of objectivity, and it is justified by the constructive role of mathematical symbols: symbols do not stand for extra-theoretical entities, but for determinate connections within a system of hypotheses.

The next chapter is devoted to the development of Cassirer's argument from *Substance and Function* (1910) to *Einstein's Theory of Relativity* (1921). The problem, after Einstein's use of Riemannian geometry and its development in the tensor calculus, was that even the generalized solution of the problem of space offered by Poincaré and advocated by Cassirer appeared to be restrictive. Cassirer's original argument deserved a substantial revision. Nevertheless, in 1921, Cassirer referred to his general perspective on measurement as formulated in 1910 to account for the physical reality of Riemann's hypotheses regarding manifolds of variable curvature, which had hitherto been considered purely mathematical speculations. By contrast, Schlick's interpretation of Poincaré's distinction between topological and metrical properties led him to separate geometric systems sharply as abstract structures from intuitive space, on the one hand, and from the empirical manifold of physical events, on the other.

References

Bauch, Bruno. 1907. Erfahrung und Geometrie in ihrem erkenntnistheoretischen Verhältnis. *Kant-Studien* 12: 213–235.

Ben-Menahem, Yemima. 2006. *Conventionalism*. Cambridge: Cambridge University Press.

Birkhoff, Garrett, and Mary Katherine Bennett. 1988. Felix Klein and his *Erlanger Programm*. In *History and philosophy of modern mathematics*, ed. William Aspray and Philip Kitcher, 144–176. Minneapolis: University of Minnesota Press.

Carnot, Lazare. 1803. *Géométrie de position*. Paris: Duprat.

Cassirer, Ernst. 1910. Substanzbegriff und Funktionsbegriff: Untersuchungen über die Grundfragen der Erkenntniskritik. Berlin: B. Cassirer. English edition in Cassirer, Ernst. 1923 *Substance and function and Einstein's theory of relativity*, ed. Marie Collins Swabey and William Curtis Swabey, 1–346. Chicago: Open Court.

Cassirer, Ernst. 1921. *Zur Einstein'schen Relativitätstheorie: Erkenntnistheoretische Betrachtungen*. Berlin: B. Cassirer. English edition in Cassirer, Ernst. 1923 *Substance and function and Einstein's theory of relativity*, ed. Marie Collins Swabey and William Curtis Swabey, 347–465. Chicago: Open Court.

Cassirer, Ernst. 1944. The concept of group and the theory of perception. *Philosophy and Phenomenological Research* 5: 1–36.

Cassirer, Ernst. 1945. The concept of group. In *Vorlesungen und Vorträge zu philosophischen Problemen der Wissenschaften 1907–1945*, ed. Jörg Fingerhut, Gerald Hartung, and Rüdiger Kramme, 183–203. Hamburg: Meiner, 2010.

Cassirer, Ernst. 1950. *The problem of knowledge: Philosophy, science, and history since Hegel*. Trans. William H. Woglom and Charles W. Hendel. New Haven: Yale University Press.

Cohen, Hermann. 1896/1984. Einleitung mit kritischem Nachtrag zur *Geschichte des Materialismus* von F. A. Lange. Repr. in *Werke*, ed. Helmut Holzhey, 5. Hildesheim: Olms.

Coffa, Alberto J. 1991. *The semantic tradition from Kant to Carnap: To the Vienna station.* Cambridge: Cambridge University Press.

Cohen, Hermann. 1888. Jubiläums-Betrachtungen. *Philosophische Monatshefte* 24: 257–291.

Couturat, Louis. 1898. Essai sur les fondements de la géométrie, revue critique de An essay on the foundations of geometry de B. Russell. *Revue de Métaphysique et de Morale* 6: 354–380.

Crocco, Gabriella. 2004. Intuition, construction et convention dans la théorie de la connaissance de Poincaré. *Philosophiques* 31: 151–177.

De Risi, Vincenzo. 2007. *Geometry and monadology: Leibniz's analysis situs and philosophy of space.* Basel: Birkhäuser.

Dedekind, Richard. 1901. *Essays on the theory of numbers.* Trans. Wooster Woodruff Beman. Chicago: Open Court.

du Bois-Reymond, Paul. 1882. *Die allgemeine Functionentheorie. Vol. 1: Metaphysik und Theorie der mathematischen Grundbegriffe: Grösse, Grenze, Argument und Function.* Tübingen: Laupp.

Folina, Janet. 1992. *Poincaré and the philosophy of mathematics.* Basingstoke: Macmillan.

Folina, Janet. 2006. Poincaré's circularity argument for mathematical intuition. In *The Kantian legacy in nineteenth-century science*, ed. Michael Friedman and Alfred Nordmann, 275–293. Cambridge, MA: The MIT Press.

Friedman, Michael. 1997. Helmholtz's *Zeichentheorie* and Schlick's *Allgemeine Erkenntnislehre*: Early logical empiricism and its nineteenth-century background. *Philosophical Topics* 25: 19–50.

Goldfarb, Warren. 1988. Poincaré against the logicists. In *History and philosophy of modern mathematics*, ed. William Aspray and Philip Kitcher, 61–81. Minneapolis: University of Minnesota Press.

Grassmann, Hermann. 1844. *Die lineale Ausdehnungslehre, ein neuer Zweig der Mathematik, dargestellt und durch Anwendungen auf die übrigen Zweige der Mathematik, wie auch die Statistik, Mechanik, die Lehre vom Magnetismus und die Krystallonomie erläutert.* Leipzig: Wigand.

Grassmann, Hermann. 1847. *Geometrische Analyse geknüpft an die von Leibniz erfundene geometrische Charakteristik.* Leipzig: Weidmann.

Gray, Jeremy J. 2008. *Plato's ghost: The modernist transformation of mathematics.* Princeton: Princeton University Press.

Gray, Jeremy J. 2013. *Henri Poincaré: A scientific biography.* Princeton: Princeton University Press.

Griffin, Nicholas. 1991. *Russell's idealist apprenticeship.* Oxford: Clarendon.

Hatfield, Gary. 1990. *The natural and the normative: Theories of spatial perception from Kant to Helmholtz.* Cambridge, MA: MIT Press.

Hawkins, Thomas. 1984. The *Erlanger Programm* of Felix Klein: Reflections on its place in the history of mathematics. *Historia Mathematica* 11: 442–470.

Heinzmann, Gerhard. 1985. *Entre intuition et analyse: Poincaré et le concept de prédicativité.* Paris: Blanchard.

Helmholtz, Hermann von. 1867. *Handbuch der physiologischen Optik.* Leipzig: Voss.

Helmholtz, Hermann von. 1868. Über die Tatsachen, die der Geometrie zugrunde liegen. In Helmholtz (1921): 38–55.

Helmholtz, Hermann von. 1870. Über den Ursprung und die Bedeutung der geometrischen Axiome. In Helmholtz (1921): 1–24.

Helmholtz, Hermann von. 1878a. Die Tatsachen in der Wahrnehmung. In Helmholtz (1921): 109–152.

von Helmholtz, Hermann. 1878b. The origin and meaning of geometrical axioms, Part 2. *Mind* 3: 212–225.

Helmholtz, Hermann von. 1887. Zählen und Messen, erkenntnistheoretisch betrachtet. In Helmholtz (1921): 70–97.

Helmholtz, Hermann von. 1921. *Schriften zur Erkenntnistheorie,* ed. Paul Hertz and Moritz Schlick. Berlin: Springer. English edition: Helmholtz, Hermann von. 1977. *Epistemological writings* (trans: Lowe, Malcom F., ed. Robert S. Cohen and Yehuda Elkana). Dordrecht: Reidel.

Hilbert, David. 1903. *Grundlagen der Geometrie,* 2nd ed. Leipzig: Teubner.

Hönigswald, Richard. 1906. *Beiträge zur Erkenntnistheorie und Methodenlehre.* Leipzig: Fock.

Hönigswald, Richard. 1909. Über den Unterschied und die Beziehungen der logischen und der erkenntnistheoretischen Elemente in dem kritischen Problem der Geometrie. In *Bericht über den III. Internationalen Kongress für Philosophie, 1 to 5 September 1908,* ed. Theodor Elsenhans, 887–893. Heidelberg: Winter.

Hönigswald, Richard. 1912. *Zum Streit über die Grundlagen der Mathematik.* Heidelberg: Winter.

Hölder, Otto. 1924. *Die mathematische Methode: Logisch erkenntnistheoretische Untersuchungen im Gebiete der Mathematik, Mechanik und Physik.* Berlin: Springer.

Hyder, David. 2009. *The determinate world: Kant and Helmholtz on the physical meaning of geometry.* Berlin: De Gruyter.

Jordan, Camille. 1870. *Traité des substitutions et des équations algébriques.* Paris: Gauthier-Villars.

Klein, Felix. 1871. Über die sogenannte Nicht-Euklidische Geometrie. *Mathematische Annalen* 4: 573–625.

Klein, Felix. 1872. *Vergleichende Betrachtungen über neuere geometrische Forschungen.* Erlangen: Deichert.

Klein, Felix. 1890. Zur Nicht-Euklidischen Geometrie. *Mathematische Annalen* 37: 544–572.

Klein, Felix. 1893. Vergleichende Betrachtungen über neuere geometrische Forschungen. 2nd ed. *Mathematische Annalen* 43: 63–100. English version: Klein, Felix. 1892. A comparative review of recent researches in geometry. Trans. Maella W. Haskell. *Bulletin of the New York Mathematical Society* 2 (1892–1893): 215–249.

Klein, Felix. 1898. Gutachten, betreffend den dritten Band der Theorie der Transformationsgruppen von S. Lie anlässlich der ersten Vertheilung des Lobatschewsky-Preises. *Mathematische Annalen* 50: 583–600.

Leibniz, Gottfried Wilhelm. 1904. *Hauptschriften zur Grundlegung der Philosophie,* vol. 1. Trans. Artur Buchenau. Ed. Ernst Cassirer. Leipzig: Dürr.

Lenoir, Timothy. 2006. Operationalizing Kant: Manifolds, models, and mathematics in Helmholtz's theories of perception. In *The Kantian legacy in nineteenth-century science,* ed. Michael Friedman and Alfred Nordmann, 141–210. Cambridge, MA: The MIT Press.

Lie, Sophus. 1893. *Theorie der Transformationsgruppen,* vol. 3. Leipzig: Teubner.

MacDougall, Margaret. 2010. Poincaréan intuition revisited: What can we learn from Kant and Parsons? *Studies in the History and Philosophy of Science* 41: 138–147.

Nabonnand, Philippe. 2000. La polémique entre Poincaré et Russell au sujet du statut des axiomes de la géométrie. *Revue d'histoire des mathématiques* 6: 219–269.

Ollig, Hans Ludwig. 1979. *Der Neukantianismus.* Stuttgart: Metzler.

Poincaré, Henri. 1882. Théorie des groupes Fuchsiens. *Acta Mathematica* 1: 1–62.

Poincaré, Henri. 1893. Le continu mathématique. *Revue de Métaphysique et de Morale* 1: 26–34.

Poincaré, Henri. 1898a. La mesure du temps. *Revue de Métaphysique et de Morale* 6: 1–13.

Poincaré, Henri. 1898b. On the foundations of geometry. *The Monist* 9: 1–43.

Poincaré, Henri. 1899. Des fondements de la géométrie: A propos d'un livre de M. Russell. *Revue de Métaphysique et de Morale* 7: 251–279.

Poincaré, Henri. 1902. *La science et l'hypothèse.* Paris: Flammarion. English Edition: Poincaré, Henri. 1905. *Science and hypothesis* (trans: Greenstreet, William John). London: Scott.

Poincaré, Henri. 1905. *La valeur de la science.* Flammarion, Paris. English edition in Poincaré (1913): 205–355.

Poincaré, Henri. 1906. Les mathématiques et la logique. *Revue de Métaphysique et de Morale* 14: 294–317.

Poincaré, Henri. 1913. *The Foundations of science: Science and hypothesis, the value of science, science and method.* Trans. George Bruce Halsted. New York: The Science Press.

Poincaré, Henri. 1997. In *Three supplementary essays on the discovery of Fuchsian functions*, ed. Jeremy J. Gray and Scott A. Walter. Berlin: Akademie-Verlag.

Pulte, Helmut. 2006. The space between Helmholtz and Einstein: Moritz Schlick on spatial intuition and the foundations of geometry. In *Interactions: Mathematics, physics and philosophy, 1860–1930*, ed. Vincent F. Hendricks, Klaus Frovin Jørgensen, Jesper Lützen, and Stig Andur Pedersen, 185–206. Dordrecht: Springer.

Riehl, Alois. 1904. Helmholtz in seinem Verhältnis zu Kant. *Kant-Studien* 9: 260–285.

Riemann, Bernhard. 1876. Fragment aus der Analysis Situs. In *Gesammelte mathematische Werke und wissenschaftlicher Nachlass*, ed. Heinrich Weber and Richard Dedekind, 448–451. Leipzig: Teubner.

Riemann, Bernhard. 1996. On the hypotheses which lie at the foundation of geometry. In *From Kant to Hilbert: A source book in the foundations of mathematics*, ed. William Bragg Ewald, 652–661. Oxford: Clarendon. Originally published as: Über die Hypothesen, welche der Geometrie zu Grunde liegen. *Abhandlungen der Königlichen Gesellschaft der Wissenschaften zu Göttingen* 13 (1867): 133–152.

Rowe, David E. 1983. A forgotten chapter in the history of Felix Klein's "Erlanger Programm." *Historia Mathematica* 10: 448–454.

Rowe, David E. 1989. The early geometrical works of Sophus Lie and Felix Klein. In *The history of modern mathematics. Vol. 1: Ideas and their reception*, ed. David E. Rowe and John McCleary, 209–273. Boston: Academic.

Rowe, David E. 1992. Klein, Lie, and the *Erlanger Programm*. In *1830–1930: A century of geometry, epistemology, history and mathematics*, ed. Luciano Boi, Dominique Flament, and Jean-Michel Salanskis, 45–54. Berlin: Springer.

Russell, Bertrand. 1897. *An essay on the foundations of geometry*. Cambridge: Cambridge University Press.

Ryckman, Thomas A. 1991. *Conditio sine qua non? Zuordnung* in the early epistemologies of Cassirer and Schlick. *Synthese* 88: 57–95.

Ryckman, Thomas A. 2005. *The reign of relativity: Philosophy in physics 1915–1925*. New York: Oxford University Press.

Torretti, Roberto. 1978. *Philosophy of geometry from Riemann to Poincaré*. Dordrecht: Reidel.

Wussing, Hans. 1969. *Die Genesis des abstrakten Gruppenbegriffes: Ein Beitrag zur Entstehungsgeschichte der abstrakten Gruppentheorie*. Berlin: VEB Deutscher Verlag der Wissenschaften.

Yaglom, Isaak M. 1988. *Felix Klein and Sophus Lie: Evolution of the idea of symmetry in the nineteenth century*. Trans. Sergei Sossinsky. Boston: Birkhäuser.

Chapter 7
Non-Euclidean Geometry and Einstein's General Relativity: Cassirer's View in 1921

> *In the progress of knowledge the deep words of Heraclitus hold that the way upward and the way downward are one and the same: ὁδὸς ἄνω κάτω μίη. Here, too, ascent and descent necessarily belong together: the direction of thought to the universal principles and grounds of knowledge finally proves not only compatible with its direction to the particularity of phenomena and facts, but the correlate and condition of the latter. (Cassirer 1921, p.444)*

7.1 Introduction

The previous chapters have been devoted to the debate about the foundations of geometry in the tradition that goes back to Helmholtz and the reception of Helmholtz in neo-Kantianism. The proposed solution of the problem of reconsidering the Kantian theory of space after the development of models of non-Euclidean geometry was a generalized version of Kant's form of outer intuition, including both Euclidean and non-Euclidean geometries as special cases. In the tradition considered, such a form was identified as that of a threefold extended manifold of constant, zero, positive or negative curvature. After Einstein's general relativity of 1915, even the assumption of constant curvature appeared to be restrictive. This corresponds to the fact that the space-time continuum of general relativity is not conceived of as a background structure for the identification of physical events; the curvature of space-time in general relativity is variable and depends on the distribution of mass and energy.

The present, concluding chapter gives a brief account of the debate about the foundations of geometry after general relativity, with a special focus on Cassirer's view in 1921. In the interpretation proposed, Cassirer was one of the leading figures of the said tradition. However, in 1921, he emphasized that the geometrical hypotheses of general relativity differed completely from those of Newtonian mechanics and of special relativity. Therefore, he revised his argument for the aprioricity of geometry as stated in 1910. Nevertheless, Cassirer argued for continuity across theory change with regard to the symbolic function of geometry in its correlation with the empirical manifold of physical events.

© Springer International Publishing Switzerland 2016
F. Biagioli, *Space, Number, and Geometry from Helmholtz to Cassirer*, Archimedes 46, DOI 10.1007/978-3-319-31779-3_7

The first part of this chapter offers a discussion of Cassirer's view by contrasting it with Schlick's thesis about the separation between geometry and experience after the development of the axiomatic method. This contrast played a key role in the debate about the philosophical interpretations of general relativity. Since Schlick's views were influential in the development of logical empiricism, his discussion with Cassirer on that subject sheds light on the main points of disagreement between neo-Kantianism and empiricism. More recent scholarship initiated by Coffa (1991), Ryckman (1991, 2005) and Friedman (1999) has drawn attention to the fact that Schlick's and others' sharp criticism of Kant was due, at least in part, to the need to contrast their own views against a neo-Kantian background. This should not obscure the fact that the renewal of Kant's transcendental philosophy in the neo-Kantian movement largely prepared the view that the principles of knowledge are necessary only relative to their implications and can be subject to revision for considerations of uniformity in the formulation of physical laws and empirical evidence. In other words, the neo-Kantian background played an important role in the development of what Coffa called the semantic tradition.

A more controversial issue is whether the neo-Kantian movement – especially in the Marburg version advocated by Cassirer – actually undertook a relativization of a priori synthetic judgments, which, in Kant's original view, are supposed to be necessary and universally valid. Friedman, who is one of the main proponents of such a view of the a priori in contemporary philosophy of science, rather draws back the idea of a relativized and historicized a priori to Hans Reichenbach's early work on *The Theory of Relativity and A Priori Knowledge* (1920). Friedman himself pointed out the substantial points of agreement between Cassirer and Reichenbach on the philosophical significance of general relativity (see esp. Friedman 1999, Ch.3). Meanwhile, new evidence of Cassirer's close relationship to Reichenbach and to the leading figures of logical positivism from 1920 to 1941 has been provided by the edition of the fourth volume of Cassirer's Nachlass on "Symbolic Pregnance, Expressive Phenomenon, and Vienna Circle" (2011) by Christian Möckel. However, on several occasions, in Friedman (2001, 2005), and, more recently – in response to Ferrari (2012) – in Friedman (2012), Friedman distanced himself from Cassirer's and the Marburg view of the a priori, which he considers to be purely regulative. In Friedman's view, the relativization of the a priori loses its significance for the main problems of a philosophical perspective on science, unless it refers to the constitutive dimension of the a priori.

As argued in more detail in the previous chapters, it seems to me that this reconstruction of the semantic tradition and its prehistory fails to appreciate the significance of Cohen's and Cassirer's arguments for the synthetic character of mathematics in terms of a conceptual and, therefore, hypothetical synthesis. It followed that the borderline between the constitutive and the regulative was set somewhere else than in Kant's *Critique of Pure Knowledge*. As first suggested by Cohen, the distinction between these two dimensions of the a priori is a relative one and depends ultimately on the advancement of science. As Cohen put it, the transcendental inquiry into the conditions of knowledge presupposes the "fact of science."

My suggestion, in the second part of this chapter, is that Helmholtz's fortune in neo-Kantianism was due: (1) to a substantial agreement between his and Cohen's – independently motivated – inductive approach to the principles of knowledge, and (2) to a broader perspective on the mathematical method than that which imposed itself after the development of modern axiomatics. Without calling into question the potential of the axiomatic method, the broader view is worth reconsidering to deal with the questions concerning continuities and discontinuities after theory change.

7.2 Geometry and Experience

Poincaré's group-theoretical approach offered a plausible account of the relation between geometry and experience. On the one hand, Poincaré made it clear that geometrical propositions differ from empirical judgments on account of idealization, which is characteristic of the formation of the geometrical notion of space. On the other hand, he identified the concept of group as a form of the understanding and introduced conventional criteria for making a choice among hypotheses in the empirical use of particular geometries. His conclusion was that geometrical axioms are neither empirical propositions nor synthetic a priori judgments, because they are "definitions in disguise" (Poincaré 1902, p.50).

Poincaré's argument for the conventionality of geometry and its reception in neo-Kantianism has been discussed in the previous chapter. The first part of this section deals with a similar way of considering geometrical axioms as "implicit definitions" according to David Hilbert's axiomatic method. Schlick's reading of Poincaré's argument in connection with the theory of implicit definitions led him to sharply separate geometries as abstract structures from intuitive space, on the one hand, and from the empirical contents of physical measurements, on the other. Therefore, Schlick posed the problem of coordinating geometries with observations and proposed a possible solution to this problem by generalizing the usual procedure of comparing magnitudes by superposition of a measuring rod. Schlick's view appeared to be confirmed by Einstein's revision of the principles of measurement in the theory of general relativity, as Einstein himself advocated a similar view in "Geometry and Experience" (1921).

The second section addresses the question whether general relativity presupposes the above conception of geometrical axioms. In order to discuss this question, the third section deals with Cassirer's solution to the coordination problem posed by Schlick. Whereas Cassirer agreed with Schlick and with Einstein on the conclusion that the principles of measurement can be subject to revision, his view in 1921 presupposed a different understanding of the relation between geometry and experience. After general relativity, Cassirer recognized that his 1910 argument for the aprioricity of geometry had to be reformulated. Nevertheless, he reaffirmed the view that the physical meaning of geometry was related strictly to the symbolic character of mathematics. In Cassirer's view, the heuristic function of mathematics – once more confirmed by Einstein's use of Riemannian geometry in general relativity – offered some of the clearest examples of the constructive and anticipatory role of symbols in the articulation of experience.

7.2.1 Axioms and Definitions: The Debate about Spatial Intuition and Physical Space after the Development of the Axiomatic Method

Schlick's starting point for the discussion about spatial intuition was a distinction between physical and phenomenal space. This distinction goes back to Schlick's 1916 paper on "Ideality of Space, Introjection and the Psycho-Physical Problem." Schlick addressed the following problem: How is it possible that different individuals perceive themselves and physical objects as located in the same single space? Schlick compared Kant's thesis about the ideality of the form of outer intuition to the solution offered by Richard Avenarius's empiriocriticism. According to Schlick, Kant's definition of space as a form of our intuition implies that all spatially determined objects are appearances, that is, "representations of my consciousness" (Schlick 1916, p.198). Avenarius's solution to the psychophysical problem was based on a distinction between the immediate acquaintance with sense qualities and the localization of physical objects. This distinction depends on the fact that localization presupposes an inference from the elements of sense perception to physical objects. The constructions required for such an inference are ideal and may vary, under the condition that they do not contradict the elements of knowledge. Therefore, Avenarius opposed the introjection thesis that sense qualities are localized in physical space and maintained that space is composed by environmental constituents or element-complexes in Mach's sense. Schlick identified Avenarius's environmental constituents with Kant's appearances as representations in consciousness: with different terminologies, both Kant and Avenarius argued for the view that "the sphere of sensory consciousness coincides with the space of appearance" (p.199).

   Schlick's considerations suggest that both Kantianism and empiriocriticism led to a solution of the psychophysical problem in terms of a parallel between psychic and physical processes. Schlick relied on these traditions to argue against the introjection thesis. However, he distanced himself from the view that physical space coincides with phenomenal space. Schlick pointed out that phenomenal space differs from the idea of a single physical space because of the plurality of the senses. He wrote:

> The space of sight is psychologically quite unlike that of touch or the muscular sense, and it is only because a univocal coordination prevails between them, that the concept arises of the physical space common to them all. This, in the end, is simply nothing else but an expression for the existence of that coordination and correspondence between all the spatial experiences of the different senses and individuals, and hence is a symbol for the order of things-in-themselves, which (even in Kant's view) we must hold responsible for the spatial order of experiential objects, though we may not describe it as itself a spatial order. But if this is the meaning of physical space, there follows at once from it the meaning of physical spatial objects, the objects of natural science: their world is not, as Kant wanted, to be put

on a level with that of the "appearances," the manifold of intuition, but has to be regarded as a conceptual symbol, a sign-system for the world of things-in-themselves. Between the structure of this world and the structure of that sign-system there is a univocal correspondence, which is traced out by the efforts of science; and in so far as in every genuine piece of knowledge the relation between concept and known object is simply no other than that of univocal coordination, one may confidently say that the truths of science are in fact just so many cognitions of things-in-themselves. (Schlick 1916, p.201)

As we saw in the third chapter, similar considerations led Alois Riehl to acknowledge empirical factors in the formation of spatial notions. Following Helmholtz, Riehl described sight and touch as "two languages that grasp the same meaning in completely different words. Therefore, the capacity to translate one of them into the other must be learned with effort – regardless of the anatomical connection between the sense organs. This could not be the case, if spatial representation were 'pure intuition' in Kant's sense, and, under different empirical circumstances, it could be grasped in exactly the same way: experience teaches us that the nature of the material plays a fundamental role in the formation of this representation" (Riehl 1879, p.139). Not only did Riehl point out that Kant mistakenly identified sensory space with mathematical space in his theory of the forms of pure intuition, but the physiological aspect of sensation led Riehl to call into question Kant's view that things in themselves are unknowable. Riehl defined sense knowledge as "the knowledge of the relations of things through the relations of the sensations of things" (Riehl 1887, p.42). Notwithstanding the impossibility of knowledge of things in themselves through reason alone, Riehl used this argument to advocate the view that things in themselves can be known by empirical intuition and empirical thought (p.42, note). Schlick's solution of the psychophysical problem is clearly reminiscent of Riehl's view that knowledge of things in themselves is restricted to the quantitative relations that find a numerical expression in measurement.

Arguably, Schlick bore in mind Riehl's argument for the knowability of things in themselves, because Riehl was the only philosopher in the neo-Kantian movement to advocate a realist interpretation of critical philosophy or the view that became known as critical realism.[1] Schlick's contacts with Riehl go back to 1905 when Riehl was appointed Professor at the University of Berlin. Schlick's mentions of

---

[1] On the debate about critical realism between the late nineteenth century and the early twentieth century, see Neuber (2014). On Schlick's relationship to this tradition, consider the following quote from Schlick's student Herbert Feigl: "It is interesting to note that both Schlick (from 1910 to 1925) and Russell (by 1948, at any rate) were critical realists and thus had to come to grips with the problems of transcendence. And, while they differed sharply in their views on probability and induction, they argued essentially inductively for the existence of entities beyond the scope of the narrow domain of immediate experience. Both Schlick and Russell thus liberalized the radical empiricism of Hume, namely, by asserting the existence of a world of knowable things-in-themselves – be they such objects of common life as sticks or stones, rivers or mountains, or be they the fields and particles of modern physics. It was only under the impact of Carnap's and Wittgenstein's ideas and criticisms that Schlick withdrew to what he conceived of as a neutral, non-metaphysical position" (in Neuber 2012, p.163). Although Schlick did not call himself a critical realist, Heidelberger (2007) and Neuber (2014) provided new evidence of Riehl's influence on Schlick regarding the central claims of critical realism.

Riehl in the *General Theory of Knowledge* and in the notes to the centenary edition of Helmholtz's *Epistemological Writings* show that he was familiar with Riehl's major work, *Critical Philosophy and Its Meaning for Positive Science*. Riehl's influence on Schlick in 1916 emerges in the latter's approach to the psychophysical problem, and is apparent in Schlick's definition of physical space as "a symbol for the order of things in themselves" in contrast to Kant's definition of space as pure intuition. By referring to the order of things in themselves, Schlick replaced the psychophysical parallelism advocated in Kantian and positivist traditions with the idea of a univocal coordination (*Zuordnung*)[2] of the phenomenal spaces with each other and with the idea of a single, common physical space.[3]

In the *General Theory of Knowledge* (1918), Schlick reformulated his argument by taking into account the role of the deductive method in the formation of geometrical concepts. He particularly referred to abstraction from the content of geometrical concepts in the works of Moritz Pasch and David Hilbert. In the *Lectures on Projective Geometry* (1882), Pasch characterized the deductive method by requiring that "inferences should be drawn independently of the *meaning* of geometrical concepts, as well as independently of figures; only the *relations* between geometrical concepts established by the adopted propositions or definitions ought to be considered" (Pasch 1882, p.98). Schlick interpreted Hilbert's "Foundations of Geometry" (1899) as a consequent development of the same requirement. The idea of geometrical axioms as implicit definitions[4] follows from Hilbert's way of introducing the elements of geometry. His work begins with the following remark: "We think of

---

[2] We have already noticed that the term "Zuordnung" indicates both the mathematical concept of function and a relation of coordination between theoretical and observational terms. Following the most common usage in the literature (see, e.g., Ryckman 1991), I translate "Zuordnung" into "coordination" when it is used in the latter sense. I render the same term with "correlation," when it refers primarily to the mathematical concept. It is noteworthy, however, that the two different meanings are related in the German original, especially in Cassirer's usage.

[3] For further information about Schlick's relationship to Riehl, see Heidelberger (2006). On Riehl's influence on Schlick's (1916) solution of the mind-body problem, see Heidelberger (2007).

[4] The expression "implicit definition" was introduced by Joseph-Diez Gergonne. Implicit definitions differ from ordinary definitions because in the former, the meaning of the terms to be defined is inferred from that of known terms. Therefore, implicit definitions presuppose that there are as many known terms as the terms to be defined (Gergonne 1818, p.23). Giovanni Vacca (1899, p.186) referred the same expression to Giuseppe Peano's definitions in terms of postulates in the axiomatic theory of arithmetic. Hilbert's definition became known as implicit after his meeting with the Peano School during the second International Congress of Mathematics, which was held in Paris in 1900. On that occasion, Hilbert mentioned his use of axioms in the definition of the fundamental concepts (Hilbert 1902, p.71). "Implicit definition" was introduced into Germany by Federigo Enriques, who used it in the entry "Principles of Geometry" in the *Encyclopedia of Mathematical Sciences* (Enriques 1907, p.11). Notice, however, that Hilbert himself never used it. Furthermore, definitions by using axioms differ from implicit definitions in Gergonne's sense in several respects. Firstly, the number of the axioms does not depend on the number of terms to be defined. Secondly, an axiomatic system admits different models. Thirdly, Hilbert's axioms define forms whose terms do not have a particular meaning; these forms rather indicate tuples of predicate variables (see Gabriel 1978, p.420).

these points, straight lines, and planes as having certain mutual relations, which we indicate by means of such words as 'are situated,' 'between,' 'parallel,' 'congruent,' 'continuous,' etc. The complete and exact description of these relations follows as a consequence of the axioms of geometry" (Hilbert 1899, p.2). Hilbert adopted a purely deductive approach towards the foundations of geometry insofar as the inquiry begins with the classification of axioms into groups and proceeds with the proof that these groups form a consistent system. At the same time, Hilbert believed that the choice of the axioms and the investigation of their relations to one another provide us with "a logical analysis of our intuition of space" (Hilbert 1899, Preface). Therefore, Hilbert (ibid.) referred to Kant's view that "all human cognition begins with intuition, goes from there to concepts, and ends with ideas" (Kant 1787, p.730).

Schlick's theory of implicit definitions led him to adopt a more critical stance towards the classical definition of geometry as the science of space. Schlick highlighted the problem of a geometrical characterization of space as follows:

> [G]eometry as a solid edifice of rigorously exact truths is not truly a science of space. The spatial figures serve simply as intuitive examples in which the relations set up *in abstracto* by the geometrical propositions are realized. As to the converse – whether geometry in so far as it does aim to be a science of space can be regarded as a firmly joined structure of absolutely rigorous truths – this is a question for the epistemology of mathematics. We shall not try to resolve it here, since our concern for the present is only with the general problems of knowledge. However, it should be clear enough from what has been said that we cannot take for granted that the answer is the affirmative, as one might otherwise suppose. For it was precisely the misgivings about the absolute rigor of propositions dealing with intuitive spatial forms that led to defining concepts not through intuitions but through systems of postulates. (Schlick 1974, pp.36–37)

Schlick's view follows from his interpretation of implicit definition as the outcome of a process of abstraction that begins with the use of polyvalent symbols. The interpretation of symbols is not relevant to the purpose of calculation. Schlick breaks with the nineteenth-century mathematical tradition by claiming that "the construction of a strict deductible science has only the significance of a game with symbols" (Schlick 1974, pp.36–37; cf. Helmholtz's view of symbols as described in 4.2.3). This claim corresponds to the fact that the logical rigor of mathematical theories is compatible with the possibility of a variety of interpretations. On the one hand, the consideration of geometry as a "stable system of absolutely rigorous truths" presupposes "a radical separation between concept and intuition, thought and reality. While we do relate the two spheres to one another, they seem not to be joined together at all. The bridges between them are down" (Schlick 1974, p.38).[5] On the

---

[5] Similarly, Freudenthal (1962, p.618) maintained that the beginning of Hilbert's inquiry into the foundations of geometry "cuts the bridge with reality." Freudenthal argued that Hilbert inaugurated a new approach to geometrical axioms, even when compared to Pasch. Whereas the debate between Kantians and empiricists focused on the origin of geometrical axioms, Hilbert introduced conventional definitions of geometrical concepts and included geometry in pure mathematics. This reading, which is clearly influenced by Schlick, can be called into question if one considers Hilbert's reference to Kant in the Preface to the "Foundations of Geometry" and Hilbert's program of axiomatizing physics in analogy to geometry (see Majer 2001, p.215; Pulte 2006, p.192). On Hilbert's relationship to Kant and to the neo-Friesian School of neo-Kantianism, see Peckhaus (1990).

other hand, Schlick ruled out Kant's assumption of apodictically true geometrical foundations of natural science. Given the use of symbols and the possibility of different interpretations in the characterization of physical space, the coordination between geometry and physics is hypothetical and presupposes both formal and empirical conditions.

Schlick's explanation of how the abstract systems of geometry are endowed with a physical meaning is found in the section of his book entitled "Quantitative and Qualitative Knowledge" (Schlick 1974, pp.272–289). Schlick's starting point was his distinction between immediate acquaintance, on the one hand, and proper knowledge, on the other: "We become acquainted with things through intuition, since everything that is given to us from the world is given to us in intuition. But we come to know things only through thinking, for the ordering and coordinating needed for cognition is precisely what we designate as thinking" (p.83). In a footnote (p.83), Schlick draws the same distinction back to Riehl (1879, p.1). Furthermore, Schlick agrees with Riehl that the ordering and coordinating needed for cognition presupposes a reduction of the heterogeneous qualities of sense perception to a common measure, namely, the presence or absence of a singularity. If, while looking at a pencil, for example, I touch its point with my finger, a singularity occurs simultaneously in both my visual space and my tactile space. These two experiences, although entirely disparate, are now correlated with one and the same point, namely, the point of contact between my finger and the pencil, which Schlick identifies as a single point of transcendent space (Schlick 1974, p.273). In other words, the transcendent ordering is revealed by abstracting from sensuous qualities and concentrating solely on the topological properties of different sensory fields. Schlick's method of coincidences consists of identifying the same way of proceeding as a necessary presupposition for scientific measurement: "In the final analysis, every precise measurement consists always and exclusively in comparing two bodies with one another, that is, in laying a measuring rod alongside the object to be measured so that certain marks on the rod (lines on a scale) are made to coincide with specific points on the object" (p.275).

We have already mentioned that Schlick attributed a similar approach to Helmholtz when dealing with the physical interpretation of the relation of congruence. As pointed out by Friedman (1997), the problematic aspect of Schlick's reading of Helmholtz's theory of signs lies in the fact that Helmholtz interpreted the localization of objects in space as a conceptual construction of space itself. There is no place for a separation between the qualitative-intuitive concept of space and physico-geometrical space in Helmholtz's view. His conditions of measurement, beginning with the free mobility of rigid bodies, work more as general rules which – once induced by observation and experiment – provide us with the necessary presuppositions for ordering or appearances. Insofar as the same rules apply to any perceptual content, Helmholtz identified the idea of a rigid geometrical structure as a (generalized) equivalent for the form of outer intuition in Kant's sense.

Another problem with Schlick's reading is that Helmholtz's definition of rigidity appears to be circular when it is supposed to be induced by acquaintance with the behavior of solid bodies: the geometrical notion of rigidity is tacitly presupposed in the identification of measuring rods as rigid bodies. Schlick's suggestion is that, in order to avoid circularity, the notion of rigidity and the coordinating principles linking uninterpreted systems of implicit definitions with actual measurements have to be stipulated. Therefore, Schlick advocated Poincaré's geometrical conventionalism as the only possible solution. As pointed out by Ryckman (2005, Ch.3), Schlick thereby turned Helmholtz's view into his own holist conventionalism. By contrast, Ryckman proposed interpreting Helmholtz's considerations regarding the notion of rigidity as an "inherently Kantian" theory of space, according to which the free mobility of rigid bodies provides us with a necessary precondition for the possibility of measurement. Helmholtz's view differs from a "strictly" Kantian view, because all of the axioms that are compatible with such a rule (i.e., all of the three classical cases of manifolds of constant curvature) can be confirmed or refuted by experience.

Although I agree with the remarks above, I wonder whether it suffices to indicate the methods of implicit definitions and of coincidences as turning points in the debate about spatial intuition and physical space. More recent scholarship initiated by Heidelberger (2007) and Neuber (2012) has drawn attention to the fact that Schlick's motivations for advocating a non-naïve form of causal realism informed by an empiricist epistemology are found not least in his commitment to the psychophysical problem. My main concern in the present work is with the received picture of the nineteenth-century mathematical tradition. With the sole exception of Riemann – who developed his theory of numbers analytically, namely, as a theory of continuous number manifolds, and sharply distinguished it from the hypotheses about the metrical structure of the physical world – Friedman includes the leading figures of this debate in the same tradition, which he characterizes as follows:

> [T]he nineteenth-century mathematical tradition in projective geometry and group theory that culminated in the work of Klein and Poincaré did not draw the now familiar distinction between pure or axiomatic geometry and applied or interpreted geometry associated with Hilbert and his followers. Mathematical geometry, in this earlier tradition, continued to be a (fully interpreted) theory of space – the very same space in which we live, move, and perceive. It was in precisely this way, in fact, that the mathematical science of geometry differed from analysis and number theory, the much more abstract sciences of "magnitude" in general. Geometry studies a particular object of thought given to us through perception or intuition, whereas number theory and analysis study all objects of thought in any way countable or measurable. So the latter sciences, unlike geometry, are not limited by the nature of our sensible intuition. (Friedman 2002, p.201)

I argued in the previous chapters that, quite on the contrary, the idea of a generalization of the Kantian form of spatial intuition can be traced back to Helmholtz's remarks about the use of analytic geometry in measurement and had its origins in the so-called arithmetization of mathematics. In my reconstruction, what distinguishes the said tradition from Schlick's view of geometry is the idea that a more general understanding of spatial order in structural terms (i.e., without limitations imposed by our sensible intuition) was necessary in order to pose the problems concerning measurement correctly. Although geometry was interpreted and not detached from the idea of space in the manner required by Schlick, the transformation of geometry in the nineteenth century implied a transformation of the concept of space itself. That space thus conceived did not differ substantially from numerical manifolds and algebraic structures seems to me to be confirmed by the fact that the restriction of the hypotheses under consideration to the manifolds of constant curvature in the earlier phase of the debate did not appear to be a limitation until after Einstein's use of Riemannian geometry in general relativity. In the following, I reconsider Cassirer's view in 1921 as a plausible account of the relation between geometry and experience in continuity with the earlier tradition.

7.2.2   Schlick and Einstein (1921)

There is evidence that Schlick influenced Einstein's epistemological views, especially if one considers the development of Einstein's views from 1915 to 1921. This section focuses particularly on Einstein's 1921 lecture on "Geometry and Experience."[6] In the extended version of the lecture, which was published in the same year by Springer, Einstein referred to Schlick's characterization of geometrical axiom as implicit definitions to make the following consideration about the epistemic value of geometry:

> As far as the laws of mathematics refer to reality, they are not certain; and as far as they are certain, they do not refer to reality. It seems to me that complete clearness as to this state of things first became common property through that new departure in mathematics which is known by the name of mathematical logic or "Axiomatics." The progress achieved by axiomatics consists in its having neatly separated the logical-formal from its objective or intuitive content; according to axiomatics the logical-formal alone forms the subject-matter of mathematics, which is not concerned with the intuitive or other content associated with the logical-formal. (Einstein 1921, pp.28–29)

Einstein's remark clearly reflects Schlick's distinction between geometric systems as abstract structures and geometry as a science of space. Einstein used this distinction to pose the following problem: "How can it be that mathematics, being after all a product of human thought which is independent of experience, is so admirably appropriate to the objects of reality?" (1921, pp.28–29). His solution was that, notwithstanding the conception of geometrical axioms as free products of the human mind, the same axioms can be provided with a physical interpretation insofar as they can be coordinated with empirical propositions in a univocal manner. Therefore, he introduced the idea of a "practical geometry," which, using the analogy of Helmholtz's physical geometry, is conceived of as a branch of physics. Einstein, for example, referred to Helmholtz's assumption that rigid bodies can be displaced without changes in shape and size in the manner of figures in three-dimensional Euclidean space. Helmholtz's free mobility of rigid bodies provides us with a coordinating principle linking the structure of Euclidean space to metrical relations in observation and experiment.

---

[6] For further evidence of Schlick's influence on Einstein based on their correspondence from 1915 to 1933, see Howard (1984).

Einstein maintained that this conception of geometry played an important role in the development of general relativity. He particularly referred to the fact that the laws of disposition of rigid bodies in a system of reference rotating relatively to an inertial system do not correspond to the rules of Euclidean geometry on account of the Lorentz contraction. Therefore, Einstein pointed out that if we admit non-inertial systems on an equal footing according to his generalized principle of equivalence of inertial and gravitational fields, we must abandon Euclidean geometry. He maintained that the physical interpretation of geometry by Helmholtz served as a "stepping-stone" in the transition to generally covariant field equations (Einstein 1921, p.33). At the same time, Einstein's approach offered a possible solution to the problem posed by Poincaré: the interpretation of mathematical structures requires some postulates. In the case of the free mobility of rigid bodies, this corresponds to the fact that geometrical notions apply only approximately to solid bodies. Poincaré concluded that the equivalence between solid bodies and figures in three-dimensional Euclidean space is a matter of convention. According to Poincaré, we choose Euclidean geometry among the set of possible hypotheses, because only the Euclidean group has the translations as a normal subgroup. Therefore, Poincaré maintained that Euclidean geometry is simpler than non-Euclidean geometry. In Schlick's view, which presupposes the distinction above between abstract systems and the theory of space, the criterion of simplicity applies to the structures of physical geometry and singles out the metrical geometry that leads us to the simplest system of natural laws. Schlick maintained that the geometrical foundations of physics can be subject to revision, because even the simplest representation of nature presupposes some arbitrary convention, if only those which underlie measurement.

Similarly, Einstein maintained that Poincaré's argument holds true *sub specie aeterni*. However, Einstein argued for the objective meaning of physical interpretations insofar as an exact equivalence between mathematical and empirical concepts is not available. Such coordinating principles as the free mobility of rigid bodies play a fundamental role in practical geometry, provided that the coordination proves to be sufficiently univocal. In this sense, Einstein compared the role of "rigid" bodies in classical mechanics to that of the ideas of the measuring rod and of the clock coordinated with it in the theory of relativity. Both the assumptions of Euclidean geometry and of Riemannian geometry – which in the sense of practical geometry includes practical Euclidean geometry as a special case – presuppose that: "If two tracts are found to be equal once and anywhere, they are equal always and everywhere" (Einstein 1921, p.37). Einstein's conclusion differed from Poincaré's for the following reason: "The question whether the structure of this continuum is Euclidean, or in accordance with Riemann's general scheme, or otherwise, is, according to the view which is here being advocated, properly speaking a physical question which must be answered by experience, and not a question of a mere convention to be selected on practical grounds" (Einstein 1921, p.39).

Summing up, Einstein posited himself between Helmholtz and Poincaré by arguing that geometry in the second of the meanings above (i.e., as a theory of space) is a branch of physics. However, Einstein's physical geometry is inconsistent with the free mobility of rigid bodies. The coordinating principle introduced by Einstein is the principle of equivalence. Whereas the free mobility of rigid bodies is equally compatible with different specific geometries of constant curvature, the principle of equivalence determines a univocal coordination between a single geometrical structure and physical reality. The structure is that of a four-dimensional semi-Riemannian manifold of variable curvature. The principle states that bodies affected only by gravitation follow geodesic trajectories in such a structure. The presence of a gravitational field is represented by a nonvanishing four-dimensional curvature.

Given the striking differences between the latter structure and those of the former theories, the question arises: Why did Einstein attach so much importance to Helmholtz's view of geometry and to the example of the rotating system in special relativity for the development of general relativity? Friedman's (2002) suggestion is to reconsider the heuristic role of this example in Einstein's line of thought. The rotating system offered a concrete example of how the behavior of measuring rods motivates the introduction of non-Euclidean geometry. Einstein was then confronted with the fact that non-Euclidean geometries cannot be described by Cartesian coordinates, but require more general Gaussian coordinates. The difference lies in the fact that the coordinates, in the more general system, no longer have a direct physical meaning. According to Friedman, the same idea, generalized to four dimensions, became the requirement of general covariance.

Friedman's reconstruction enables him to make the following point. From our contemporary perspective, the example of the rotating system in (flat) Minkowski space-time has nothing to do with the variably curved space-time we now use to represent the gravitational field. At the time when Einstein was articulating a generalized equivalence between gravitational and inertial phenomena, however, the four-dimensional space-time geometry we now identify with his name was yet to be discovered. The use of non-Euclidean geometry in that case appeared to him to be an indispensable application of the principle of equivalence and, therefore, played a crucial role in the articulation of his view. As Friedman puts it, "Einstein's own appeal to rigid bodies, in the spirit of Helmholtz, [...] provides an especially striking example of how an older perspective on geometrical constitutive principles can be subtly transformed into a radically new perspective on such principles that is actually inconsistent with the old" (Friedman 2002, p.218).

Although I agree with Friedman's characterization of the discontinuity between Einstein and earlier views, I believe that more can be said about the continuity in the transformation of the concepts of space and of space-time from the old to the new view inaugurated by Einstein. This will require us to reconsider the rather different view of geometry advocated at that time by Cassirer. Cassirer agreed with Schlick that the advance of Hilbert's axiomatic method over the synthetic method of Euclidean geometry depended on the separation of geometrical from intuitive contents. In contrast to Euclidean definitions, which take such concepts as point and straight line as immediate data of intuition, Hilbert derived the properties of geometrical objects as the consequences of general rules of connection. His work began with a group of axioms, which we assume, and whose compatibility has to be proven. Cassirer interpreted this result as follows:

Intuition seems to grasp the content as an isolated self-contained existence; but as soon as we go on to characterize this existence in judgment, it resolves into a web of related structures which reciprocally support each other. Concept and judgment know the individual only as a member, as a point in a systematic manifold; here as in arithmetic, the manifold, as opposed to all particular structures, appears as the real logical *prius*. (Cassirer 1910, p.94)

Cassirer relied on Hilbert to clarify the relationship between definitions and geometrical axioms. From a psychological viewpoint, the definition of the elements of geometry in terms of axioms might seem to imply a vicious circle: the axioms themselves seem to presuppose some concepts, because, in a psychological sense, the meaning of some relations can only be conceived in connection with given terms. By contrast, the starting point of mathematics is given by hypothetical terms, whose closer determinations are obtained by inserting them into various relational complexes. Cassirer agreed with Schlick that the object of mathematical investigations such as Hilbert's consists not so much of particular elements, as of the relational structure as such. Nevertheless, Cassirer emphasized that objectivity is defined in terms of such a structure.

To sum up, Cassirer's view differed from Schlick's, and from Einstein's – insofar as he relied upon Schlick's theory of implicit definitions – because the separation of concepts and intuitions in Cassirer's sense did not entail a separation of form and content: contents themselves are redefined from the more general viewpoint of relational structures. Therefore, Cassirer maintained that the formal study of mathematical structures was a necessary condition for the definition of physical objects. His point, in 1921, was that the solution to the problems concerning measurement in general relativity offered an example of his original view, insofar as such a solution was due to immanent developments in the theory of manifolds first introduced by Riemann.

7.2.3 Cassirer's Argument about the Coordination between Geometry and Physical Reality in General Relativity

Cassirer's book on *Einstein's Theory of Relativity* (1921) offered a possible solution to the problems of a Kantian perspective on science after radical changes in the forms considered a priori by Kant. In particular, Cassirer addressed the following problem:

Kant believed that he possessed in Newton's fundamental work, in the *Philosophiae Naturalis Principia Mathematica*, a fixed code of physical "truth" and believed that he could definitively ground philosophical knowledge on the "factum" of mathematical natural science as he here found it; but the relation between philosophy and exact science has since changed fundamentally. Ever more clearly, ever more compellingly do we realize today that the Archimedean point on which Kant supported himself and from which he undertook to raise the whole system of knowledge, as if by a lever, no longer offers an unconditionally fixed basis. (Cassirer 1921, pp.352–353)

Cassirer referred to the role played by Euclidean geometry in classical mechanics. Is there a place for a similar role of geometry in general relativity? Cassirer's answer presupposed a dynamical interpretation of Kant's forms of intuition. Cassirer considered his interpretation a consistent development of Kant's conception of space and time, not as things (*Dinge*), but as conditions (*Bedingungen*) for observation and experiment. The Kantian theory of pure sensibility rules out the possibility of identifying space and time as any fixed content: as conditions of experience, space and time cannot themselves be experienced. Cassirer borrowed the idea from Cohen that the core of the Kantian theory consists not so much in the assumption of pure intuitions, as in the corresponding distinction of levels in the architectonic of knowledge. The appearances present themselves in a spatiotemporal order on account of the forms of intuition. As we saw in Chap. 3, the formation of geometrical concepts is a further development, which presupposes the application of the concept of number to that of magnitude and requires further specifications concerning the metric of space-time.

Given the difference of levels above, Cassirer pointed out that the Kantian theory of space cannot be contradicted by a physical theory: physical theories contain a doctrine of empirical space and time, not of pure space and time (Cassirer 1921, p.409). At the same time, Cassirer emphasized the philosophical relevance of Einstein's theory insofar as it presupposed a revision of the principles of measurement. The space-time structure of general relativity is characterized by the fact that the measure of curvature depends on physical factors. Therefore, Einstein replaced the Euclidean form of the line element with the general form: $ds^2 = \sum_1^4 g_{\mu\nu} \, dx_\mu \, dx_\nu$, and adopted Riemannian geometry for the expression of the measure of curvature of space-time in its correlation with matter. Regarding Einstein's use of Riemannian geometry, Cassirer maintained that the relation between Euclidean and non-Euclidean geometry appeared in a new light. He wrote:

> The real superiority of Euclidean geometry seems at first glance to consist in its concrete and intuitive determinateness in the face of which all "pseudo-geometries" fade into logical "possibilities." These possibilities exist only for thought, not for "being;" they seem analytic plays with concepts, which can be left unconsidered when we are concerned with experience and with "nature," with the synthetic unity of objective knowledge. [...] [T]his view must undergo a peculiar and paradoxical reversal. (Cassirer 1921, pp.442–443)

As we saw in the previous chapter, Cassirer's argument in 1910 was borrowed from Poincaré and stated that the choice among different but physically equivalent geometries would be guided by simplicity; "simplicity" in the case of manifolds of constant curvature being determined by the property of having a normal subgroup. The choice of Euclidean geometry, hence, depended on the fact that only the Euclidean group has the translations as a normal subgroup. In 1921, Cassirer revised the argument as follows. On the one hand, the view that geometry is independent of experience was confirmed by the fact that deviations from the law of homogeneity required a physical explanation: the value of $g_{\mu\nu}$, which indicates the deviation, depends on the gravitational field, and can be neglected at the infinitesimal level and in the other cases in which the same results are obtained according to special relativity. On the other hand, there is a reversal in what appeared to be "abstract" and "concrete." Relatively complex expressions now have a physical meaning and Euclidean geometry is considered a limiting case. Cassirer had to reconsider his argument that the priority of homogeneity over heterogeneity in the formation of geometrical concepts speaks for the choice of Euclidean geometry: general relativity shows that because Euclidean space represents the logically simplest form of spatial construction, Euclidean geometry is not adequate to solve the problem of the heterogeneous.

Another aspect of the line of argument above that seems to be problematic is the idea of ordering metrical geometries from the simplest to the most complex in the first place. Nevertheless, the said reversal lends plausibility to Cassirer's core idea that there is no fixed hierarchy of concepts from concrete to abstract. Therefore, he reaffirmed the view that the heuristic function of mathematics in physics depends on the complete freedom of pure mathematics. This view, and the related view of a priori notions as a range of conceptual hypotheses, goes back to Cohen. In the *Principle of the Infinitesimal Method*, Cohen characterized the relation between pure and applied mathematics as follows:

> What is conceived of by pure mathematics, thereby possess the character of *applicability*; for pure is only that which under given circumstances can be applicable. However, the question whether such conditions are given or not does not affect what is pure. Therefore, the new way of calling mathematics *free* appears to be appropriate. Pure mathematics is not created by mathematical ingenuity or imagination according to some game rules; rather, spatial and numerical constructions are created in the economy of our means of knowledge, without presupposing their application to natural objects. But the less the pure mathematician looks for such an application, the more he can rest assured that what he conceives of pure – namely, according to laws – precisely because of that, is applicable, whatever time it may take, until such application is implemented. The conic sections belonged to pure mathematics for Kepler as well as for Greek geometers; and even they attached to the conic sections the value of being referable to nature. [...] Without the construction of the conic sections, we would not know about the natural process of planetary motions; the same process would manifest itself at most as a problem, like that of the epicycles. This construction of pure geometry is therefore applied, not to the nature of planetary motions per se, but in the *construction* of that nature, which without those pure and constructive means would manifest itself only as an open question, not as a known natural process. (Cohen 1883, pp.131–132)

Cohen referred Georg Cantor's usage of "free" instead of "pure mathematics" to point out that the existence of a mathematical concept depends solely on its being well defined, free from contradiction, and entering into fixed relations with previously accredited concepts. By contrast, the reality of empirical concepts depends on their representing processes and relations in the external world (Cantor 1883, 181). Cantor used this argument to advocate the existence of the transfinite numbers introduced by him. However, it is hard to see how the same argument can support Cohen's view of mathematics. It might be helpful to notice that Cohen's understanding of freedom is reminiscent of the Kantian notion of autonomy or self-determination (rather than freedom from something), as is the case with Cassirer's understanding of Dedekind's definition of numbers as free creations of the human mind. On the one hand, Cohen himself emphasizes that mathematical constructions are independent of their applications to the natural world. On the other hand, the constructive character of the mathematical reasoning is the very condition for the applicability of mathematical laws. The relation of pure and applied mathematics postulated by Cohen is meant to explain the role of mathematics in the solution of the problems that occur in physics. Such a relation offered an argument for critical idealism, because Cohen believed that the solution of physical problems by means of pure mathematics provides us with a construction and an exact definition of natural processes.

In Cassirer's view, the same relation corresponds to the fact that the highest abstractions of mathematics play a fundamental role in the definition of physical objects. Cassirer compared Cohen's example from astronomy to the application of Riemannian geometry to the problem of the heterogeneous in general relativity. He wrote:

> When the concept of the special three-dimensional manifold with a curvature 0 is broadened here to the thought of a system of manifolds with different constant or variable curvatures, a new ideal means is discovered for the mastery of complex manifolds; new conceptual symbols are created, not as expressions of things, but of possible relations according to law. Whether these relations are realized within phenomena at any place only experience can decide. But it is not experience that grounds the content of the geometrical concepts; rather these concepts foreshadow it as methodological anticipations, just as the form of the ellipse was anticipated as a conic section long before it attained concrete application and significance in the courses of the planets. When they first appeared, the systems of non-Euclidean geometry seemed lacking in all empirical meaning, but there was expressed in them the intellectual preparation for problems and tasks, to which experience was to lead later. (Cassirer 1921, p.443)

Cassirer's view of geometrical concepts as methodological anticipations enabled him to recognize that general relativity presupposed completely different geometrical hypotheses from those of classical mechanics and special relativity. At the same time, he drew attention to the fact that the hypotheses required for the solution of the problem of the heterogeneous had been elaborated in such a highly abstract branch of mathematics as the theory of manifolds. This example suggests that theory change confirms that there is an interaction between pure and applied mathematics in Cohen's sense. What we can learn from this is that "the possibility of such an application must be held open for all, even the most remote constructions of pure mathematics and especially of non-Euclidean geometry. For it has always shown in the history of mathematics that its complete freedom contains the guarantee and condition of its fruitfulness" (p.443).

In order to justify such a requirement, Cassirer integrated Cohen's considerations about the construction of nature with the view that mathematical symbols stand for a system of connections freely established by the mind in the articulation of experience rather than for external entities. This view corresponds to Cassirer's account of physical reality in terms of laws and relations. Therefore, Cassirer rejected the view that geometries can have physical reality in the sense that they capture the features of space. The question is not whether non-Euclidean geometry can be a correct description of space, but whether an exact correlation between the symbolic language of non-Euclidean geometry and the empirical manifold of spatiotemporal events can be established. Since relativity theory answered this question affirmatively, Cassirer reformulated his former argument for the aprioricity of geometry by identifying the a priori of space not as a definite structure of space in itself, but as the general function of spatiality that finds expression in the line element of spacetime (Cassirer 1921, p.433).

Cassirer used his argument to provide a Kantian interpretation of Einstein's claim that the requirement of general covariance "takes away from space and time the last remnant of physical objectivity" (Einstein 1916, p.117). The requirement is that the form of natural laws remains unchanged under arbitrary changes of spacetime values. Einstein deemed this requirement a condition for the implementation of the general principle of relativity: the laws of physics are to be such that they apply to systems of reference in any kind of motion. The classical objection against this interpretation of the principle of relativity goes back to Erich Kretschmann (1917) and it is based on the possibility of giving a covariant formulation of both classical mechanics and special relativity. Thus, the fulfillment of general covariance, as a formal property, is not a distinctive characteristic of general relativity. Nevertheless, it is clear from Cassirer's interpretation that general covariance is not restricted to the formal aspect, insofar as it relates to the idea of a univocal coordination. According to Cassirer, the advance of Einstein's theory of 1915 over the former theories depends on the possibility of univocally identifying space-time coincidences regardless of the choice of privileged systems of reference, such as inertial systems (Cassirer 1921, pp.382–384). This corresponds to the fact that general relativity, unlike classical mechanics and special relativity, must have a covariant formulation.

More recent scholarship initiated by Stachel (1980) reconsiders Einstein's remarks about general covariance and the principle of relativity as follows. The missing premise of Einstein's remarks from 1916 goes back to his 1912 attempt to give a generally covariant weak field approximation to Newtonian gravity. In this connection, he formulated the so-called "Hole Argument" (*Lochbetrachtung*) about the possibility of considering a "hole" in the space-time continuum or "empty space." The problem is that the principle of univocal coordination does not apply to the supposed points inside the hole. Einstein later developed the insight that bare mathematical points inside the hole have no physically defining properties by assuming that these points are unobservable in principle and should have no place in physical theory.[7]

---

[7] On the reconstruction of the Hole Argument, see Norton (2011). On the connection with the principle of univocal coordination, see Ryckman (1999, p.593): "For Einstein, the fundamental meaning of general covariance may be expressed thus: there can be no principled distinction between the structure of space-time and its 'contents'; in brief, 'no metric, no space-time.' Thus space and time have lost 'the last remnants of physical objectivity.' In this broadened sense, then, the *physical meaning* of general covariance encompasses the purely formal *requirement* of general covariance (freedom to make 'arbitrary' (including nonlinear) transformations of the coordinates); the former, the field-theoretic programmatic framework within which the distasteful notion of an inertial sys-

In Cassirer's interpretation, the philosophical aspect of Einstein's theory lies in its consequences for the idea of physical objectivity and of reality in general. Cassirer maintained that Einstein's theory lent plausibility to Pierre Duhem's view that in order to provide the basis for the mathematical development of physical theory, empirical facts have to be transformed and put into a "symbolic form" (Duhem 1906, p.322). Cassirer (1921, p.427, and note) borrowed this expression from Duhem to indicate the fact that the interpretation of measurements presupposes theoretical principles and in the latter, a general function of coordination between the principles and the empirical manifold. According to Cassirer, Einstein revised the principles of measurement in order for such a coordination to be univocal. Therefore, Cassirer maintained that: "The postulate of relativity may be the purest, most universal and sharpest expression of the physical concept of objectivity, but this concept of the physical object does not coincide, from the standpoint of the general criticism of knowledge, with reality absolutely" (Cassirer 1921, p.446). The reference to particular facts does not provide a plausible criterion of objectivity, because the exclusion of arbitrary (i.e., psychological or anthropomorphic) factors from our images of nature presupposes mathematical abstraction. As Cassirer put it, "'relativization,' the resolution of the natural object into pure relations of measurement constitutes the kernel of physical *procedure*, the fundamental cognitive function of physics" (p.446). Considered the advancement of general relativity in the process of "resolution" of natural objects into observer-independent relations of measure, Cassirer contrasted physical objectivity with the naïve realist tendency to hypostatize a single concept of reality and to set it up as a norm and pattern for all the others. It was in this connection that Cassirer first posed the problem of accounting for different symbolic forms:

> It is the task of systematic philosophy, which extends far beyond the theory of knowledge, to free the idea of the world from this one-sidedness. It has to grasp the whole system of symbolic forms, the application of which produces for us the concept of an ordered reality, and by virtue of which subject and object, ego and world are separated and opposed to each other in definite form, and it must refer each individual in this totality to its fixed place. (Cassirer 1921, p.447)

Summing up, the idea of univocal coordination played a central role in Cassirer's argument in two senses. Firstly, it provided the construction and the exact definition of series of elements, as in the case of Dedekind's definitions of numerical domains. Secondly, univocal correlation between different series was the guiding idea for the definition of physical objectivity in Cassirer's interpretation of the history of science in terms of the logic of the concept of function. Correlation in this second sense or coordination includes the relation between mathematical symbols and the empirical manifold. This view has more recently been reconsidered in the literature about early philosophical interpretations of general relativity. In particular, although the discussion about univocal coordination goes back to Joseph Petzoldt's 1895 paper

---

tem can finally be dismantled, has the latter as an implication, and so any theory in which there is no principled distinction between space-time structure and the 'contents' of space-time must be given a generally covariant formulation."

"The Law of Uniqueness," Don Howard attributes one of the first model-theoretic versions of the principle to Cassirer: "Cassirer not only asserts the *Eindeutigkeit* of the real, but also the *Eindeutigkeit* of those objects implicitly constructed or defined within a mathematical theory by the accumulation of sufficiently many axioms or principles intended to characterize the essential properties of the objects supposed to fall within the domain of the theory, in effect, the objects constituting a model for the theory" (Howard 1988, p.199). According to Howard, the principle, in this formulation, amounts to the requirement that an acceptable theory be categorical in the later terminology of model theory. In this sense, Howard (1988, p.158) distinguishes model-theoretic *Eindeutigkeit* from the classical, metaphysical version of the principle, according to which the world described by our theories is unique. Given Cassirer's commitment to the view that mathematics is synthetic, however, Howard charges him with blurring the distinction between model and reality. Therefore, Howard attributes a metaphysical version of the principle to Cassirer as well.

Cassirer's argument for the physical significance of general covariance has been reconsidered by Ryckman (1999). Similar to Howard, Ryckman relies upon the consideration that the relation of isomorphism is implicit in Cassirer's definition of coordination (see especially Ryckman 1991, p.69). With regard to Cassirer's conception of reality, Ryckman considers Cassirer's analysis of the significance of general covariance to be one of his main motivations for a further "relativization" of knowledge besides the relativized a priori, that is, for the transition from the critique of knowledge to the philosophy of symbolic forms. With the requirement of general covariance, the general theory of relativity has shown the purely symbolic nature of the physical concepts of space and time, which have nothing to do with the meanings of these concepts in such cultural domains as art and myth.[8] However, there seems to be a paradox between the widening of Cassirer's perspective to other symbolic forms and the demand of abstraction from anthropomorphic factors in the physical conception of objectivity. On the one hand, the philosophy of symbolic forms enables Cassirer to account for different dimensions of reality in non-reductionist terms. On the other hand, he seems to reintroduce anthropomorphism in the philosophical account of physical science, insofar as he announces the view that theoretical knowledge (in particular exact science) provides only a single illustration of the more general notion of symbolic form. As Ryckman puts it, "the inevitability of anthropomorphism – a tenet of any idealism – is tied up with this initial announcement of the project of a philosophy of symbolic forms" (Ryckman 1999, p.613).

The idealist aspect of Cassirer's philosophical project has been criticized especially by Friedman. In contrast to Schlick, Cassirer did not clearly distinguish mathematical structure from empirical contents. According to Friedman, Cassirer's idealized conception of physical reality, therefore, opens the door to an idealistic system of symbolic forms, in which any consistent construction of metaphysics – however arbitrary – is on a level with scientific theories (Friedman 1999, p.27, and

---

[8] Cassirer articulated the idea of different although equally relevant concepts for the spatial ordering of appearance in his 1931 essay on "Mythical, Aesthetic, and Theoretical Space."

note). For the same reason, Friedman maintains that Cassirer's view of physical reality precluded him from posing the problems concerning the coordinating principles linking mathematical structures to empirical concepts correctly. This corresponds to the fact that Friedman advocates a classical reading of the requirement of general covariance as a formal property and identifies the coordinating principle of general relativity as the principle of equivalence. However, the main objection concerns the fact that Cassirer – in line with the interpretation of Kant in the Marburg School of neo-Kantianism – denied the assumption of pure intuition as a source of mathematical knowledge and as a mediating term between mathematics and reality. Friedman's objection is that the mathematical concept of function cannot fulfill the same role. Therefore, he denies that Cassirer's logic of knowledge can account for the stratification of the conditions of knowledge which is characteristic of Kant's architectonic. According to Friedman, Cassirer's view qualifies more as a formalistic variant of holism, according to which theoretical statements can be subject to revision by including the mathematical structures of the previous theories in the structure of the new theory. However, such a view would fail to appreciate discontinuities across theory change in such cases as general relativity, as the introduction of new coordinating principles contributed to the discovery of the new structure itself. The advantage of Friedman's relativized conception of a priori knowledge over Cassirer's resides in the possibility of accounting for both continuities and discontinuities across theory change (Friedman 2005).

To sum up, different objections against Cassirer's argument focused on the fact that he, by elevating the mathematical concept of function to an epistemological principle, was faced with the shortcomings of a formalistic view of scientific theories and with the paradoxes of an idealistic conception of reality. My reconstruction offers a partial defense of Cassirer's view, insofar as the kind of continuity he was concerned with in his considerations about the applicability of non-Euclidean geometry was not one of mathematical structures in the formal sense, but depended more on the heuristic function of the mathematical method. In order to appreciate this aspect of Cassirer's argument, my suggestion is to reconsider his broader perspective on mathematical method in contrast to Schlick's. Despite the fact that the latter view imposed itself over the past century, Cassirer's view was based on a well-documented account of the nineteenth-century mathematical tradition and of how this prepared Einstein's revision of the principles of measurement. The next section provides evidence of the fact that Cassirer thereby profited largely from arguments drawn from the empiricist tradition, especially if one considers his references to Helmholtz in support of the view above concerning the relation between pure and applied mathematics. Whereas Cassirer's neo-Kantian perspective appears to be at odds with an empiricist epistemology in his dispute with Schlick, his reception of earlier empiricist views sheds some light on the aspects of Cassirer's epistemology that cannot be reduced to a formalistic approach.[9]

---

[9] On Helmholtz's influence on Cassirer's idea of univocal coordination as a necessary requirement for physics, cf. Ryckman 1991. The proposed interpretation is largely indebted to Ryckman's article. However, it offers a different account of Helmholtz's legacy insofar as Ryckman suggests

Regarding Friedman's more specific objection against Cassirer's unKantian architectonic of knowledge, I argued in the previous chapters that Cassirer's logic of the mathematical concept of function, when reconsidered in context, provides an equivalent differentiation of the conditions of knowledge based on Cassirer's reconstruction of the interaction between nineteenth-century mathematics and physics. In order to show that Cassirer's perspective implies a relativized conception of the a priori, the next section compares Cassirer's account of continuities and discontinuities across theory change in the case of general relativity to Reichenbach's account of 1920.

7.3 Kantianism and Empiricism

Schlick's discussion of Cassirer's philosophical interpretation of general relativity is found in a paper published in 1921 in *Kant-Studien* under the title "Critical or Empiricist Interpretation of Modern Physics? Remarks on Ernst Cassirer's *Einstein's Theory of Relativity*." This paper was influential in the development of logical empiricism in opposition to Kantianism and neo-Kantianism, as it contains a classical objection against Kant's assumption of a priori synthetic judgments. Schlick relied on the theory of implicit definitions to argue for the view that mathematics is analytic. He borrowed the view from Poincaré that geometrical axioms are neither a priori synthetic judgments nor empirical statements, because they are conventions. According to Schlick, Poincaré's view was confirmed by Einstein's revision of the principles of measurement. Therefore, Schlick contrasted the logical empiricist interpretation of general relativity with both Cassirer's view and the view advocated by Hans Reichenbach in *The Theory of Relativity and A Priori Knowledge* (1920). In both cases, Schlick drew attention to the problems of any attempt to indicate a priori synthetic judgments in Kant's sense after general relativity.

This section provides a brief account of the points of agreement between Cassirer and Reichenbach and of Cassirer's discussion with Schlick. Although Schlick's argumentative strategy proved to be more successful, I suggest that Schlick's focus on the status of specific judgments obscured the possibility of reinterpreting the notion of synthetic a priori in terms of a range of hypotheses. Such a possibility, as envisioned by Cohen and explored especially by Cassirer, enables us to reconsider Cassirer's view of the relation between mathematics and physics. Cassirer traced his view back to Cohen and to the role of mathematical symbols in the empiricist epistemologies of such scientists such Helmholtz and Heinrich Hertz. My suggestion is that Cassirer's synthesis between neo-Kantianism and Helmholtz's empiricism led him to develop a relativized conception of a priori independently of the more stan-

---

that a model-theoretic approach is also implicit in Helmholtz's theory of signs (Ryckman 1991, p.69). Given the problems of this reading, I emphasize that the interpretation of Helmholtz's theory of perception in terms of mathematical structures is justified only insofar as it reflects the fact that there are both a top-down and a bottom-up direction in the conceptual articulation of experience.

dard view advocated by Friedman and drawn back by him to Reichenbach (1920). Cassirer's view differs from Reichenbach's because it is rooted in Cassirer's account of mathematical method as one of the highest expressions of rationality and as a tool of discovery.

7.3.1   Reichenbach and Cassirer

In his first published book, *The Theory of Relativity and A Priori Knowledge* (1920), Reichenbach distinguished two meanings of the notion of a priori in Kant's philosophy. On the one hand, a priori principles are supposed to be valid for all time. On the other hand, they are constitutive of the objects of experience insofar as they provide nonempirical presuppositions for the definition of empirical concepts. The second meaning corresponds to the fact that, for Kant, "the object of knowledge is not immediately given but constructed, and that it contains conceptual elements not contained in pure perception" (Reichenbach 1920, p.49).

Kant identified the a priori part of scientific knowledge as Euclidean geometry and Newton's laws of motion. He distinguished between the foundations of Newtonian physics and specific laws of nature, such as universal gravitation (Kant 1786, p.469). The foundations are a priori insofar as they do not depend on observation and experiment, in the manner that empirical generalizations do, and provide us, at the same time, with necessary presuppositions for the formulation of empirical judgments.

After radical changes in the foundations of physical theory, Reichenbach's distinction suggests that the first meaning of Kant's notion of a priori was disproved. Reichenbach argued for a reconsideration of the notion of a priori in the second sense. Therefore, he introduced another distinction between axioms of coordination and axioms of connection. The axioms of coordination differ from those of connection "in that they do not connect certain variables of state with others but contain general rules according to which connections take place" (Reichenbach 1920, p.54). Reichenbach mentioned, for example, the principle of probability. This may serve as an example of coordinating principle, insofar as it defines when a class of measured values is to be regarded as pertaining to the same constants. Insofar as this kind of principles defines the individual elements of reality, Reichenbach considered the axioms of coordination to be constitutive of the real object. The axioms of connection, on the other hand, correspond to the individual laws of physics. Kant's example was Newton's law of gravitation. In the context of general relativity, the axioms of connection include Einstein's equations of gravitation and metrical geometry.

Regarding space-time theories, Reichenbach's view implies a relativized conception of a priori knowledge according to the invariants of the groups of transformations under consideration. In special relativity, for example, the structure of Newtonian space-time is replaced with the structure of Minkowski space-time. According to Reichenbach's definition, both structures provide axioms of coordination relative to specific theories. In the case of general relativity, a priori knowledge includes only the infinitesimal Lorentzian manifold structure, whereas the particular Riemannian metric realized within this framework depends on the distribution of mass and energy. Einstein's use of Riemannian geometry suggests that geometrical hypotheses differ in their physical meaning. The revolutionary aspect of Einstein's general relativity lies in the fact that the specific principles of metrical geometry provide us with axioms of connection in Reichenbach's sense and, therefore, belong to the empirical part of the theory. For the same reason, Reichenbach criticized Poincaré's argument for the equivalence of Euclidean and non-Euclidean geometry, and especially Poincaré's exclusion of Riemannian geometry from the hypotheses considered: "If [Poincaré] had known that it would be this geometry that physics would choose, he would not have been able to assert the arbitrariness of geometry" (Reichenbach 1920, p.109).

Arbitrariness for Reichenbach occurs in the formulation of the axioms of coordination. He wrote:

> It is [...] not possible, as Kant believed, to single out in the concept of object a component that reason regards as necessary. It is experience that decides which elements are necessary. The idea that the concept of object has its origin in reason can manifest itself only in the fact that this concept contains elements for which *no* selection is prescribed, that is, elements that are independent of the nature of reality. The arbitrariness of these elements shows that they owe their occurrence in the concept of knowledge altogether to reason. *The contribution of reason is not expressed by the fact that the system of coordination contains unchanging elements, but in the fact that arbitrary elements occur in the system.* This interpretation represents an essential modification compared to Kant's conception of the contribution of reason. The theory of relativity has given an adequate presentation of this modification. (Reichenbach 1920, pp.88–89)

Arbitrariness in the formulation of the axioms of coordination enabled Reichenbach to vindicate their constitutive character. At the same time, the constitutive role of the axioms of coordination in general relativity presupposed a modification of Kant's view about the status of the a priori components in their relationship to the empirical part of physical theory: whereas empirical elements are necessary, arbitrariness affects a priori elements.[10]

Although Reichenbach did not refer to earlier attempts to disentangle the notion of a priori from that of necessity, his distinction of the two meanings of the a priori is clearly reminiscent of Cohen's distinction between the metaphysical and the transcendental a priori in Kant's work.[11] A historicized conception of the a priori clearly emerges, furthermore, from Cohen's identification of the transcendental method as the analytic method adopted by Kant in the *Prolegomena*. Whereas Kant's analytic method presupposed the synthetic method of the *Critique of Pure Reason*, Cohen defended the autonomy of the first method, which, according to him, depends solely on the fact of science and consists of the conceptual reconstruction of its presuppositions.

---

[10] For the interpretation of Reichenbach's view as a relativization of the notion of a priori, see Friedman 1999, pp.59–62.

[11] Cohen's distinction is discussed in Chap. 2.

Regarding the relativization of the a priori in relativistic physics, it is noteworthy that there are substantial points of agreement between Reichenbach and Cassirer. Not only did Cassirer (1910) emphasize the inductive aspect of his project of a universal invariant theory of experience relative to scientific knowledge in its historical development, but in 1921, he also fundamentally agreed with Reichenbach as to the revolutionary and philosophical aspect of Einstein's general relativity. On the one hand, Cassirer identified the a priori of space as the line element in its general form; on the other hand, he acknowledged the correlation between space, time and matter in the determination of specific metrical properties. Such an agreement is confirmed by the fact that Reichenbach and Cassirer read and expressed appreciation for each other's works. Reichenbach reported that he read Cassirer's book while his own was in press. Although he was not able to deal with Cassirer in *The Theory of Relativity and A Priori Knowledge*, Reichenbach referred both to Cassirer's *Substance and Function* (1910) and to *Einstein's Theory of Relativity* (1921) as the most suitable starting point in neo-Kantianism for initiating a debate between scientists and philosophers about the philosophical significance of general relativity (Reichenbach 1920, p.72, note 20). Cassirer read Reichenbach's work in manuscript. Since Cassirer's work was being printed, he wrote in a bibliographical note:

> I can here only refer to this thorough and penetrating work, which has much in common with the present essay in its way of stating the problem; I cannot, however, completely agree with its results, especially with regard to the relation of the theory of relativity to the Kantian critique of cognition. (Cassirer 1921, p.460)

Friedman draws attention to Reichenbach's relationship to Cassirer (Friedman 1999, p.63). Nevertheless, in his historical reconstruction of the idea of a relativized a priori, Friedman refers not so much to Cassirer, as to Reichenbach's view in 1920. Friedman's objection to Cassirer is that there is no place for axioms of coordination in Cassirer's project of a universal invariant theory of experience insofar as Cassirer defends Duhem's view that empirical facts occur in physics only after their transposition into a symbolic form. It follows that the empirical part of a physical theory cannot be considered regardless of theoretical, higher level assumptions.

However, Cassirer's relativized conception of the a priori emerged clearly from his consideration of the epistemological significance of Riemann's 1854 habilitation lecture "On the Hypotheses Which Lie at the Foundation of Geometry" (1867), especially if one considers the following quote from the fourth volume of *The Problem of Knowledge*. Cassirer wrote:

> Even the very title of this work suggests the revolution in thinking that had come about in mathematics, for Riemann speaks of "hypotheses," where his predecessors had spoken of "axioms." Where absolute and self-evident propositions had been envisioned he sees "hypothetical" truths that are dependent upon the validity of certain assumptions, and no longer expects a decision on this validity from logic or mathematics but from physics. [...] The whole character of mathematics appeared radically changed by this view, and axioms that had been regarded for centuries as the supreme example of eternal truth now seemed to belong to an entirely different kind of knowledge. In the words of Leibniz, the "eternal verities" had apparently become merely "truths of fact." (Cassirer 1950, pp.21–22)

Cassirer maintained that Riemann's ideas revolutionized the traditional way of thinking about mathematical truth and truth in general. Arguably, in 1921, Cassirer did not emphasize this aspect, because he was mainly concerned with continuities in the anticipatory role of the mathematical method. The quotation above shows that, on the other hand, he fundamentally agreed with Reichenbach regarding the discontinuity. Cassirer's disagreement with Reichenbach does not concerns so much the physical meaning of Riemannian geometry in general relativity, but Reichenbach's interpretation of general covariance and his conclusions about the arbitrariness of reason. The coordination between geometry and empirical reality for Cassirer rather presupposes the freedom of mathematical thinking in Cohen's sense: freedom or spontaneity is an essential characteristic of reason because, contrary to arbitrariness, it enables us to account for the positive role of mathematics in the formulation of new hypotheses. It is because of the anticipatory role of the mathematical method in the formulation of hypotheses – and not with regard to a purely formal consideration of mathematical structures – that Cassirer defended the view that mathematics is synthetic. As the status of mathematics plays a central part in the debate under consideration, it may be helpful to turn back to this point after a brief account of Cassirer's discussion with Schlick on this point.

7.3.2 Cassirer's Discussion with Schlick

In "Critical or Empiricist Interpretation of Modern Physics?" (1921), Schlick formulated two main objections against Cassirer's philosophical interpretation of Einstein's general relativity. The first concerns the conception of critical philosophy. According to Schlick, critical philosophy has its roots in its relationship to Newtonian physics: the goal of Kant's transcendental philosophy is to prove that the constitutive principles of physics are a priori synthetic judgments; where a priori knowledge is characterized by its being apodictic (i.e., universal and necessarily valid). Schlick's objection regarding the possibility of indicating the a priori synthetic judgments of other physical theories (in particular of general relativity) is that this would presuppose an indiscriminate broadening of the Kantian perspective. For Schlick, the logical idealism of the Marburg School of neo-Kantianism "runs the risk of losing its bold colouring, and hence its philosophical value; the most heterogeneous opinions could be brought in under it" (Schlick 1921, p.323).

The second objection concerns, more specifically, Cassirer's revision of the Kantian theory of space. As a consequence of his reliance upon the literal meaning of a priori knowledge as introduced by Kant (i.e., what Cohen called the metaphysical a priori), Schlick considered the Kantian theory of space committed essentially to the view that a priori synthetic judgments are grounded in the form of outer intuition and, therefore, admit a unique formulation. Pure intuition is supposed to provide necessary preconditions for the use of ideal constructions in the definition of empirical objects. Cassirer's identification of the a priori of space as the general form of the line element is problematic. On the one hand, even the supposed system of axioms, including, for example, the axiom of continuity, might not be invariant relative to the transformation groups of other physical theories (e.g., of quantum mechanics). In other words, even a generalized form of spatial intuition would be too restrictive a commitment to account for every physical theory, and hence, to work as a universal invariant of experience in Cassirer's sense. On the other hand, the substantial advance of general relativity depends not so much on the system of axioms under consideration, as on the correlation between space, time and matter established by the use of Riemannian geometry in the expression of specific metrical properties. In this regard, Cassirer's argument does not seem to be specific enough to account for the empirical status of geometry in general relativity. Schlick maintained that a more promising philosophical approach to the problem of measurement is given by the principle that "differences in reality may be assumed only where there are differences that can, in principle, be experienced" (Schlick 1921, p.330). According to Schlick, this principle goes back to Leibniz's principle of observability, and it was assumed in this form by Cassirer as well (Cassirer 1921, p.376); but it was only in the epistemology of Ernst Mach that the formulation of the principle as an empiricist demand played a fundamental role in the development of the principle of relativity from Mach to Einstein.

Schlick believed that the Kantian theory of space lost its significance altogether on account of the particular metrical structure of general relativity:

> The spatial and the temporal retain a meaning in which they can no longer be regarded merely as "forms" in the usual sense, but now belong to the physical determinants of bodies; the "metric" does not just mean a mathematical measuring of the physically real, but itself gives expression to the presence of this. (Schlick 1921, p.329)

But the main objection depends on the said distinction between the conceptions of geometry as a theory of space, on the one hand, and as an axiomatic system, on the other. The use of Riemannian geometry in general relativity shows that geometry in the former sense is not independent of experience. However, Schlick ruled out the assumption of pure intuition, mostly because he believed that Kant's problem of indicating a source of mathematical certainty had been solved by the method of implicit definitions. Accordingly, there can be no mediating term between thought and reality, concepts and intuitions: the method of implicit definitions attains exactness precisely because of the unbridgeable gap which subsists between pure concepts and psychological factors. In this regard, Schlick distanced himself from both Kantianism and Mach's positivism, and adopted Poincaré's definition of geometrical axioms as conventions. Schlick apparently overlooked Poincaré's idea of a topological intuition of space. Nevertheless, it is worth noting that Schlick's distinction between intuitive and mathematical space corresponds to Poincaré's distinction between topological and metrical properties. As we saw in the previous chapter, it is because of the need of conventions for the introduction of the concept of measure in the topological continuum that Poincaré emphasized the conventional aspect of the formation of geometrical concepts. On the one hand, Schlick used the distinction between intuitive and geometric space against the Kantian assumption of pure intuition; on the other hand, he recognized the possibility of exploring a psychological counterpart of Kant's forms of intuition (Schlick 1921, p.332).

Schlick's conclusion is that constitutive principles are either hypotheses or conventions: "In the first case they are not a priori (since they lack apodicticity), and in the second they are not synthetic" (Schlick 1921, p.324). For similar reasons, in the concluding part of Schlick (1921), he distanced himself from Reichenbach's relativization of the notion of a priori. For Schlick, Reichenbach's attempt to disentangle the constitutive meaning of the a priori from its apodictic validity leads to the abandonment of critical philosophy altogether. However, Schlick attributed his disagreement with Reichenbach to a choice of terminology. In order to make it clear that Reichenbach's constitutive principles can be subject to revision, Schlick's suggestion was to call them conventions in Poincaré's sense (Schlick 1921, p.333).

Following Schlick's suggestion, in a later paper, "The Present State of the Discussion on Relativity" (1922), Reichenbach classed himself with Schlick, Poincaré and Einstein as representative of the relativistic conception of physical geometry and contrasted the latter conception with Cassirer's neo-Kantianism, on the one hand, and with Mach's and Petzoldt's positivism, on the other.[12]

Friedman has made it plausible that Reichenbach's change of terminology involves, in fact, a substantive change in doctrine. This change particularly rules out the possibility of theory-specific a priori principles. In order to reconsider this aspect of Reichenbach's original view, Friedman draws attention to what he considers to be the most important result of Reichenbach's approach of 1920: "In the context of general relativity, physical geometry (the metric of physical space) is *no longer* constitutive" (Friedman 1999, p.66). As Reichenbach put it, the principles of physical geometry have been transformed from axioms of coordination into axioms of connection. It was for that reason that Reichenbach distanced himself from Poincaré's geometrical conventionalism: Reichebach's approach enabled him to attribute a univocal meaning to Riemannian geometry in its correlation with space, time and matter. To call the constitutive principles conventions now amounts to an advocation of the holistic view that geometrical hypotheses – as any hypothesis in the context of a physical theory – cannot be put to the test in isolation, but only the total system of hypotheses has empirical consequences. By contrast, the relativized perspective on the a priori advocated by Friedman presupposes a distinction between constitutive and empirical elements relative to specific theories and entails that constitutive elements can be subject to revision by introducing new coordinating principles (i.e., in the case of general relativity, the equivalence principle and the principle of the invariance of the velocity of light).

Friedman's objection to Cassirer is strictly related to his reconstruction of this debate, because Friedman's starting point is Schlick's argument against the assumption of pure intuition. Friedman writes:

> The Kantian bridge between thought and reality – namely, pure intuition – has indeed been demolished. Hence, if we persist in a "holistic" and "formalistic" account of knowledge and

---

[12] For further evidence of Schlick's influence on Reichenbach in this regard, see Parrini (1993), Friedman (1999, pp.63–68) and Ryckman (2005, Ch.3).

judgment, we are driven toward idealism and the coherence theory of truth. In particular, we will have a hard time distinguishing physical or empirical knowledge from pure mathematics, on the one hand, and from arbitrary coherent systems of metaphysics or myth, on the other. Either way it will be difficult to maintain the Kantian commitment to mathematical physics as a paradigm of knowledge. (Friedman 1999, p.27)

Friedman makes it clear in a footnote that this view was the path taken by Cassirer and the Marburg School of neo-Kantianism with the denial of pure intuition. Since Friedman attributes a coherence theory of truth to Cassirer, he puts Cassirer's revision of the Kantian theory of space in connection with the later development of the philosophy of symbolic forms. Notwithstanding the connection between Cassirer's interpretation of general relativity and the idea that there are different forms of spatiality and temporality, this reading overlooks the fact that it is problematic to attribute a coherence theory of truth in the current sense to Cassirer. The functional dependence that subsists between the elements of a theory for Cassirer differs from the coherence of a proposition with some specified set of propositions on account of Cassirer's reliance on Kant's transcendental logic. Both Cassirer's logic of objective knowledge and transcendental logic ruled out what Cassirer called a "copy" theory of knowledge – namely, the view that the truth conditions of knowledge depend on a relation of similarity with an external object. However, from such a Kantian perspective of knowledge as Cassirer's, the objects of knowledge are constituted according to the conditions of experience, and not merely thought without contradiction. Cassirer's conditions differ from Kant's, because in Cassirer's view, the Kantian bridge between thought and reality (i.e., pure intuition) corresponds to the logic of the mathematical concept of function in the history of mathematics and physics.

Furthermore, Friedman's description of Cassirer's view presupposes Schlick's argument against the assumption of pure intuition. Cassirer's account of theory change then appears to imply a holistic and formalistic account of knowledge rather than a relativized conception of the a priori.[13] However, we have already drawn attention to the fact that Schlick's conception of geometry as completely detached from reality differed significantly from Cassirer's. Whereas Schlick argued for a sharp separation between geometry as a description of space and as an axiomatic system, Cassirer referred to the axiomatic method as one of the clearest expressions of the tendency to substitute concepts of substance with concepts of function. It is because of this tendency that nineteenth-century geometry (in particular Riemannian geometry) provided the conceptual foundations for a solution of the problem of physical space. Even in 1910, Cassirer did not identify the form of space generally as a specific axiomatic structure, but as a system of hypotheses, whose specification in terms of metrical geometry is a matter for empirical investigation. In 1921, Cassirer fundamentally agreed with Schlick and Reichenbach about the physical meaning of Riemannian metric: the revolutionary aspect of Einstein's theory and its philosophical significance depend on a shift of metrical geometry from the purely

---

[13] On Friedman's contrast between his own conception of the a priori after general relativity and Cassirer's, see esp. Friedman (2005).

mathematical to the empirical part of the theory. At the same time, Cassirer's emphasis is on the continuity with the nineteenth-century mathematical tradition. The use of Riemannian geometry in Einstein's equations confirmed the conviction, expressed by mathematicians such as Riemann and Klein, that the introduction of abstract concepts for the classification of geometrical hypotheses was necessary for a better understanding of the connection of geometry with different branches of mathematics and with physics. Cassirer's goal was to find a justification of such a conviction in the Kantian view that mathematics is synthetic. Therefore, he proposed a revision of the Kantian theory of space regarding the assumption that synthetic a priori judgments are grounded in pure intuition. Cassirer replaced Kant's forms of intuition with those rules of coordination that enable a comparison between mathematical structures. The same rules of coordination provide a mediating term between mathematical structures and observations. Mediation is possible, because even at the level of the formation of mathematical concepts, the meaning of mathematical symbols depends on conceptual relations and mathematical reasoning. Symbols occur in the definition of physical objects insofar as complexes of empirical and conceptual conditions are sufficiently determined to make the interpretation of the phenomena univocal.

Considering Cassirer's conception of symbol, it appears inappropriate to attribute to him a formalistic approach to mathematics or even to science. For the same reason, Cassirer's interpretation of Duhemian holism clearly presupposes a stratification of levels in the conditions of knowledge. The difference with Kant lies in the fact that Cassirer acknowledged the possibility of shifts from the mathematical to the empirical level in line with Cohen's historicized conception of the a priori. In this regard, Cassirer clearly distanced himself from Schlick's conventionalism. This is confirmed by the fact that, whereas Reichenbach adopted Schlick's view in 1922, Cassirer defended a neo-Kantian view of the a priori in response to Schlick's critique.[14]

In order to highlight this point, the concluding section provides a brief account of Cassirer's reply to Schlick about the possibility of reconsidering the notion of a priori after general relativity. To conclude, I compare the neo-Kantian view with Helmholtz's empiricist approach to the relation between mathematics and empirical reality. Not only did neo-Kantian and empiricist methodologies agree fundamentally on the idea of an interaction in the development of mathematics and empirical knowledge, but Cassirer especially agreed with Helmholtz that a shift from the mathematical to the empirical level, and vice versa, was justified by the requirement of the comprehensibility of nature. Regarding Schlick's objection against the philosophical relevance of Cassirer's work on Einstein's general relativity, it will be helpful to consider the fact that Cassirer's connection with Kantianism and empiricism goes back to his early writings. In 1921, Cassirer drew the connection in the same terms and reaffirmed his commitment to a neo-Kantian conception of knowledge.

---

[14] On Cassirer's discussion with Schlick, see Ferrari (1991).

7.3.3   Kantian and Neo-Kantian Conceptions of the A Priori

In a letter to Schlick dated October 23, 1920, Cassirer replied to Schlick's comments on the manuscript of *Einstein's Theory of Relativity* as follows. Cassirer maintained that his interpretation of critical philosophy and Schlick's empiricism largely agreed on the approach towards the epistemological issues at stake concerning general relativity. According to Cassirer, his disagreement with Schlick depended mainly on the conception of the a priori. Whereas Schlick relied upon Kant's definition of a priori in terms of apodictic knowledge, Cassirer deemed the a priori was "not a constant, once and for all fixed component of 'intuitions' or concepts, but a function that is determined according to laws, and therefore remains identical with itself in its *direction* and form. However, with the advancement of knowledge, the same function can assume the most diverse characteristics as for the content" (Cassirer 2009, pp.50–51). Therefore, Cassirer attributed apriority to the idea of unity of nature or the principle of univocal coordination alone. In a manner which is reminiscent of Helmholtz's distinction between general and narrower characteristics of the form of space, Cassirer distinguished the idea of univocal coordination from its specifications. He urged a revision of the Kantian theory of space regarding the meaning of such a distinction, insofar as Einstein's general relativity showed that specific coordinating principles can be subject to revision. Cassirer agreed with Schlick and with Reichenbach about the physical meaning of Riemannian geometry in general relativity. In contrast to empiricist views, Cassirer maintained that the principle of univocal coordination itself can be neither a convention nor an inductive generalization. Cassirer called it an "expression of reason" insofar as the same principle is necessary for empirical facts to be defined.

This letter was written before the publication of Schlick's review of Cassirer's book in *Kant-Studien*. Therefore, in the concluding part of the letter, Cassirer expressed his wish that Schlick's paper could initiate a dialogue between philosophers and physicists. On this occasion, Cassirer recognized that Schlick's standpoint was closer to the views of the physicists than his own. Cassirer's worry about his book was that the physicists might have perceived it as written in a "foreign language" (Cassirer 2009, p.51). What Cassirer might not have expected was that Schlick had no less influence on the philosophical aspects of the debate. Not only did Schlick rule out a relativized conception of the a priori as it emerged from the works of Cassirer and of Reichenbach, but he emphasized the opposition between Kantianism and empiricism. Schlick pointed out an impasse in the debate in order to argue for his integration of the empiricist view with a realist view of scientific knowledge, on the one hand, and with geometrical conventionalism, on the other.

Without going into the details of Schlick's view and its reception in the twentieth-century philosophy of science, in the concluding part of this chapter, I limit myself to pointing out that, by contrast, according to Cassirer, there were substantial points of agreement between neo-Kantian and empiricist approaches to the principles of knowledge. Cassirer's connection with empiricist views goes back to his early writings. In a 1907 lecture held in Berlin under the title "Substance and Function" and recently published in the eighth volume of Cassirer's Nachlass (2010), Cassirer made the same claim about the philosophical interpretation of energetics.[15] According to Cassirer, the shift from the concept of force to that of energy offered a solution to the problems concerning the application of geometrical concepts to empirical concepts. Cassirer particularly referred to Boscovich's problem of correlating mathematical and physical space: whereas mathematical space is continuous, physical space – as composed out of a finite number of non-extended force points – is discrete. Since the conservation of energy is an integral law, the physical processes under consideration need not have been reduced to infinitesimal quantities previously. Cassirer's remark is that the mathematical characterization of such a process as a whole does not require the mediation of an intuitive schema: "The variations that occur in experiment can be referred to and compared with one another immediately, without having been previously transformed in such a way that they have become homogenous and commensurable in the sense of intuition" (Cassirer 2010, pp.8–9).

Cassirer used the example of energy to avoid reference to external entities in the definition of fundamental concepts, and instead to argue for a functionalist interpretation of the foundations of physics. From a substantialist viewpoint, relations of measure are accidental. However, in the case of energy, it is clear that the assumption of an external bearer would be a metaphysical dogma. The definition of energy depends on a coordinating principle relating physical processes to one another, namely, the principle of the conservation of energy. According to the latter principle, energy is the numerical proportion that indicates the equivalence between different physical processes. In Cassirer's view, this way of proceeding shed light onto the relation between thought and experience. The concepts thereby introduced owe their mediating role in the interpretation of sense perception to their functional character: their definition refers not to the intuitive representation of the phenomena considered in their singularity, but to their relations to one another. Given the purely conceptual nature of the systems under consideration, Cassirer distanced himself implicitly from Kant's assumption of mediating schemas grounded in the forms of intuition. Nevertheless, Cassirer (2010, p.9) traced the role of the concept of function in the definition of physical concepts back to Kant's claim that "whatever we can cognize only in matter is pure relations [...]; but there are among these some self-sufficient and persistent ones, through which a determinate object is given to us" (Kant 1787, p.341). Cassirer referred to Kant regarding the view that the mathematical interrelation of the phenomena is not a fact, but rather a presupposition for the interpretation of the phenomena. Therefore, Cassirer compared empiricist and critical approaches to the problem of measurement as follows:

> Even though the particular contents of the functional law depend on experiment, the general *demand* for such laws is a condition and a guiding idea of empirical research. The phenomena could never assume a conceptual and especially a mathematical order *by themselves*, if

---

[15] The term "energetics," first introduced by William Rankine (1855), indicated the idea of a generalized thermodynamics as pursued in various ways by physicists such as Wilhelm Ostwald and Pierre Duhem.

it were not for the fact that this order has been imposed on them from the outset on account of the scientific formulation of the problem, which we posed. It appears that, in this regard, there is a clear methodological agreement between empiricism and critical philosophy. Both of them agree on the fundamental view that all of human knowledge does not attain absolute things, but culminates in a system of *relations*. Therefore, both views indicate in experience the domain and the limit of any contents of knowledge. (Cassirer 2010, p.13)

Cassirer recognized that the contents of mathematical structures depend solely on functional connections. The functionality of mathematical structures lies at the foundation of the coordination between these structures and the phenomena. Consider, for example, Helmholtz's theory of measurement, as discussed in Chap. 4. Critical philosophy in Cassirer's sense and Helmholtz's empiricism agreed that such a correlation is necessary for the definition of physical magnitudes. Therefore, empiricism and critical philosophy agreed, furthermore, on the identification of the conditions of accessibility as conditions of the objects of experience. For the same reason, Friedman (1997) interpreted Helmholtz's conditions as constitutive principles in the Kantian sense and contrasted Helmholtz's view with causal realism. Cassirer argued for critical idealism as a consequent development of the same view, insofar as this provides us with a clear distinction between physical objectivity and the idea of an absolute reality.

To sum up, Cassirer in 1907 relied on both neo-Kantian and empiricist approaches to scientific knowledge in the tradition inaugurated by Helmholtz, with one noteworthy exception. Cassirer distanced himself from Mach's empiricism, insofar as knowledge for Mach depends solely on observation, on the one hand, and on the conventional demand of the economy of thought, on the other. Mach's view presupposes the assumption of a mind-independent reality, which lies at the foundation of knowledge. Since, at the same time, Mach recognized the relational character of physical laws, the assumption of absolute elements called into question the possibility of objective knowledge itself. In order to avoid skeptical consequences, Cassirer argued for the constitutive role of conceptual relations in the determination of empirical contents.

In *Substance and Function* (1910), Cassirer dealt with the same issue by relying on Duhem's view that there is a tendency in the history of science to abstract from empirical knowledge in the common sense insofar as the role of theoretical assumptions in the interpretation of empirical facts increases. Duhem distanced himself from Mach by saying: "The theory is not solely an economical representation of experimental laws; it is also a classification of these laws" (Duhem 1906, p.32).

Cassirer's example was once again the following argument for a functionalist interpretation of the concept of energy: "Energy is able to institute an order among the totality of phenomena, because it itself is on the same plane with no one of them; because, lacking all concrete existence, energy only expresses a pure relation of mutual dependency" (Cassirer 1910, p.200). The same example played a key role in Duhem's views about the advancement of science, as he emphasized that, according to the law of conservation, such expressions as "quantity of heat" can be defined without borrowing anything from the specific perceptions of warmth and cold. Therefore, Duhem deemed energetics a "physics of qualities." As Duhem put it, "the advantages sought by past physicists when they substituted a hypothetical quantity for the qualitative property revealed to the senses, and measured the magnitude of that quantity, can be obtained without employing the hypothetical quantity simply by the choice of a suitable scale" (Duhem 1906, p.191).

Cassirer borrowed the idea of a physics of qualities from Duhem and characterized the advantage of energetics over the mechanistic conception of nature as follows:

> The conflict between these two conceptions can ultimately only be decided by the history of physics itself; for only history can show which of the two views can finally be most adequate to the concrete tasks and problems. Abstracting from this, however, energism is in any case of preeminent epistemological interest in so far as the attempt is made to establish the minimum of conditions, under which we can still speak of a "measurability" of phenomena in general. Only those principles and rules are truly general, on which rests the numerical determination of any particular process whatever and its numerical comparison with any other process. The comparison itself, however, does not presuppose that we have already discovered any unity of "essence" – for example, between heat and motion; but, on the contrary, mathematical physics begins by establishing an exact numerical relation, on the basis of which it also maintains the homogeneity of such processes as can in no way be sensuously reduced to each other. (Cassirer 1910, pp.202–203)

The example of the definition of energy showed that translation into the language of the abstract numerical concepts – no less than translation into the language of spatial concepts – presupposes a theoretical transformation of the empirical material of perception. Thereby, Cassirer emphasized the importance of distinguishing clearly between the mathematical principles, on the one hand, and those specific hypotheses which serve in the treatment of a particular field, on the other (p.203).

Although Cassirer did not mention Helmholtz in this connection, his way of proceeding clearly is reminiscent of Helmholtz's theory of measurement as presented in "Counting and Measuring" (1887). As we saw in Chap. 4, Helmholtz analyzed the concepts of number and of sum as given only in the form of inner intuition in order to formulate general additive principles suitable for a variety of interpretations in mathematical and physical domains. Specific hypotheses are required for any method of measurement. These are necessary, but not sufficient conditions of measurement in Helmholtz's sense, because the definition of physical magnitudes and the general validity of the results of measurement depend on the method of addition.

Furthermore, it is noteworthy that Duhem presented his argument for a physics of qualities as a generalized solution to the question: Under what conditions can a physical attribute be signified by a numerical symbol? Similar to Helmholtz, Duhem identified such conditions as the defining conditions of physical magnitudes. Duhem (1906, pp.173–177) assumed that: (1) if two magnitudes are both equal to a third magnitude, they are equal amongst themselves, and (2) if a first magnitude overcomes a second magnitude and this overcomes a third magnitude, the first magnitude overcomes the third as well. These assumptions enable the adoption of a measuring standard and, therefore, provide an interpretation of the symbols "=," ">" and "<," respectively. Additive principles are required for the interpretation of "+." Such principles are: (1) independence from order in the composition of magnitudes of the same kind, and (2) the equivalence of magnitudes of the same kind with their sum. The use of an arithmetic sum in the representation of the composition of magnitudes is justified by the possibility, under these conditions, to apply the laws of addition (i.e., the commutative law and the associative law, respectively).[16]

According to Duhem, the difference between quantities and qualities depends on the lack of conditions of the second kind in the case of the intensity of a quality. In other words, only the symbols "=," ">" and "<" apply to qualities, but not "+." This corresponds to the fact that the intensity of a quality does not result from a composition of many qualities of the same kind: qualities are heterogeneous. Nevertheless, Duhem emphasized the analogy between qualities and quantities with regard to the interpretation of numerical symbols. He wrote:

> As the definition of a magnitude is given by an abstract number only when conjoined with concrete knowledge about a measuring standard, the intensity of a quality is not entirely represented by a numerical symbol; this symbol must be given in conjunction with a concrete procedure for obtaining the scale of these intensities. Only the knowledge of this scale allows one to give a physical meaning to the algebraic propositions which we state concerning the numbers representing the different intensities of the quality studied. (Duhem 1906, p.190)

Insofar as algebraic propositions assume a physical meaning in both cases, Duhem argues for an extension of the scope of physics as follows: "In order to make physics a part of a universal arithmetic, as Descartes suggested, it is not necessary to imitate him and to exclude all qualities, because the language of algebra enables us to consider the different intensities of a quality as well as the different magnitudes of a quantity" (p.193).

Similarly, in the chapter of *Substance and Function* dedicated to "The Concept of Reality," Cassirer referred to Helmholtz, not as a representative of the classical view of mechanics, but as an epistemologist. Cassirer particularly emphasized the significance of Helmholtz's theory of signs for the modern conception of physical reality as follows:

> [N]atural science, even where it retains the concept of the absolute object, can find no other expression for its import than the purely formal relations at the basis of experience. This fact appears in significant form in Helmholtz theory of signs, which is a typical formulation of the general theory of knowledge held by natural science. Our sensations and representations are signs, not copies of objects. From a copy we demand some sort of similarity with the object copied, but we can never be sure of this in the case of our representations. The sign, on the contrary, does not require any actual similarity in the elements, but only a functional correspondence of the two structures. What is retained in it is not the special charac-

---

[16] Although Duhem did not refer to Helmholtz (1887), the influence of Helmholtz emerges clearly from Duhem's definition of quantity in terms of relations of equality and inequality established by a concrete operation of addition. He departed from Helmholtz (and from Maxwell's consideration of temperature measurement) by interpreting the latters' distinction between directly and indirectly measurable quantities as a rigid distinction. Whereas for Helmholtz and for Maxwell a property was a quality in the temporary lack of a concrete procedure of addition, for Duhem, a property identified as a quality had to remain a quality for all time. On Duhem's relationship to Helmholtz, see Darriol (2003, pp.567–569).

ter of the signified thing, but the objective relations, in which it stands to others like it. (Cassirer 1910, p.304)

To sum up, Cassirer's connection with empiricism in the individuation of increasing levels of generality in the classification of physical laws goes back to his early writings and is related strictly to his conception of symbols as expressions of objective relations. Cassirer referred to the ideas of such physicists as Duhem and Helmholtz about the interpretation of mathematical symbols in physical theories to support his reformulation of the Kantian schematism of the pure concepts of the understanding in terms of the logic of the concept of function. Insofar as Einstein's general relativity can be interpreted as a further generalization in the functionalist conception of physical reality, Cassirer pointed out that the leading ideas of *Substance and Function* were confirmed. Therefore, in 1921, Cassirer proposed a neo-Kantian conception of knowledge again and described the connection between critical idealism and empiricism as follows:

> It is indeed at first glance strange and paradoxical that the most diverse epistemological standpoints, that radical empiricism and positivism as well as critical idealism have all appealed to the theory of relativity in support of their fundamental views. But this is satisfactorily explained by the facts that empiricism and idealism meet in certain presuppositions with regard to the doctrine of empirical space and of empirical time, and that the theory of relativity sets up just such a doctrine. Both here grant to experience the decisive role, and both teach that every exact measurement presupposes universal empirical laws. (Cassirer 1921, p.426)

In 1907, Cassirer indicated the same points of agreement between neo-Kantianism and empiricism regarding the philosophical significance of energetics. Therefore, it is misleading to claim – as Schlick did – that it is in the interpretation of general relativity that Cassirer broadened his original view to the point of triviality. Furthermore, it is worth noting that in both 1907 and 1921, Cassirer referred specifically to methodological points of agreement, although we know from Chap. 4 that Helmholtz and Marburg neo-Kantians disagreed about the ultimate justification of knowledge. In the proposed interpretation, the agreement depends on a similar approach to the problem of giving a physical interpretation of mathematical symbols and on the assumption of general coordinating principles. Cassirer agreed with Helmholtz that there is both a top-down direction from mathematical laws to empirical interpretations and a bottom-up direction from the problems concerning empirical measurements to the development of generalized mathematical models for their solution in the conceptual articulation of experience. It was in these terms that Cassirer interpreted the use of Riemannian geometry for the solution of the problem of the heterogeneous in general relativity.

In 1921, Cassirer particularly referred to Helmholtz's (1870) mention of the possibility of considering the concept of a rigid geometrical figure as a transcendental concept because of its role in the definition of rigid bodies. At the same time, similar to Cohen before him, Cassirer distanced himself from Helmholtz's conclusion that a transcendental concept in this sense would imply analytic judgments. According to Cassirer, the coordinating principles establish a synthetic kind of unity: "Assumptions of this sort refer to the object in so far as in their totality they 'constitute' the object and render possible knowledge of it; but none of them, taken for itself, can be understood as an assertion concerning things or relations of things" (Cassirer 1921, p.438). Regarding the physical meaning of Riemannian geometry in general relativity, Cassirer added:

> That the elements, to which we must ascribe, methodologically, a certain "simplicity" must be adequate for the interpretation of the laws of nature, cannot be demanded a priori. But even so, thought does not simply give itself over passively to the mere *material* of experience, but it develops out of itself new and more complex *forms* to satisfy the demands of the empirical manifold. (Cassirer 1921, p.438)

These considerations suggest that the idea of constitutive principles relative to scientific theories emerged both from the possibility of giving a physical interpretation of non-Euclidean geometry as considered in different ways by Riemann and Helmholtz and from the historical perspective on the mathematical method in Marburg neo-Kantianism. Such a perspective led Cohen and especially Cassirer to give a plausible account of the view that mathematics is synthetic in terms of an interaction between immanent mathematical developments and the history of physics.

More importantly, also in connection with the current debate about theory change, Cassirer's neo-Kantian perspective on the a priori led him to argue for continuity across theory change in a way that differs significantly from the more well-known logical positivist argument. This is the argument that, even after such a radical change as general relativity, the consequences of the former theories can be derived as limiting special cases of the new theory. As pointed out by Thomas Kuhn, the problem of this view is that continuity can be showed only retrospectively, from the viewpoint of the latter theories. Although the laws derived in this way are special cases of relativistic physics, they are not Newtonian laws. "Or at least they are not unless those laws are reinterpreted in a way that would have been impossible until after Einstein's work" (Kuhn 1962, p.101). Cassirer did not restrict his consideration to the retrospective view, because he looked at the example of mathematics to explore the connection between symbolic thinking and future experience. It did not suffice to point out the limiting cases; Cassirer's goal was to show that even the new formulation of natural laws had been foreshadowed in the form of mathematical hypotheses. It is because of the symbolic character of mathematics that the system of experience, in the sense of the transcendental philosophy, is always capable of further generalizations.

In the *Dynamics of Reason* (2001), Friedman refers to Cassirer's project of a universal invariant theory of experience in order to address the same problem. However, Friedman believes that a solution to this problem requires us to regard a convergent sequence of successive frameworks of "paradigms" – to borrow Kuhn's expression – as "approximating at the limit (but never actually reaching) an ideal state of maximally comprehensive communicative rationality in which all participants in the ideal community of inquiry agree on a common set of truly-universal, trans-historical constitutive principles" (Friedman 2001, p.67). In other words, Friedman's suggestion of a positive characterization of such an ideal is to look for a more comprehensive idea of rationality outside the domain of the exact sciences.

Therefore, in the quotation above, he borrows the idea from Habermas of a communicative kind of rationality as opposed to an instrumental one.

The present study explored different approaches to the same problem based on a more comprehensive view of the mathematical method itself. Cassirer's argument offered one of the clearest expressions of a tradition that looked at mathematics, especially in its transformation in the nineteenth century from a science of quantities to the study of mathematical structures, as a source of discovery.

References

Cantor, Georg. 1883. *Grundlagen einer allgemeinen Mannigfaltigkeitslehre: Ein mathematisch-philosophischer Versuch in der Lehre des Unendlichen.* Leipzig: Teubner.

Cassirer, Ernst. 1910. *Substanzbegriff und Funktionsbegriff: Untersuchungen über die Grundfragen der Erkenntniskritik.* Berlin: B. Cassirer. English edition in Cassirer (1923): 1–346.

Cassirer, Ernst. 1921. *Zur Einstein'schen Relativitätstheorie: Erkenntnistheoretische Betrachtungen.* Berlin: B. Cassirer. English edition in Cassirer (1923): 347–465.

Cassirer, Ernst. 1923. *Substance and function and Einstein's theory of relativity.* Trans. Marie Collins Swabey and William Curtis Swabey. Chicago: Open Court.

Cassirer, Ernst. 1931. Mythischer, ästhetischer und theoretischer Raum. *Zeitschrift für Ästhetik und allgemeine Kunstwissenschaft* 25: 21–36.

Cassirer, Ernst. 1950. *The problem of knowledge: Philosophy, science, and history since Hegel.* Trans. William H. Woglom, and Charles W. Hendel. New Haven: Yale University Press.

Cassirer Ernst. 2009. *Nachgelassene Manuskripte und Texte.* Vol. 18: *Ausgewählter wissenschaftlicher Briefwechsel*, ed. John Michael Krois, Marion Lauschke, Claus Rosenkranz, and Marcel Simon-Gadhof. Hamburg: Meiner.

Cassirer, Ernst. 2010. *Nachgelassene Manuskripte und Texte.* Vol. 8: *Vorlesungen und Vorträge zu philosophischen Problemen der Wissenschaften 1907–1945*, ed. Jörg Fingerhut, Gerald Hartung, and Rüdiger Kramme. Hamburg: Meiner.

Cassirer, Ernst. 2011. *Symbolische Prägnanz, Ausdrucksphänomen und "Wiener Kreis*, ed. Christian Möckel. Hamburg: Meiner.

Coffa, Alberto. 1991. *The semantic tradition from Kant to Carnap: To the Vienna station.* Cambridge: Cambridge University Press.

Cohen, Hermann. 1883. *Das Princip der Infinitesimal-Methode und seine Geschichte: Ein Kapitel zur Grundlegung der Erkenntniskritik.* Berlin: Dümmler.

Darriol, Olivier. 2003. Number and measure: Hermann von Helmholtz at the crossroads of mathematics, physics, and psychology. *Studies in History and Philosophy of Science* 34: 515–573.

Duhem, Pierre. 1906. *La théorie physique: son objet et sa structure.* Paris: Chevalier et Rivière. English translation of the 2nd edition of 1914: Duhem, Pierre. 1954. *The aim and structure of physical theory* (trans: Wiener, Philip P.). Princeton: Princeton University Press.

Einstein, Albert. 1916. Die Grundlagen der allgemeinen Relativitätstheorie. *Annalen der Physik* 49: 769–822. English edition: Einstein, Albert. 1923. The foundation of the general theory of relativity. Trans. George Barker Jeffrey and Wilfrid Perrett. In *The principle of relativity*, ed. Hendrik A. Lorentz, Albert Einstein, Hermann Minkowski, and Hermann Weyl, 111–164. London: Methuen.

Einstein, Albert. 1921. Geometrie und Erfahrung. In *The collected papers of Albert Einstein. Vol. 7: The Berlin years: Writings, 1918–1921*, ed. Michel Janssen, Robert Schulmann, Jószef Illy, Christoph Lehner, and Diana Kormos Buchwald, 383–405. Princeton: Princeton University Press, 2002. English edition: Einstein, A. 1922. Geometry and experience. In *Sidelights on relativity* (trans: Jeffrey, George Barker and Perrett, Wilfrid), 27–55. London: Methuen.

Enriques, Federigo. 1907. Prinzipien der Geometrie. In *Enzyklopädie der mathematischen Wissenschaften*, 3a.1b: 1–129.

Ferrari, Massimo. 1991. Cassirer, Schlick e l'interpretazione "kantiana" della teoria della relatività. *Rivista di filosofia* 82: 243–278.

Ferrari, Massimo. 2012. Between Cassirer and Kuhn. Some remarks on Friedman's relativized a priori. *Studies in History and Philosophy of Science* 43: 18–26.

Freudenthal, Hans. 1962. The main trends in the foundations of geometry in the 19th century. In *Logic, methodology and philosophy of science*, ed. Ernest Nagel, Patrick Suppes, and Alfred Tarski, 613–621. Stanford: Stanford University Press.

Friedman, Michael. 1997. Helmholtz's Zeichentheorie and Schlick's *Allgemeine Erkenntnislehre*: Early logical empiricism and its nineteenth-century background. *Philosophical Topics* 25: 19–50.

Friedman, Michael. 1999. *Reconsidering logical positivism*. Cambridge: Cambridge University Press.

Friedman, Michael. 2001. *Dynamics of reason: The 1999 Kant lectures at Stanford University*. Stanford: CSLI Publications.

Friedman, Michael. 2002. Geometry as a branch of physics: Background and context for Einstein's 'Geometry and experience'. In *Reading natural philosophy: Essays in the history and philosophy of science and mathematics*, ed. David B. Malament, 193–229. Chicago: Open Court.

Friedman, Michael. 2005. Ernst Cassirer and contemporary philosophy of science. *Angelaki* 10: 119–128.

Friedman, Michael. 2012. Reconsidering the dynamics of reason: Response to Ferrari, Mormann, Nordmann, and Uebel. *Studies in History and Philosophy of Science* 43: 47–53.

Gabriel, Gottfried. 1978. Implizite Definitionen: Eine Verwechslungsgeschichte. *Annals of Science* 35: 419–423.

Gergonne, Joseph-Diez. 1818. Essai sur la théorie des définitions. *Annales de mathématiques pures et appliquées* 9: 1–35.

Heidelberger, Michael. 2006. Kantianism and realism: Alois Riehl (and Moritz Schlick). In *The Kantian legacy in nineteenth-century science*, ed. Michael Friedman and Alfred Nordmann, 227–247. Cambridge, MA: The MIT Press.

Heidelberger, Michael. 2007. From neo-Kantianism to critical realism: Space and the mind-body problem in Riehl and Schlick. *Perspectives on Science* 15: 26–47.

Helmholtz, Hermann von. 1870. Über den Ursprung und die Bedeutung der geometrischen Axiome. In Helmholtz (1921): 1–24.

Helmholtz, Hermann von. 1887. Zählen und Messen, erkenntnistheoretisch betrachtet. In Helmholtz (1921): 70–97.

Helmholtz, Hermann von. 1921. *Schriften zur Erkenntnistheorie*, ed. Paul Hertz and Moritz Schlick. Berlin: Springer. English edition: Helmholtz, Hermann von. 1977. *Epistemological writings* (trans: Lowe, Malcom F., ed. Robert S. Cohen and Yehuda Elkana). Dordrecht: Reidel.

Hilbert, David. 1899. Grundlagen der Geometrie. In *Festschrift zur Feier der Enthüllung des Gauss-Weber-Denkmals in Göttingen*, 1–92. Leipzig: Teubner. English edition: Hilbert, David. 1902. *The foundations of geometry* (trans: Townsend, Edgar Jerome). Chicago: Open Court.

Hilbert, David. 1902. Sur les problèmes futurs des mathématiques. In *Compte rendu du deuxième Congrès International des Mathématiciens, held in Paris, from 6 to 12 August 1900*, ed. Ernest Duporcq, 58–114. Paris: Gauthier-Villars.

Howard, Don. 1984. Realism and conventionalism in Einstein's philosophy of science: The Einstein-Schlick correspondence. *Philosophia Naturalis* 21: 616–630.

Howard, Don. 1988. Einstein and *Eindeutigkeit*: A neglected theme in the philosophical background to general relativity. In *Studies in the history of general relativity: Based on the Proceedings of the 2nd international conference on the history of general relativity, Luminy, France, 1988*, ed. Jean Eisenstaedt, and A.J. Kox, 154–243. Boston: Birkhäuser, 1992.

Kant, Immanuel. 1786. *Metaphysische Anfangsgründe der Naturwissenschaft*. Riga: Hartknoch. Repr. in *Akademie-Ausgabe*. Berlin: Reimer, 4: 465–565.

Kant, Immanuel. 1787. *Critik der reinen Vernunft*. 2nd ed. Riga: Hartknoch. Repr. in *Akademie-Ausgabe*. Berlin: Reimer, 3. English edition: Kant, Immanuel. 1998. *Critique of Pure Reason* (trans: Guyer, Paul and Wood, Allen W.). Cambridge: Cambridge University Press.

Kretschmann, Erich. 1917. Über den physikalischen Sinn der Relativitätspostulate: A. Einsteins neue und seine ursprüngliche Relativitätstheorie. *Annalen der Physik* 53: 575–614.

Kuhn, Thomas S. 1962. *The structure of scientific revolutions: Foundations of the unity of science*. Chicago: University of Chicago Press.

Majer, Ulrich. 2001. Hilbert's program to axiomatize physics (in analogy to geometry) and its impact on Schlick, Carnap and other members of the Vienna Circle. In *History of philosophy of science. New trends and perspectives*, ed. Michael Heidelberger and Friedrich Stadler, 213–224. Dordrecht: Springer.

Neuber, Matthias. 2012. Helmholtz's theory of space and its significance for Schlick. *British Journal for the History of Philosophy* 20: 163–180.

Neuber, Matthias. 2014. Critical realism in perspective: Remarks on a neglected current in neo-Kantian epistemology. In *The philosophy of science in a European perspective: New directions in the philosophy of science*, ed. Maria Carla Galavotti, Dennis Dieks, Wenceslao J. Gonzales, Stephan Hartmann, Thomas Uebel, and Marcel Weber, 657–673. Cham: Springer.

Norton, John D. 2011. *The hole argument*. http://plato.stanford.edu/entries/spacetime-holearg/. Accessed 28 Jan 2016.

Parrini, Paolo. 1993. Origini e sviluppi dell'empirismo logico nei suoi rapporti con la "filosofia continentale". Alcuni testi inediti. *Rivista di storia della filosofia* 48: 121–146.

Pasch, Moritz. 1882. *Vorlesungen über neuere Geometrie*. Leipzig: Teubner.

Peckhaus, Volker. 1990. *Hilberts Programm und kritische Philosophie: Das Göttinger Modell interdisziplinärer Zusammenarbeit zwischen Mathematik und Philosophie*. Göttingen: Vandenhoeck und Ruprecht.

Petzoldt, Joseph. 1895. Das Gesetz der Eindeutigkeit. *Vierteljahrsschrift für wissenschaftliche Philosophie und Soziologie* 19: 146–203.

Poincaré, Henri. 1902. *La science et l'hypothèse*. Paris: Flammarion. English Edition: Poincaré, Henri. 1905. *Science and hypothesis* (trans: Greenstreet, William John). London: Scott.

Pulte, Helmut. 2006. The space between Helmholtz and Einstein: Moritz Schlick on spatial intuition and the foundations of geometry. In *Interactions: Mathematics, physics and philosophy, 1860–1930*, ed. Vincent F. Hendricks, Klaus Frovin Jørgensen, Jesper Lützen, and Stig Andur Pedersen, 185–206. Dordrecht: Springer.

Rankine, William John Macquorn. 1855. Outlines of the science of energetics. *Glasgow Philosophical Society Proceedings* 3: 121–141.

Reichenbach, Hans. 1920. *Relativitätstheorie und Erkenntnis apriori*. Berlin: Springer. English edition: Reichenbach, Hans. 1965. *The Theory of Relativity and A Priori Knowledge* (trans: Reichenbach, Maria). Los Angeles: University of California Press.

Reichenbach, Hans. 1922. Der gegenwärtige Stand der Relativitätsdiskussion. *Logos* 10: 316–378. English edition: Reichenbach, Hans. 1978. The present state of the discussion on relativity. In *Hans Reichenbach: Selected writings, 1909–1953*, vol. 2, ed. Robert S. Cohen and Maria Reichenbach, 3–47. Dordrecht: Reidel.

Riehl, Alois. 1879. *Der Philosophische Kriticismus. Vol 2: Die sinnlichen und logischen Grundlagen der Erkenntnis*. Leipzig: Engelmann.

Riehl, Alois. 1887. *Der Philosophische Kriticismus. Vol. 3: Zur Wissenschaftstheorie und Metaphysik*. Leipzig: Engelmann. English edition: Riehl, Alois. 1894. *The principles of the critical philosophy: Introduction to the theory of science and metaphysics* (trans: Fairbanks, Arthur). London: Paul, Trench, Trübner.

Riemann, Bernhard. 1867. Über die Hypothesen, welche der Geometrie zu Grunde liegen. *Abhandlungen der Königlichen Gesellschaft der Wissenschaften zu Göttingen* 13: 133–152.

Ryckman, Thomas A. 1991. *Conditio sine qua non? Zuordnung* in the early epistemologies of Cassirer and Schlick. *Synthese* 88: 57–95.

Ryckman, Thomas A. 1999. Einstein, Cassirer, and general covariance – then and now. *Science in Context* 12: 585–619.

Ryckman, Thomas A. 2005. *The reign of relativity: Philosophy in physics 1915–1925*. New York: Oxford University Press.

Schlick, Moritz. 1916. Idealität des Raumes: Introjektion und psychophysisches Problem. *Vierteljahrsschrift für wissenschaftliche Philosophie und Soziologie* 40: 230–254. English edition: Schlick, Moritz. 1979. Ideality of space, introjection and the psycho-physical problem. In *Moritz Schlick: Philosophical papers*, vol. 1, ed. Henk L. Mulder, and Barbara F.B. van de Velde, 190–206. Dordrecht: Reidel.

Schlick, Moritz. 1918. *Allgemeine Erkenntnislehre*. Berlin: Springer.

Schlick, Moritz. 1921. Kritizistische oder empiristische Deutung der neuen Physik? Bemerkungen zu Ernst Cassirers Buch *Zur Einstein'schen Relativitätstheorie*. *Kant-Studien* 26: 96–111. English edition: Schlick, Moritz. 1979. Critical or empiricist interpretation of modern physics? In *Moritz Schlick: Philosophical papers*, vol. 1, ed. Henk L. Mulder, and Barbara F.B. van de Velde, 322–334. Dordrecht: Reidel.

Schlick, Moritz. 1974. *General theory of knowledge*. Trans. Albert E. Blumberg. Originally published as *Allgemeine Erkenntnislehre* (1925). New York: Springer.

Stachel, John. 1980. Einstein's search for general covariance. In *Einstein and the history of general relativity*, ed. Don Howard and John Stachel, 63–100. Boston: Birkhäuser, 1989.

Vacca, Giovanni. 1899. Sui precursori della logica matematica, II: J.D. Gergonne. *Rivista di matematica* 6: 183–186.

Index

A

Adickes, Erich, 40
Analysis situs, 155, 156, 158
Appearance(s)
    form(s) of, 12, 24
    form *vs.* matter of, 32, 36
    manifold of, 24
    object of, 47
    order of, 44, 160
    outer, 17, 18, 23
    sensuous, 112
    space of, 192
A priori
    constitutive *vs.* regulative, 7, 46, 145, 190
    metaphysical *vs.* transcendental, 34, 71, 76, 211
    notion of, 28, 30, 34, 51, 75, 119, 152, 163, 210, 211, 215, 217
    psychological, 11, 139
    relativized, 15, 18, 19, 35, 65, 76, 77, 135, 145, 190, 207, 209, 210, 212, 216, 218
    of space, 205, 212, 213
    theory of, 18, 19, 24, 30–37, 51, 65, 66, 71, 75, 76
    transcendental, 35, 46, 71, 74, 75
Avenarius, Richard, 192
Avigad, Jeremy, 134
Axiom(s)
    of addition, 95
    Archimedean, 122, 126
    arithmetical, 83
    vs. canons, 40
    of comprehension, 171
    of congruence, 90, 122, 126, 161
    of connection, 210, 211, 215
    and conventions, 214

of coordination, 210, 212
of coordination *vs.* connection, 215
(Dedekind's) axiom of continuity, 134, 135, 139, 143
and definitions, 51–77, 173, 192–198
Euclidean, 72, 85, 86, 90, 108
Euclid's first, 169, 172
and formulas, 166
geometrical, 9, 14, 15, 18, 23, 24, 30, 48, 51, 52, 57, 60, 61, 63–66, 75, 76, 84–88, 91, 92, 117, 141, 151, 156, 166, 173–176, 191, 194, 195, 198, 201, 209, 214
Grassmann's, 94
and hypotheses, 51–77
of intuitions, 74
mechanical, 91
notion of, 140, 173
of order, 109
parallel, 131
of parallel lines, 122, 126
Peano, 107
of quantity, 97
of separation, 171
synthetic a priori, 161
Axiomatics, 191, 198

B

Banks, Erik C., 57
Bauch, Bruno, 52, 70, 152, 167, 176–181
Beiser, Frederick C., 29, 31, 32
Beltrami, Eugenio, 58, 63, 129, 154, 160
Ben-Menahem, Yemima, 167, 175, 180
Benis-Sinaceur, Houria, 135
Bennett, Mary Katherine, 120, 155
Bessel, Friedrich Wilhelm, 53

© Springer International Publishing Switzerland 2016
F. Biagioli, *Space, Number, and Geometry from Helmholtz to Cassirer*,
Archimedes 46, DOI 10.1007/978-3-319-31779-3

The manufacturer's authorised representative in the EU is Springer
Nature Customer Service Centre GmbH, Europaplatz 3, 69115 Heidelberg,
Germany. If you have any concerns regarding our products, please
contact ProductSafety@springernature.com

Printed and bound by CPI Group (UK) Ltd, Croydon, CR0 4YY
27/04/2026
02097562-0005